COLLEGE ALGEBRA

COLLEGE ALGEBRA

BERNARD KOLMAN
Drexel University

ARNOLD SHAPIRO
Pennsylvania State University
Ogontz Campus

ACADEMIC PRESS

New York / London / Toronto / Sydney / San Francisco
A subsidiary of Harcourt Brace Jovanovich, Publishers

Cover art: *Lyra* by Vasarely
courtesy of Vasarely Center,
New York

Academic Press, Inc.
111 Fifth Avenue, New York, New York   10003

United Kingdom Edition published by
Academic Press, Inc. (London) Ltd.
24/28 Oval Road, London NW1 7DX

ISBN: 0-12-417884-7
Library of Congress Catalog Card Number: 79-50492

Printed in the United States of America

To the memory of my mother Eva
B.K.
To my mother Helen
A.S.

CONTENTS

PREFACE

This book is a complete and self-contained presentation of the fundamentals of algebra. Our objective has been *to write a textbook that will help the instructor teach the material and that the student will find readable.* Moreover, we believe that it is almost impossible to oversimplify an idea in the fundamentals of mathematics. To this end we have provided the following helpful features.

FEATURES

- [ ] **Presentation.** We have adopted an informal, supportive style to encourage the student to read the text and to develop confidence under its guidance. Concepts are introduced gradually with accompanying diagrams and illustrations that aid the student to grasp intuitively the "reasonableness" of results. Many algebraic procedures are described with the aid of a "split-screen" that simultaneously displays both the steps of an algorithm and a worked out example.

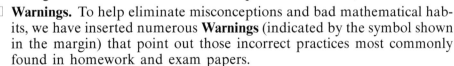

- [ ] **Warnings.** To help eliminate misconceptions and bad mathematical habits, we have inserted numerous **Warnings** (indicated by the symbol shown in the margin) that point out those incorrect practices most commonly found in homework and exam papers.

- [ ] **Progress Checks.** At carefully selected places, problems similar to those worked in the text have been inserted (with answers) to enable the student to test his or her understanding of the material just discussed.

- [ ] **End-of-Chapter.** Every chapter contains a summary that includes the following:

Terms and Symbols with appropriate page references.
Key Ideas for Review to stress the concepts.
Review Exercises to provide additional practice.
Progress Tests to provide self-evaluation and reinforcement.
The answers to all **Review Exercises** and **Progress Tests** appear in the back of the book.

☐ **Exercises.** Over 3,000 carefully graded exercises provide practice in both the mechanical and conceptual aspects of algebra. Exercises requiring the use of a calculator are indicated by the symbol shown in the margin. Answers to all odd-numbered exercises appear at the back of the book. Answers to all even-numbered exercises appear in the Instructor's Manual that is available to instructors upon request. The Instructor's Manual also contains an extensive Test Bank with solutions.

☐ **Supplement.** A Study Guide containing solutions to about one-third of the exercises as well as additional chapter tests can be purchased from the publisher at a nominal price.

☐ **Pedagogic Approach**
  • Section 1.1 presents an axiomatic development of the foundations of algebra. If desired, the material in this section can be omitted from classroom presentation and the student can be asked to read the section.
  • Set theory is not stressed. A brief introduction to set theory is presented in Chapter One and set notation is used throughout the book.
  • Complex numbers are defined in Chapter One and are used in Chapter Two to find the roots of *any* quadratic equation. Advanced concepts involving complex numbers are presented later in the chapter on roots of polynomials.
  • Functions and function notation are introduced in Chapter Three and are used as a unifying concept throughout the rest of the book.
  • Computations with logarithms are treated briefly in Section 4.4 and is an optional topic.
  • An entire chapter is devoted to analytic geometry and the conic sections. The use of a coordinate system to prove theorems from plane geometry is demonstrated and the equations of the conics are derived by applying the methods of analytic geometry.
  • The material on matrices includes matrix operations and matrix inverses. For a "softer" approach, Sections 2 and 3 of the chapter "Matrices and Determinants" as well as the Gauss–Jordan method can be omitted.

ACKNOWLEDGMENTS

We thank the following for their review of the manuscript and their helpful comments: Professor Jacqueline Peterson at Arizona State University; Professor Monty Strauss at Texas Tech University; Professor Gerald Bradley at Claremont Men's College; Professor Norman Mittman at Northeastern Illinois University; Professor Stanley Lukawecki at Clemson University; and Professor James Snow at Whatcom Community College.

Our thanks to Susan Pagano for typing much of the manuscript. We are grateful to Harry Spector for his contributions as copy editor. We know of no other copy editor who checks the worked-out examples!

TO THE STUDENT

This book was written for you. It gives you every possible chance to succeed—if you use it properly.

We would like to have you think of mathematics as a challenging game—but not as a spectator sport. Which leads to our primary rule: *Read this textbook with pencil and paper handy.* Every new idea or technique is illustrated by fully worked examples. As you read the text, carefully follow the worked-out examples and then do the **Progress Checks.** The key to success in a math course is working problems, and the **Progress Checks** are there to provide immediate practice with the material you have just learned.

Your instructor will assign homework from the extensive selection of exercises that follow each section in the book. *Do the assignments regularly, thoroughly, and independently.* By doing lots of problems you will develop the necessary skills in algebra and your confidence will grow. Since algebraic techniques and concepts build upon previous results, you can't afford to skip any of the work.

To help you eliminate improper habits and to help you avoid those errors that we see each semester as we grade papers, we have interspersed **Warnings** throughout the book. The **Warnings** point out common errors and show you the proper method.

There is other important review material at the end of each chapter. The **Terms and Symbols** should all be familiar by the time you reach them. If your understanding of a term or symbol is hazy, use the page reference to find the place in the text where it is introduced. Go back and read the definition.

It is possible to become so involved with the details of techniques that you may lose track of the broader concepts. The list of **Key Ideas for Review** at the end of each chapter will help you focus on the principal ideas.

The **Review Exercises** at the end of each chapter can be used as part of your preparation for examinations. You are then ready to try **Progress Test A.** You will soon pinpoint your weak spots and can go back for further review and more exercises in those areas. Then, and only then, should you proceed to **Progress Test B.**

The authors believe that the eventual "payoff" in studying mathematics is an improved ability to tackle practical problems in your chosen field of interest. To that end, this book places special emphasis on word problems, which recent surveys show are often troublesome to students. Since algebra is the basic language of most mathematical techniques as used in virtually all fields, the mastery of algebra is well worth your effort.

COLLEGE ALGEBRA

CHAPTER ONE
THE FOUNDATIONS OF ALGEBRA

No one would debate that "2 + 2 = 4" or that "5 + 3 = 3 + 5." The significance of the statement "2 + 2 = 4" lies in the recognition that it is true whether the objects under discussion are apples or ants, cradles or cars. Further, the statement "5 + 3 = 3 + 5" indicates that the order of addition is immaterial, and this is true for any pair of integers.

These simple examples illustrate the fundamental task of algebra: to abstract those properties that apply to a number system. Of course, the properties will depend upon the type of numbers we choose to deal with. We therefore begin with a discussion of the *real number system* and its properties since much of our work in algebra will involve this number system. We will then indicate a correspondence between the real numbers and the points on a real number line that permits a graphical presentation of many of our results.

The remainder of this chapter is devoted to a review of some fundamentals of algebra: the meaning and use of variables; algebraic expressions and polynomial forms; factoring; and operations with rational expressions or algebraic fractions.

1.1
THE REAL NUMBER SYSTEM
SETS

We will need to use the notation and terminology of sets from time to time. Recall that a set is a collection of objects or numbers that are called the **elements** or **members** of the set. The elements of a set are written within braces, so that

$$A = \{4, 5, 6\}$$

tells us that the set A consists of the numbers 4, 5, and 6. The set

$$B = \{\text{Exxon, Ford, Honeywell}\}$$

consists of the names of these three corporations. We also write $4 \in A$, which we read as "4 is a member of the set A." Similarly, Ford $\in B$ is read as "Ford is a member of the set B," and IBM $\notin B$ is read as "IBM is not a member of the set B."

If every element of a set A is also a member of a set B, then A is a **subset** of B. For example, the set of all robins is a subset of the set of all birds.

EXAMPLE 1
The set C consists of the names of all coins whose denominations are less than 50 cents. We may write C in set notation as

$$C = \{\text{penny, nickel, dime, quarter}\}$$

We see that dime $\in C$ but half dollar $\notin C$. Further, the set $H = \{\text{nickel, dime}\}$ is a subset of C.

PROGRESS CHECK
The set V consists of the vowels in the English alphabet.

(a) Write V in set notation.
(b) Is the letter k a member of V?
(c) Is the letter u a member of V?
(d) List the subsets of V having four elements.

Answers
(a) $V = \{a, e, i, o, u\}$ *(b) No*   *(c) Yes*   *(d)* $\{a, e, i, o\}, \{e, i, o, u\},$
$\{a, i, o, u\}, \{a, e, o, u\}, \{a, e, i, u\}$

THE REAL NUMBER SYSTEM

Much of our work in algebra deals with the real number system and we now review the composition of this number system.

The numbers 1, 2, 3, . . . used for counting form the set of **natural numbers.** If we had only these numbers to use to show the profit earned by a company, we would have no way to indicate that the company has no profit or has a loss. To indicate no profit we introduce 0 and for losses we need to introduce negative numbers. The numbers

$$\ldots, -2, -1, 0, 1, 2, \ldots$$

form the set of **integers.** Thus, every natural number is an integer, and the set of natural numbers is seen to be a subset of the set of integers.

When we try to divide two apples equally among four people we find no number in the set of integers that will express how many apples each person should get. We need to introduce the set of **rational**

numbers, which are numbers that can be written as a ratio of two integers,

$$\frac{p}{q}, \quad \text{with } q \text{ not equal to zero}$$

Examples of rational numbers are

$$0, \quad \frac{2}{3}, \quad -4, \quad \frac{7}{5}, \quad \frac{-3}{4}$$

Thus, when we divide two apples equally among four people, each person gets $\frac{1}{2}$ apple. Since every integer n can be written as $n/1$, we see that every integer is a rational number. The decimal number 1.3 is also a rational number since $1.3 = 13/10$.

We have now seen three fundamental number systems: the natural number system, the system of integers, and the rational number system. Each later system includes the previous system(s) and is more complicated. However, the rational number system is still inadequate for mature uses of mathematics since there exist numbers which are not rational, that is, numbers that cannot be written as the ratio of two integers. It can be shown that the number a that satisfies $a \cdot a = 2$ is such a number. The number π, which is the ratio of the circumference of a circle to its diameter, is also such a number. These are called **irrational numbers.** The decimal form of a rational number will either terminate, as

$$\frac{3}{4} = 0.75; \quad -\frac{4}{5} = -0.8$$

or will form a repeating pattern, as

$$\frac{2}{3} = 0.666\ldots; \qquad \frac{1}{11} = 0.090909\ldots; \qquad \frac{1}{7} = 0.1428571\ldots$$

Remarkably, the decimal form of an irrational number *never* forms a repeating pattern.

The rational and irrational numbers together comprise the **real number system** (Figure 1).

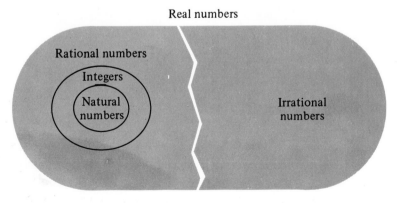

FIGURE 1

With respect to the operations of addition and multiplication, the real number system has properties that are fundamental to algebra. The letters a, b, and c will denote real numbers.

Closure

Property 1. The sum of a and b, denoted by $a + b$, is a real number.

Property 2. The product of a and b, denoted by $a \cdot b$ or ab, is a real number.

We say that the set of real numbers is **closed** with respect to the operations of addition and multiplication since the sum and product of two real numbers are also real numbers and are therefore members of the set.

Commutativity

Property 3. $a + b = b + a$ **Commutative law of addition**

Property 4. $ab = ba$ **Commutative law of multiplication**

That is, we may add or multiply real numbers in any order.

Associativity

Property 5. $(a + b) + c = a + (b + c)$ **Associative law of addition**

Property 6. $(ab)c = a(bc)$ **Associative law of multiplication**

That is, when adding or multiplying real numbers we may group them in any order.

Identities

Property 7. There is a unique real number, denoted by 0, such that $a + 0 = 0 + a = a$ for every real number a.

Property 8. There is a unique real number, denoted by 1, such that $a \cdot 1 = 1 \cdot a = a$ for every real number a.

The real number 0 of Property 7 is called the **additive identity**; the real number 1 of Property 8 is called the **multiplicative identity.**

Inverses

Property 9. For every real number a, there is a unique real number, denoted by $-a$, such that

$$a + (-a) = (-a) + a = 0$$

Property 10. For every real number $a \neq 0$, there is a unique real number, denoted by $1/a$, such that

$$a\left(\frac{1}{a}\right) = \left(\frac{1}{a}\right)a = 1$$

The number $-a$ of Property 9 is called the **negative** or **additive inverse** of a. The number $1/a$ of Property 10 is called the **reciprocal** or **multiplicative inverse** of a.

Distributive Laws

Property 11. $a(b + c) = ab + ac$

Property 12. $(a + b)c = ac + bc$

EXAMPLE 2

Specify the appropriate properties in each of the following.

(a) $2 + 3 = 3 + 2$ Commutative law of addition

(b) $(2 \cdot 3) \cdot 4 = 2 \cdot (3 \cdot 4)$ Associative law of multiplication

(c) $2 \cdot \dfrac{1}{2} = 1$ Multiplicative inverse

(d) $2(3 + 5) = 2 \cdot 3 + 2 \cdot 5$ Distributive law

Properties 1–12 are called **field axioms** because any set of elements together with two operations that satisfy these properties is called a **field.** Thus, the real numbers form a field. There are also other important examples of fields. Some of these play key roles in computers and in the coding of information for transmission.

When we say that two numbers are **equal,** we mean that they are identical. Thus, when we write

$$a = b$$

(read "*a* equals *b*"), we mean that a and b represent the same number. For example, $2 + 5$ and $4 + 3$ are different ways of writing the number 7, so we can write

$$2 + 5 = 4 + 3$$

Equality satisfies four basic properties.

Properties of Equality

Let a, b, and c be elements of a set.

1. $a = a$ **Reflexive property**
2. If $a = b$ then $b = a$.   **Symmetric property**
3. If $a = b$ and $b = c$, then $a = c$.   **Transitive property**
4. If $a = b$ then a may be replaced by b in any statement that involves a or b.   **Substitution property**

EXAMPLE 3

Specify the appropriate property in each of the following.

(a) If $5a - 2 = b$, then $b = 5a - 2$.   Symmetric property

(b) If $a = b$ and $b = 5$ then $a = 5$.   Transitive property

(c) If $3(a + 2) = 3a + 6$, and $a = b$, then $3(b + 2) = 3b + 6$.   Substitution property

Using the field axioms given in Properties 1–12, the properties of equality, and rules of logic, we can *prove* many other properties of the real numbers.

> **Theorem 1** If a, b, and c are real numbers, and $a = b$, then
> (a) $a + c = b + c$
> (b) $ac = bc$

This theorem, which will be used often in working with equations, allows us to add the same number to both sides of an equation and to multiply both sides of an equation by the same number. We will prove Theorem 1a and leave the proofs of the other theorems as exercises. By the closure property, $a + c$ is a real number. Then, by the reflexive property of equality, $a + c = a + c$. Since $a = b$, the substitution property permits us to replace a with b in the right hand side of the last equation, so that we may write $a + c = b + c$. Thus we have completed the proof of Theorem 1a and have made use of the field axioms and properties of equality to do so.

The following theorem is the converse of Theorem 1.

> **Theorem 2** Let a, b, and c be real numbers.
> (a) If $a + c = b + c$, then $a = b$.
> (b) If $ac = bc$ and $c \neq 0$, then $a = b$.

Part (b) of Theorem 2 is often called the **cancellation law of multiplication.** We can restate part (a) of Theorems 1 and 2 in this way: if a, b, and c are real numbers, then $a + c = b + c$ if and only if $a = b$. The connector "if and only if" is used to indicate that both statements are true or both statements are false.

> **Theorem 3** Let a and b be real numbers.
> (a) $a \cdot 0 = 0 \cdot a = 0$
> (b) If $ab = 0$ then $a = 0$ or $b = 0$.

The real numbers a and b are said to be **factors** of the product ab. Part (b) of Theorem 3 says that a product of two real numbers can be zero only if at least one of the factors is zero.

The next theorem gives us the usual rules of signs.

> **Theorem 4** Let a and b be real numbers. Then
>
> | | | *Example* |
> |---|---|---|
> | (a) | $-(-a) = a$ | $-(-3) = 3$ |
> | (b) | $(-a)(b) = -(ab) = a(-b)$ | $(-2)(3) = -6$ |
> | (c) | $(-1)(a) = -a$ | $(-1)(4) = -4$ |
> | (d) | $(-a)(-b) = ab$ | $(-2)(-4) = 8$ |
> | (e) | $-(a + b) = (-a) + (-b)$ | $-(2 + 5) = (-2) + (-5)$ |

It is important to note that $-a$ is not necessarily a negative number. In fact, the first example shows that $-(-3) = 3$.

We next introduce the operations of subtraction and division. If a and b are real numbers, the **difference** between a and b, denoted by $a - b$, is defined by

$$a - b = a + (-b)$$

and the operation is called **subtraction.** Thus,

$$6 - 2 = 6 + (-2) = 4 \qquad 2 - 2 = 0 \qquad 0 - 8 = -8$$

If a and b are real numbers, with $b \neq 0$, then the **quotient** of a by b, denoted by a/b, is defined by

$$\frac{a}{b} = a \cdot \frac{1}{b}$$

and the operation is called **division.** We also write a/b as $a \div b$ and speak of the **fraction** a over b.

$$3 \div 5 = \frac{3}{5} \qquad 5 \div 4 = \frac{5}{4} \qquad 1 \div 2 = \frac{1}{2} \qquad 2 \div 1 = \frac{2}{1} = 2$$

The numbers a and b are called the **numerator** and **denominator,** respectively, of the fraction a/b. Observe that we have not defined division by zero, since 0 has no reciprocal.

The following theorem summarizes the familiar properties of fractions.

Theorem 5 Let a, b, c, and d be real numbers with $b \neq 0$, $d \neq 0$. Then

Example

(a) $\dfrac{a}{b} = \dfrac{c}{d}$ if and only if $ad = bc$ $\qquad$ $\dfrac{2}{3} = \dfrac{4}{6}$ since $2 \cdot 6 = 3 \cdot 4$

(b) $\dfrac{a}{b} = \dfrac{ad}{bd}$ $\qquad$ $\dfrac{6}{12} = \dfrac{6 \cdot 3}{12 \cdot 3} = \dfrac{18}{36}$

(c) $\dfrac{a}{b} + \dfrac{c}{b} = \dfrac{a+c}{b}$ $\qquad$ $\dfrac{2}{3} + \dfrac{5}{3} = \dfrac{2+5}{3} = \dfrac{7}{3}$

(d) $\dfrac{a}{b} + \dfrac{c}{d} = \dfrac{ad+bc}{bd}$ $\qquad$ $\dfrac{2}{5} + \dfrac{3}{4} = \dfrac{2 \cdot 4 + 5 \cdot 3}{5 \cdot 4} = \dfrac{23}{20}$

(e) $\dfrac{a}{b} \cdot \dfrac{c}{d} = \dfrac{ac}{bd}$ $\qquad$ $\dfrac{2}{3} \cdot \dfrac{4}{5} = \dfrac{2 \cdot 4}{3 \cdot 5} = \dfrac{8}{15}$

(f) $\dfrac{\frac{a}{b}}{\frac{c}{d}} = \dfrac{a}{b} \cdot \dfrac{d}{c}$ (if $c \neq 0$) $\qquad$ $\dfrac{\frac{2}{3}}{\frac{5}{7}} = \dfrac{2}{3} \cdot \dfrac{5}{7} = \dfrac{2 \cdot 5}{3 \cdot 7} = \dfrac{10}{21}$

PROGRESS CHECK

Perform the indicated operations.

(a) $\dfrac{3}{5} + \dfrac{1}{4}$ $\qquad$ (b) $\dfrac{5}{2} \cdot \dfrac{4}{15}$ $\qquad$ (c) $\dfrac{2}{3} \div \dfrac{3}{7}$

Answers

(a) $\dfrac{17}{20}$ $\qquad$ (b) $\dfrac{2}{3}$ $\qquad$ (c) $\dfrac{14}{9}$

EXERCISE SET 1.1

Write each set by listing its elements within braces.

1. The set of natural numbers between 3 and 7, inclusive.
2. The set of integers between -4 and 2.
3. The set of integers between -10 and -8.
4. The set of natural numbers between -9 and 3, inclusive.
5. The subset of the set $S = \{-3, -2, -1, 0, 1, 2\}$ consisting of the positive integers in S.
6. The subset of the set $S = \left\{-\frac{2}{3}, -1.1, 3.7, 4.8\right\}$ consisting of the negative rational numbers in S.
7. The subset of all $x \in S$, $S = \{1, 3, 6, 7, 10\}$, such that x is an odd integer.
8. The subset of all $x \in S$, $S = \{2, 5, 8, 9, 10\}$, such that x is an even integer.

In Exercises 9–22 determine whether the given statement is true (T) or false (F).

9. -14 is a natural number.

10. $-\frac{4}{5}$ is a rational number.

11. $\frac{\pi}{3}$ is a rational number.

12. $\sqrt{9}$ is an irrational number.
13. -1207 is an integer.
14. 0.75 is an irrational number.

15. $\frac{4}{5}$ is a real number.

16. 3 is a rational number.
17. $\sqrt{5}$ is a real number.
18. The sum of two rational numbers is always a rational number.
19. The sum of two irrational numbers is always an irrational number.
20. The product of two rational numbers is always a rational number.
21. The product of two irrational numbers is always an irrational number.
22. The difference of two irrational numbers is always an irrational number.

In the following, the letters represent real numbers. Identify the property or properties of real numbers that justify each statement.

23. $a + x = x + a$
24. $(xy)z = x(yz)$
25. $xyz + xy = xy(z + 1)$
26. $x + y$ is a real number
27. $(a + b) + 3 = a + (b + 3)$
28. $5 + (x + y) = (x + y) + 5$
29. cx is a real number
30. $(a + 5) + b = (a + b) + 5$
31. $uv = vu$
32. $x + 0 = x$
33. $a(bc) = c(ab)$
34. $xy - xy = 0$

35. $5 \cdot \frac{1}{5} = 1$
36. $xy \cdot 1 = xy$

Find a counterexample for each of the following statements; that is, find real values for which the statement is false.

37. $a - b = b - a$
38. $\frac{a}{b} = \frac{b}{a}$

39. $a(b + c) = ab + c$
40. $(a + b)(c + d) = ac + bd$

Indicate the property or properties of equality that justify each statement.

41.   If $3x = 5$, then $5 = 3x$.
42.   If $x + y = 7$ and $y = 5$, then $x + 5 = 7$.
43.   If $2y = z$ and $z = x + 2$, then $2y = x + 2$.
44.   If $x + 2y + 3z = r + s$ and $r = x + 1$, then $x + 2y + 3z = x + 1 + s$.
45.   If $a = b$, then $ac = bc$. (Theorem 1b)

46.   If $a = b$ and $c \neq 0$, then $\dfrac{a}{c} = \dfrac{b}{c}$.

In each of the following, a, b, and c are real numbers. Use the field properties of the real numbers and the properties of equality to prove each theorem.

47.   If $a + c = b + c$, then $a = b$. (Theorem 2a)
48.   If $ac = bc$ and $c \neq 0$, then $a = b$. (Theorem 2b)
49. $a(b - c) = ab - ac$
50.   Prove that the real number 0 does not have a reciprocal.

(*Hint:* Assume $b = \dfrac{1}{0}$ is the reciprocal of 0. Supply a reason for each of the following steps.

$$1 = 0 \cdot \frac{1}{0}$$

$$= 0 \cdot b$$

$$= 0$$

Since this conclusion is a contradiction, the original assumption must be false.)

1.2
THE REAL NUMBER LINE

There is a simple and very useful geometric interpretation of the real number system. Draw a horizontal straight line; pick a point on this line, label it with the number 0, and call it the origin. Choose the **positive direction** to the right of the origin and the **negative direction** to the left of the origin.

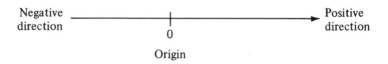

Negative direction — 0 (Origin) — Positive direction

Next, select a unit of length for measuring distance. With each positive real number r we associate the point which is r units to the right of the origin and with each negative number $-r$ we associate the point which is r units to the left of the origin. Thus, the set of real numbers is identified with all possible points on a straight line. For every point on the line there is a real number and for every real number there is a point on the line. The line is called the **real number line** and we can now show some points on this line.

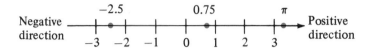

The numbers to the right of zero are called **positive,** the numbers to the left of zero are called **negative.** The positive numbers and zero together are called the **nonnegative** numbers, and the combination of zero and the negative numbers are the **nonpositive** numbers.

We will frequently turn to the real number line to help us picture the results of algebraic computations.

EXAMPLE 1
Draw a real number line and plot the points: $-\dfrac{3}{2}$, 2, $\dfrac{13}{4}$.

Solution

INEQUALITIES

If a and b are real numbers we can compare their relative positions on the real number line by defining the relations of **less than, greater than, less than or equal to,** and **greater than or equal to,** denoted by the **inequality symbols** $<$, $>$, $\leq$, and $\geq$, respectively. Table 1 describes both algebraic and geometric interpretations of the inequality symbols.

TABLE 1

Algebraic Statement	Equivalent Statement	Geometric Statement
$a > 0$	a is positive	a lies to the right of the origin
$a < 0$	a is negative	a lies to the left of the origin
$a > b$	$a - b$ is positive	a lies to the right of b
$a < b$	$a - b$ is negative	a lies to the left of b
$a \geq b$	$a - b$ is zero or positive	a coincides with b or lies to the right of b
$a \leq b$	$a - b$ is zero or negative	a coincides with b or lies to the left of b

Expressions involving inequality symbols, such as $a < b$ and $a \geq b$, are called **inequalities.** We often combine these expressions so that

$a \leq b < c$ means both $a \leq b$ and $b < c$. Thus, $-5 \leq x < 2$ is equivalent to $-5 \leq x$ and $x < 2$.

PROGRESS CHECK
Verify that the following inequalities are true by using either the "Equivalent Statement" or "Geometric Statement" of Table 1.

(a) $-1 < 3$ (b) $2 \leq 2$ (c) $-2.7 < -1.2$

(d) $-4 < -2 < 0$ (e) $-\dfrac{7}{2} < \dfrac{7}{2} < 7$

The real numbers satisfy the following useful properties of inequalities.

Properties of Inequalities

Let a, b, and c be real numbers.

(a) One and only one of the following relations holds.
$$a < b, \quad a > b, \quad a = b \quad \text{Trichotomy property}$$
(b) If $a < b$ and $b < c$, then $a < c$.   Transitive property
(c) If $a < b$, then $a + c < b + c$.
(d) If $a < b$ and $c > 0$, then $ac < bc$.   When an inequality is multiplied by a positive number, the sense of the inequality is preserved.
(e) If $a < b$ and $c < 0$, then $ac > bc$.   When an inequality is multiplied by a negative number, the sense of the inequality is reversed.

EXAMPLE 2
(a)   Since $-2 < 4$ and $4 < 5$, then $-2 < 5$.
(b)   Since $-2 < 5$, $-2 + 3 < 5 + 3$, or $1 < 8$.
(c)   Since $3 < 4$, $3 + (-5) < 4 + (-5)$, or $-2 < -1$.
(d)   Since $2 < 5, 2(3) < 5(3)$, or $6 < 15$.
(e)   Since $-3 < 2, (-3)(-2) > 2(-2)$, or $6 > -4$.

ABSOLUTE VALUE

Suppose we are interested in the *distance* between the origin and the points labeled 4 and -4 on the real number line. Each of these points is four units from the origin, that is, the *distance is independent of the direction* and is nonnegative (Figure 2).

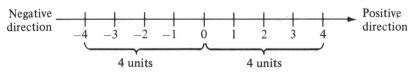

FIGURE 2

When we are interested in the size or magnitude of a number a, and don't care about the direction or sign, we use the concept of **absolute value**, which we write as $|a|$. The formal definition of absolute value is stated this way.

$$|a| = \begin{cases} a & \text{if} \quad a \geq 0 \\ -a & \text{if} \quad a < 0 \end{cases}$$

EXAMPLE 3
(a) $|5| = 5$ since $5 \geq 0$
(b) $|-5| = -(-5) = 5$ since $-5 < 0$
(c) $|0| = 0$ since $0 \geq 0$
(d)   The distance between the origin and the point labeled 3.4 on the real number line is $|3.4| = 3.4$
(e)   The distance between the origin and the point labeled -2.3 on the real number line is $|-2.3| = 2.3$

In working with the notation of absolute value, it is important first to perform the operations within the bars. Here are some examples.

EXAMPLE 4
(a) $|5 - 2| = |3| = 3$
(b) $|2 - 5| = |-3| = 3$
(c) $|3 - 5| - |8 - 6| = |-2| - |2| = 2 - 2 = 0$
(d) $\dfrac{|4 - 7|}{-6} = \dfrac{|-3|}{-6} = \dfrac{3}{-6} = -\dfrac{1}{2}$

It is possible to prove the following properties of absolute value.

> **Properties of Absolute Value**
>
> For all real numbers a and b,
> (i) $|a| \geq 0$
> (ii) $|a| = |-a|$
> (iii) $|a - b| = |b - a|$

Property (i) follows immediately from the definition of absolute value. Property (ii) is illustrated in Example 3a and 3b; it can be established formally by considering two cases. If $a \geq 0$, then $|a| = a$ and $|-a| = -(-a) = a$; if $a < 0$, then $|a| = -a$ and $|-a| = -a$. In either case, $|a| = |-a|$. Property (iii) is illustrated in Example 4a and 4b and we leave the proof as an exercise for the student with this hint: consider separately the cases where $a \geq b$ and where $a < b$.

We began by showing a use for absolute value in denoting distance from the origin without regard to direction. We will conclude by demonstrating the use of absolute value to denote the distance between *any* two points a and b on the real number line. In Figure 3, the distance between the points labeled 2 and 5 is 3 units and can be obtained by evaluating either $|5 - 2|$ or $|2 - 5|$. Similarly, the distance between the points labeled -1 and 4 is given by either $|4 - (-1)| = 5$ or $|-1 - 4| = 5$. Using the notation $\overline{AB}$ to denote the distance between the points A and B, we provide the following definition.

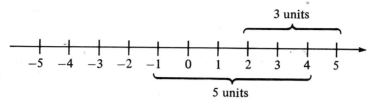

FIGURE 3

The distance $\overline{AB}$ between points A and B on the real number line whose coordinates are a and b, respectively, is given by

$$\overline{AB} = |b - a|$$

Property (iii) then tells us that $\overline{AB} = |b - a| = |a - b|$. Viewed another way, Property (iii) states that the distance between any two points on the real number line is independent of the direction.

EXAMPLE 5
Let points A, B, and C have coordinates -4, -1, and 3, respectively, on the real number line. Find the following distances.

(a) $\overline{AB}$ (b) $\overline{CB}$ (c) $\overline{OB}$

Solution
Using the definition, we have

(a) $\overline{AB} = |-1 - (-4)| = |-1 + 4| = |3| = 3$
(b) $\overline{CB} = |-1 - 3| = |-4| = 4$
(c) $\overline{OB} = |-1 - 0| = |-1| = 1$

PROGRESS CHECK
Let the coordinates of points P, Q, and R be -6, 4, and 6, respectively. Find the following distances.

(a) $\overline{PR}$ (b) $\overline{QP}$ (c) $\overline{PQ}$

Answers
(a) 12   (b) 10   (c) 10

EXERCISE SET 1.2
1. Draw a real number line and plot the following points.

(a) 4   (b) -2 (c) $\dfrac{5}{2}$ (d) -3.5 (e) 0

2. Draw a real number line and plot the following points.

(a) -5 (b) 4   (c) -3.5 (d) $\dfrac{7}{2}$ (e) -4

3.   On the following real number line give the real number associated with the points A, B, C, D, Q, and E.

4.   Represent the following by real numbers.
   (a)   a profit of $10
   (b)   a loss of $20
   (c)   a temperature of 20° above zero
   (d)   a temperature of 5° below zero

In the following exercises indicate which of the two given numbers appears first, viewed from left to right, on the real number line.

5.   4, 6

6. $\frac{1}{2}$, 0

7. -2, $\frac{3}{4}$

8.   0, -4

9. -5, $-\frac{2}{3}$

10.   4, -5

Indicate the following sets of numbers on a real number line.

11.   The natural numbers less than 8.
12.   The natural numbers greater than 4 and less than 10.
13.   The integers that are greater than 2 and less than 7.
14.   The integers that are greater than -5 and less than or equal to 1.

Express each statement as an inequality.

15.   10 is greater than 9.99.
16. -6 is less than -2.
17. a is nonnegative.
18. b is negative.
19. x is positive.
20. a is between 3 and 7.
21. b is less than or equal to -4.

22. a is between $\frac{1}{2}$ and $\frac{1}{4}$

23. b is greater than or equal to 5.
24. x is negative.

State a property of inequalities which justifies each of the following statements.

25.   Since $-3 < 1$, then $-1 < 3$.
26.   Since $-5 < -1$ and $-1 < 4$, then $-5 < 4$.
27.   Since $14 > 9$, then $-14 < -9$.
28.   Since $5 > 3$, then $5 \neq 3$.
29.   Since $-1 < 6$, then $-3 < 18$.
30.   Since $6 > -1$, then 7 is a positive number.

Find the value of each of the following.

31. $|2|$

32. $\left|-\dfrac{2}{3}\right|$

33. $|1.5|$

34. $|-0.8|$

35. $-|2|$

36. $-\left|-\dfrac{2}{5}\right|$

37. $|2-3|$

38. $|2-2|$

39. $|2-(-2)|$

40. $|2|+|-3|$

41. $\dfrac{|14-8|}{|-3|}$

42. $\dfrac{|2-12|}{|1-6|}$

43. $\dfrac{|3|-|2|}{|3|+|2|}$

44. $\dfrac{|3-2|}{|3+2|}$

The coordinates of points A and B are given in each of the following. Find $\overline{AB}$.

45. $2, 5$

46. $-3, 6$

47. $-3, -1$

48. $-4, \dfrac{11}{2}$

49. $-\dfrac{4}{5}, \dfrac{4}{5}$

50. $2, 2$

1.3
ALGEBRAIC EXPRESSIONS; POLYNOMIALS

In Section 1.1 we saw that a rational number is one that can be written as p/q, where p and q are integers (and q is not zero). This was the first time we used symbols which can take on one or more distinct values. For example, when $p = 5$ and $q = 7$ we have the rational number 5/7; when $p = -3$ and $q = 2$ we have the rational number $-3/2$. The symbols p and q are called **variables** since various values can be assigned to them.

If we invest P dollars at an annual interest rate of 6%, then we will earn $0.06P$ dollars interest per year, and we will have $P + 0.06P$ dollars at the end of the year. We call $P + 0.06P$ an **algebraic expression.** Note that an algebraic expression involves **variables** (in this case P), **constants** (such as 0.06), and **algebraic operations** (such as $+, -, \times, \div$). Virtually everything we do in algebra involves algebraic expressions, sometimes as simple as our example and sometimes very involved.

An algebraic expression takes on a **value** when we assign a specific number to each variable in the expression. Thus, the expression

$$\frac{3m + 4n}{m + n}$$

is **evaluated** when $m = 3$ and $n = 2$ by substituting these values for m and n:

$$\frac{3(3) + 4(2)}{3 + 2} = \frac{9 + 8}{5} = \frac{17}{5}$$

We often need to write algebraic expressions in which a variable multiplies itself repeatedly. We use the notation of exponents to indicate such repeated multiplication.

$$a^1 = a$$
$$a^2 = a \cdot a$$
$$\vdots \qquad \vdots$$
$$a^n = \underbrace{a \cdot a \cdot \ldots \cdot a}_{n \text{ factors}}$$

where n is a natural number and a is a real number. We call a the **base** and n the **exponent** and say that a^n is the nth **power** of a. When $n = 1$, we simply write a rather than a^1.

EXAMPLE 1
Write without using exponents.

(a) $\left(\dfrac{1}{2}\right)^3 = \dfrac{1}{2} \cdot \dfrac{1}{2} \cdot \dfrac{1}{2} = \dfrac{1}{8}$

(b) $2x^3 = 2 \cdot x \cdot x \cdot x$

(c) $(2x)^3 = 2x \cdot 2x \cdot 2x = 8 \cdot x \cdot x \cdot x$

(d) $-3x^2y^3 = -3 \cdot x \cdot x \cdot y \cdot y \cdot y$

WARNING ───────────────────────────────

Note the difference between

$$(-3)^2 = (-3)(-3) = 9$$

and

$$-3^2 = -(3 \cdot 3) = -9$$

───────────────────────────────────────

Later in this chapter we shall need an important rule of exponents. Observe that if m and n are natural numbers and a is any real number then

$$a^m \cdot a^n = \underbrace{a \cdot a \cdot \ldots \cdot a}_{m \text{ factors}} \cdot \underbrace{a \cdot a \cdot \ldots \cdot a}_{n \text{ factors}}$$

Since there are a total of $m + n$ factors on the right side, we conclude that

$$a^m a^n = a^{m+n}$$

EXAMPLE 2

(a) $x^2 \cdot x^3 = x^{2+3} = x^5$

(b) $(3x)(4x^4) = 3 \cdot 4 \cdot x \cdot x^4 = 12x^{1+4} = 12x^5$

PROGRESS CHECK
Multiply.

(a) $x^5 \cdot x^2$ (b) $(2x^6)(-2x^4)$

Answers
(a) x^7 (b) $-4x^{10}$

POLYNOMIALS

A polynomial is an algebraic expression of a certain form. Polynomials play an important role in the study of algebra since many word problems translate into equations or inequalities which involve polynomials. We first study the manipulative and mechanical aspects of polynomials; this will serve as background for dealing with their applications in later chapters.

Let x denote a variable and let n be a nonnegative integer. The expression cx^n, where c is a constant real number, is called a **monomial in** x. A **polynomial in** x is an expression that is a sum of monomials. Thus, a polynomial in x is an expression of the form

$$P = c_n x^n + c_{n-1} x^{n-1} + \cdots + c_1 x + c_0 \qquad c_n \neq 0 \qquad (1)$$

Each of the monomials in Equation (1) is called a **term** of P, and $c_0, c_1, \ldots,$ c_n are constant real numbers that are called the **coefficients** of the terms of P. Note that a polynomial may consist of just one term, that is, a monomial is also considered to be a polynomial.

EXAMPLE 3
The following expressions are polynomials in x.

$$3x^4 + 2x + 5, \qquad 2x^3 + 5x^2 - 2x + 1, \qquad \frac{3}{2} x^3$$

Notice that we write $2x^3 + 5x^2 + (-2)x + 1$ as $2x^3 + 5x^2 - 2x + 1$. The following expressions are not polynomials in x. Why not?

$$2x^{1/2} + 5, \qquad 3 - \frac{4}{x}, \qquad \frac{2x - 1}{x - 2}, \qquad x^2 + 2xy + y^2$$

The **degree of a monomial in** x is the exponent of x. Thus, the degree of $5x^3$ is 3. A monomial in which the exponent of x is 0 is called a **constant term** and is said to be of **degree zero**. The coefficient c_n of the term in P with highest degree is called the **leading coefficient** of P and we say that P is a **polynomial of degree** n. A special case is the polynomial all of whose coefficients are zero. Such a polynomial is called the **zero polynomial**, is denoted by 0, and is said to have no degree.

EXAMPLE 4
Given the polynomial

$$P = 2x^4 - 3x^2 + \frac{4}{3} x - 1$$

The terms of P are $2x^4, \quad 0x^3, \quad -3x^2, \quad \frac{4}{3}x, \quad -1.$

The coefficients of the terms are $2, \quad 0, \quad -3, \quad \frac{4}{3}, \quad -1.$

The degree of P is 4 and the leading coefficient is 2.

A **monomial in the variables** x **and** y is an expression of the form $ax^m y^n$, where a is a constant and m and n are nonnegative integers. The number a is called the **coefficient** of the monomial. The **degree of a monomial in** x **and** y is the sum of the exponents of x and y. Thus, the degree of $2x^3 y^2$ is $3 + 2 = 5$. A **polynomial in** x **and** y is an expression which is a sum of monomials. The **degree of a polynomial in** x **and** y is the degree of the monomial with nonzero coefficient that has the highest degree.

EXAMPLE 5
The following are polynomials in x and y.

$2x^2 y + y^2 - 3xy + 1$ Degree is 3.

xy Degree is 2.

$3x^4 + xy - y^2$ Degree is 4.

OPERATIONS WITH POLYNOMIALS

Note that the coefficients of a polynomial are real numbers and that the variable x is assigned real values. It then follows that the properties of the real number system will be applicable in defining addition of polynomials. For example,

$(3x^2 - 2x + 5) + (x^2 + 4x - 9)$

$= (3x^2 + x^2) + (-2x + 4x) + (5 - 9)$ Commutative and associative laws

$= (3 + 1)x^2 + (-2 + 4)x + (5 - 9)$ Distributive law

$= 4x^2 + 2x - 4$

We can restate the rule for adding polynomials by introducing the following terminology. The terms of a polynomial which differ only in their coefficients are called **like terms**. For example, $4x^2$ and $-3x^2$ are like terms; $-5xy^3$ and $-17/2xy^3$ are like terms. It is clear then that we add polynomials by adding the coefficients of like terms.

Subtraction of polynomials is accomplished by proper removal of parentheses followed by addition. To handle

$(3x^2 - xy - 5) - (x^2 - 3xy + 1)$

we first observe that

$-(x^2 - 3xy + 1) = -x^2 + 3xy - 1$ Distributive law

We then see that

$$(3x^2 - xy - 5) - (x^2 - 3xy + 1)$$
$$= 3x^2 - xy - 5 - x^2 + 3xy - 1$$
$$= 3x^2 - x^2 - xy + 3xy - 5 - 1$$
$$= (3 - 1)x^2 + (-1 + 3)xy + (-5 - 1) \qquad \text{Distributive law}$$
$$= 2x^2 + 2xy - 6$$

Thus, the difference of two polynomials is found by subtracting the coefficients of like terms.

WARNING

$$(x + 5) - (x + 2) \neq x + 5 - x + 2.$$

The coefficient -1 must multiply each term in the parentheses. Thus,

$$-(x + 2) = -x - 2$$

and

$$(x + 5) - (x + 2) = x + 5 - x - 2$$
$$= 3$$

EXAMPLE 6

(a) Add $2x^3 + 2x^2 - 3$ and $x^3 - x^2 + x + 2$.

(b) Subtract $2x^3 + x^2 - x + 1$ from $3x^3 - 2x^2 + 2x$.

Solution

(a) $(2x^3 + 2x^2 - 3) + (x^3 - x^2 + x + 2) = 3x^3 + x^2 + x - 1$

(b) $(3x^3 - 2x^2 + 2x) - (2x^3 + x^2 - x + 1) = x^3 - 3x^2 + 3x - 1$

Multiplication of polynomials is based upon the rule for exponents developed earlier in this section,

$$a^m a^n = a^{m+n}$$

and upon the distributive laws

$$a(b + c) = ab + ac$$
$$(a + b)c = ac + bc$$

EXAMPLE 7

Multiply $3x^3(2x^3 - 6x^2 + 5)$.

Solution

$3x^3 (2x^3 - 6x^2 + 5)$

$\qquad = (3x^3) (2x^3) + (3x^3) (-6x^2) + (3x^3) (5) \qquad$ Distributive law

$\qquad = (3) (2)x^{3+3} + (3) (-6)x^{3+2} + (3) (5)x^3 \qquad a^m a^n = a^{m+n}$

$\qquad = 6x^6 - 18x^5 + 15x^3$

EXAMPLE 8

Multiply $(x + 2) (3x^2 - x + 5)$.

Solution

$(x + 2) (3x^2 - x + 5)$

$\qquad = x(3x^2 - x + 5) + 2(3x^2 - x + 5) \qquad$ Distributive law

$\qquad = 3x^3 - x^2 + 5x + 6x^2 - 2x + 10 \qquad$ Distributive law and $a^m a^n = a^{m+n}$

$\qquad = 3x^3 + 5x^2 + 3x + 10$

PROGRESS CHECK

Multiply.

(a) $(x^2 + 2) (x^2 - 3x + 1)$ (b) $(x^2 - 2xy + y) (2x + y)$

Answers

(a) $x^4 - 3x^3 + 3x^2 - 6x + 2$ *(b)* $2x^3 - x^2y + 2xy - 2xy^2 + y^2$

The multiplication in Example 8 can be carried out in "long form" as follows.

$$
\begin{array}{ll}
3x^2 - x + 5 & \\
\underline{x + 2} & \\
3x^3 - x^2 + 5x & = x(3x^2 - x + 5) \\
\underline{6x^2 - 2x + 10} & = 2(3x^2 - x + 5) \\
3x^3 + 5x^2 + 3x + 10 & = \text{sum of above lines}
\end{array}
$$

Products of the form $(2x + 3) (5x - 2)$ or $(2x + y) (3x - 2y)$ occur often and we can handle them mentally as follows:

$$
\begin{array}{c}
10x^2 \quad -6 \\
(2x + 3) (5x - 2) \qquad = 10x^2 + 11x - 6 \\
15x \\
-4x \\
\text{Sum} = 11x
\end{array}
$$

PROGRESS CHECK

(a) Multiply $(2x^2 - xy + y^2) (3x + y)$ in long form.
(b) Multiply $(2x - 3) (3x - 2)$ mentally.

Answers

(a) $6x^3 - x^2y + 2xy^2 + y^3$ *(b)* $6x^2 - 13x + 6$

A number of special products occur often and it is worthwhile knowing them. Three such products are

$$(x + y)^2 = (x + y)(x + y) = x^2 + 2xy + y^2$$
$$(x - y)^2 = (x - y)(x - y) = x^2 - 2xy + y^2$$
$$(x + y)(x - y) = x^2 - y^2$$

EXAMPLE 9
Multiply mentally.

(a) $(x + 2)^2 = (x + 2)(x + 2) = x^2 + 4x + 4$

(b) $(x - 3)^2 = (x - 3)(x - 3) = x^2 - 6x + 9$

(c) $(x + 4)(x - 4) = x^2 - 16$

EXERCISE SET 1.3

Evaluate the given expression in Exercises 1–8 when $r = 2$, $s = 3$, and $t = 4$.

1. $r + 2s + t$

2. rst

3. $\dfrac{rst}{r + s + t}$

4. $(r + s)t$

5. $\dfrac{r + s}{rt}$

6. $\dfrac{r + s + t}{t}$

7. Evaluate $2\pi r$ when $r = 3$ (recall that π is approximately 3.14).

8. Evaluate $\dfrac{9}{5}C + 32$ when $C = 37$.

9. If P dollars are invested at a simple interest rate of r percent per year for t years, the amount on hand at the end of t years is $P + Prt$. Suppose \$2000 is invested at 8% per year ($r = 0.08$). How much money is on hand after
    (a) one year?   (b) half a year?   (c) 8 months?

10. The perimeter of a rectangle is given by the formula $P = 2(L + W)$, where L is the length and W is the width of the rectangle. Find the perimeter if

    (a) $L = 2$ feet, $W = 3$ feet   (b) $L = \dfrac{1}{2}$ meter, $W = \dfrac{1}{4}$ meter

11. Evaluate $0.02r + 0.314st + 2.25t$ when $r = 2.5$, $s = 3.4$, and $t = 2.81$.

12. Evaluate $10.421x + 0.821y + 2.34xyz$ when $x = 3.21$, $y = 2.42$, and $z = 1.23$.

Evaluate the given expression in Exercises 13–18.

13. $|x| - |x \cdot y|$ when $x = -3$, $y = 4$

14. $|x + y| + |x - y|$ when $x = -3$, $y = 2$

15. $\dfrac{|a - 2b|}{2a}$ when $a = 1$, $b = 2$

16. $\dfrac{|x| + |y|}{|x| - |y|}$ when $x = -3$, $y = 4$

17. $\dfrac{-|a - 2b|}{|a + b|}$ when $a = -2$, $b = -1$

18. $\dfrac{|a - b| - 2|c - a|}{|a - b + c|}$ when $a = -2$, $b = 3$, $c = -5$

Carry out the indicated operations in Exercises 19–24.

19. $b^5 \cdot b^2$

20. $x^3 \cdot x^5$

21. $(4y^3)(-5y^6)$

22. $(-6x^4)(-4x^7)$

23. $\left(\dfrac{3}{2} x^3\right)(-2x)$

24. $\left(-\dfrac{5}{3} x^6\right)\left(-\dfrac{3}{10} x^3\right)$

25. Which of the following expressions are *not* polynomials?

    (a) $-3x^2 + 2x + 5$

    (b) $-3x^2 y$

    (c) $-3x^{2/3} + 2xy + 5$

    (d) $-2x^{-4} + 2xy^3 + 5$

26. Which of the following expressions are not polynomials?

    (a) $4x^5 - x^{1/2} + 6$

    (b) $\dfrac{2}{5} x^3 + \dfrac{4}{3} x - 2$

    (c) $4x^5 y$

    (d) $x^{4/3} y + 2x - 3$

Indicate the leading coefficient and the degree of each given polynomial in Exercises 27–30.

27. $2x^3 + 3x^2 - 5$

28. $-4x^5 - 8x^2 + x + 3$

29. $\dfrac{3}{5} x^4 + 2x^2 - x - 1$

30. $0.75x^7 + 7x^3 - 1.5$

Find the degree of each given polynomial in Exercises 31–34.

31. $3x^2 y - 4x^2 - 2y + 4$

32. $4xy^3 + xy^2 - y^2 + y$

33. $2xy^3 - y^3 + 3x^2 - 2$

34. $\dfrac{1}{2} x^3 y^3 - 2$

35. Find the value of the polynomial $3x^2 y^2 + 2xy - x + 2y + 7$ when $x = 2$ and $y = -1$.

36. Find the value of the polynomial $0.02x^2 + 0.3x - 0.5$ when $x = 0.3$.

37. Find the value of the polynomial $2.1x^3 + 3.3x^2 - 4.1x - 7.2$ when $x = 4.1$.

38. Write a polynomial giving the area of a circle of radius r.

39. Write a polynomial giving the area of a triangle of base b and height h.

40. A field consists of a rectangle and a square arranged as shown in Figure 4.

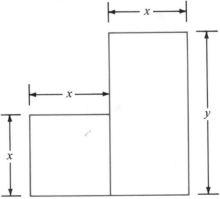

FIGURE 4

What does each of the following polynomials represent?
(a) $x^2 + xy$ (b) $2x + 2y$ (c) $4x$ (d) $4x + 2y$

41. An investor buys x shares of G.E. stock at \$55 per share, y shares of Exxon stock at \$45 per share, and z shares of A.T.&T. stock at \$60 per share. What does the polynomial $55x + 45y + 60z$ represent?

Perform the indicated operations in Exercises 42–60.

42. $(4x^2 + 3x + 2) + (3x^2 - 2x - 5)$
43. $(2x^2 + 3x + 8) - (5 - 2x + 2x^2)$
44. $4xy^2 + 2xy + 2x + 3 - (-2xy^2 + xy - y + 2)$
45. $(2s^2t^3 - st^2 + st - s + t) - (3s^2t^2 - 2s^2t - 4st^2 - t + 3)$
46. $3xy^2z - 4x^2yz + xy + 3 - (2xy^2z + x^2yz - yz + x - 2)$
47. $a^2bc + ab^2c + 2ab^3 - 3a^2bc - 4ab^3 + 3$
48. $(x + 1)(x^2 + 2x - 3)$ 49. $(2 - x)(2x^3 + x - 2)$
50. $(2s - 3)(s^3 - s + 2)$ 51. $(-3s + 2)(-2s^2 - s + 3)$
52. $(x^2 + 3)(2x^2 - x + 2)$ 53. $(2y^2 + y)(-2y^3 + y - 3)$
54. $(x^2 + 2x - 1)(2x^2 - 3x + 2)$ 55. $(a^2 - 4a + 3)(4a^3 + 2a + 5)$
56. $(2a^2 + ab + b^2)(3a - b^2 + 1)$ 57. $(-3a + ab + b^2)(3b^2 + 2b + 2)$
58. $5(2x - 3)^2$ 59. $2(3x - 2)(3 - x)$
60. $(x - 1)(x + 2)(x + 3)$

61. An investor buys x shares of IBM stock at \$260 per share at Thursday's opening of the stock market. Later in the day, he sells y shares of G&W stock at \$13 per share and z shares of Holiday Inn stock at \$17 per share. Write a polynomial that expresses the money transactions for the day.

62. An artist takes a rectangular piece of cardboard whose sides are x and y and cuts out a square of side $x/2$ (Figure 5) to obtain a mat for a painting. Write a polynomial giving the area of the mat.

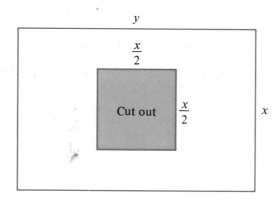

FIGURE 5

In Exercises 63–76, perform the multiplication mentally.

63. $(x - 1)(x + 3)$ 64. $(x + 2)(x + 3)$
65. $(2x + 1)(2x + 3)$ 66. $(3x - 1)(x + 5)$
67. $(3x - 2)(x - 1)$ 68. $(x + 4)(2x - 1)$
69. $(x + y)^2$ 70. $(x - 4)^2$
71. $(3x - 1)^2$ 72. $(x + 2)(x - 2)$
73. $(2x + 1)(2x - 1)$ 74. $(3a + 2b)^2$
75. $(x^2 + y^2)^2$ 76. $(x - y)^2$

1.4
FACTORING

Now that we can find the product of two polynomials, let's consider the reverse problem: Given a polynomial, can we find factors whose product will yield the given polynomial? This process is known as **factoring**. Factoring is one of the very basic tools of algebra. We will approach factoring by learning to recognize the situations in which factoring is possible.

Consider the polynomial

$$x^2 + x$$

Since the factor x is common to each term, we can write

$$x^2 + x = x(x + 1)$$

EXAMPLE 1
Factor.

(a) $15x^3 - 10x^2$

Both 5 and x^2 are common to *both* terms. Therefore

$$15x^3 - 10x^2 = 5x^2(3x - 2)$$

(b) $4x^2y - 8xy^2 + 6xy$

Here we see that 2, x, and y are common to each term. Therefore

$$4x^2y - 8xy^2 + 6xy = 2xy(2x - 4y + 3)$$

(c) $2ab + b + 2ac + c$

We group those terms containing b and those terms containing c.

$$
\begin{aligned}
2ab + b + 2ac + c &= (2ab + b) + (2ac + c) &&\text{Grouping} \\
&= b(2a + 1) + c(2a + 1) &&\text{Common factors } b, c \\
&= (2a + 1)(b + c) &&\text{Common factor } 2a + 1
\end{aligned}
$$

PROGRESS CHECK
Factor.

(a) $4x^2 - x$ (b) $3x^4 - 9x^2$

(c) $5x^2y - 10xy^2 + 5x^2y^2$ (d) $2ab - 6b + ac - 3c$

Answers
(a) $x(4x - 1)$ (b) $3x^2(x^2 - 3)$

(c) $5xy(x - 2y + xy)$ (d) $(a - 3)(2b + c)$

In this chapter we establish the following conventions: A polynomial with *integer* coefficients is to be factored as a product of polynomials of lower degree with *integer* coefficients; a polynomial with *rational* coefficients is to be factored as a product of polynomials of lower degree with *rational* coefficients. Thus, consider the polynomial

$$x^2 + 5x + 6$$

We see that the term x^2 can have come only from $x \cdot x$, so we can write two incomplete factors like this.

$$x^2 + 5x + 6 = (x \qquad) (x \qquad)$$

The constant term $+6$ can be the product of either two positive numbers or two negative numbers. Since the middle term $+5x$ is the sum of two other terms, both signs must be positive. Thus,

$$x^2 + 5x + 6 = (x + \qquad) (x + \qquad)$$

Finally, the number 6 can be written as the product of two integers in only two ways: $1 \cdot 6$ or $2 \cdot 3$. The first pair gives a middle term of $7x$. The second pair gives the middle term, $5x$.

$$x^2 + 5x + 6 = (x + 2) (x + 3)$$

EXAMPLE 2
Factor.

(a) $x^2 - 7x + 10$ (b) $x^2 - 3x - 4$

Solution

(a)   Since the constant term is positive and the middle term is negative, we must have two negative signs. The integer factors of 10 are 1 and 10, and 2 and 5. We find that

$$x^2 - 7x + 10 = (x - 2) (x - 5)$$

(b)   Since the constant term is negative, we must have opposite signs. The integer factors of 4 are 1 and 4, and 2 and 2. We find that

$$x^2 - 3x - 4 = (x + 1) (x - 4)$$

Consider now a polynomial of the form $ax^2 + bx + c$, where a, b, and c are integers, with $a \neq 1$. If we write

$$ax^2 + bx + c = (rx + u) (sx + v)$$

where r, s, u, and v are integers, then we must have $rs = a$, $uv = c$, and $rv + su = b$. These three equations give candidates for r, s, u, and v. The final choices from among the candidates are determined by trial and error, which is made easier by using mental multiplication.

EXAMPLE 3
Factor $2x^2 - x - 6$.

Solution

The term $2x^2$ can result only from the factors $2x$ and x. We can write the constant term -6 as a product of factors as follows.

$$(-1) (6) \qquad (1) (-6) \qquad (6) (-1) \qquad (-6) (1)$$
$$(-2) (3) \qquad (2) (-3) \qquad (3) (-2) \qquad (-3) (2)$$

Using mental multiplication we find that the correct middle term is obtained by using the factors (3) and (-2). Thus

$$2x^2 - x - 6 = (2x + 3) (x - 2)$$

PROGRESS CHECK
Factor.

(a) $3x^2 - 16x + 21$ (b) $2x^2 + 3x - 9$

Answers
(a) $(3x - 7)(x - 3)$ (b) $(2x - 3)(x + 3)$

There is a special case of the second-degree polynomial which occurs frequently and factors easily. Consider the polynomial $x^2 - 9$. Factoring gives

$$x^2 - 9 = (x - 3)(x + 3)$$

In fact, the difference of two perfect squares always factors immediately.

$$a^2 - b^2 = (a + b)(a - b)$$

EXAMPLE 4
Factor.

(a) $4x^2 - 25$ (b) $9x^2 - 16y^2$

Solution

(a) $4x^2 - 25 = (2x + 5)(2x - 5)$
(b) $9x^2 - 16y^2 = (3x + 4y)(3x - 4y)$

PROGRESS CHECK
Factor.

(a) $x^2 - 49$ (b) $16x^2 - 9$ (c) $25x^2 - y^2$

Answers
(a) $(x + 7)(x - 7)$ (b) $(4x + 3)(4x - 3)$ (c) $(5x + y)(5x - y)$

A good rule to follow in factoring is the following.

Always remove common factors before attempting any other factoring techniques.

EXAMPLE 5
Factor.

(a) $x^3 - 6x^2 + 8x$

Following our rule, we first remove the common factor.

$$x^3 - 6x^2 + 8x = x(x^2 - 6x + 8)$$
$$= x(x - 2)(x - 4)$$

(b) $2x^3 - 18xy^2$

Removing the common factor of $2x$,

$$2x^3 - 18xy^2 = 2x(x^2 - 9y^2)$$

We then recognize that $x^2 - 9y^2$ has the form $a^2 - b^2$ so that

$$2x(x^2 - 9y^2) = 2x(x + 3y)(x - 3y)$$

PROGRESS CHECK
Factor.

(a) $x^3 + 5x^2 - 6x$ (b) $5x^2 - 20$ (c) $2x^3 - 2x^2y - 4xy^2$

Answers
(a) $x(x + 6)(x - 1)$ (b) $5(x - 2)(x + 2)$ (c) $2x(x + y)(x - 2y)$

Are there polynomials that cannot be written as a product of polynomials of lower degree with integer coefficients? The answer is yes. An example is the polynomial $x^2 + x + 1$. A polynomial is said to be **prime** or **irreducible** if it cannot be written as a product of two polynomials each of positive degree. Thus, $x^2 + x + 1$ is irreducible.

WARNING ——————————————————————————————————

The polynomial $x^2 - 6x$ can be written as

$$x^2 - 6x = x(x - 6)$$

and is therefore a product of two polynomials of positive degree. Students often fail to consider the polynomial x to be a "true" factor.

EXERCISE SET 1.4
Factor completely.

1. $5x - 15$

2. $\frac{1}{2}x + \frac{1}{4}y$

3. $-2x - 8y$

4. $3x - 6y + 15$

5. $5bc + 25b$

6. $2x^4 + x^2$

7. $-3y^2 - 4y^5$

8. $3abc + 12\,bc$

9. $3x^2 + 6x^2y - 9x^2z$

10. $9a^3b^3 + 12a^2b - 15ab^2$

11. $x^2 + 4x + 3$

12. $x^2 + 2x - 8$

13. $y^2 - 8y + 15$

14. $y^2 + 7y - 8$

15. $a^2 - 7ab + 12b^2$

16. $x^2 - 49$

17. $y^2 - \frac{1}{9}$

18. $a^2 - 7a + 10$

19. $9 - x^2$

20. $4b^2 - a^2$

21. $x^2 - 5x - 14$

22. $x^2y^2 - 9$

23. $\frac{1}{16} - y^2$

24. $4a^2 - b^2$

25. $x^2 - 6x + 9$

26. $a^2b^2 - \frac{1}{9}$

27. $x^2 - 12x + 20$

28. $x^2 - 8x - 20$

29. $x^2 + 11x + 24$

30. $y^2 - \frac{9}{16}$

31. $2x^2 - 3x - 2$

32. $2x^2 + 7x + 6$

33. $3a^2 - 11a + 6$

34. $4x^2 - 9x + 2$

35. $6x^2 + 13x + 6$

36. $4y^2 - 9$

37. $8m^2 - 6m - 9$
38. $9x^2 + 24x + 16$
39. $10x^2 - 13x - 3$
40. $9y^2 - 16x^2$
41. $6a^2 - 5ab - 6b^2$
42. $4x^2 + 20x + 25$
43. $10r^2s^2 + 9rst + 2t^2$
44. $x^{12} - 1$
45. $16 - 9x^2y^2$
46. $6 + 5x - 4x^2$
47. $8n^2 - 18n - 5$
48. $15 + 4x - 4x^2$
49. $2x^2 - 2x - 12$
50. $3y^2 + 6y - 45$
51. $30x^2 - 35x + 10$
52. $x^4y^4 - x^2y^2$
53. $18x^2m + 33xm + 9m$
54. $25m^2n^3 - 5m^2n$
55. $12x^2 - 22x^3 - 20x^4$
56. $10r^2 - 5rs - 15s^2$
57. $x^4 - y^4$
58. $a^4 - 16$
59. $b^4 + 2b^2 - 8$
60. $4b^4 + 20b^2 + 25$
61. $6b^4 + 7b^2 - 3$
62. $4(x + 1)(y + 2) - 8(y + 2)$
63. $2(x + 1)(x - 1) + 5(x - 1)$
64. $3(x + 2)^2 (x - 1) - 4(x + 2)^2 (2x + 7)$
65. $4(2x - 1)^2 (x + 2)^3 (x + 1) - 3(2x - 1)^5 (x + 2)^2 (x + 3)$

1.5
RATIONAL EXPRESSIONS

Much of the terminology and many of the techniques of arithmetic fractions carry over to **algebraic fractions**, which are the quotients of algebraic expressions. In particular, we refer to a quotient of two polynomials as a **rational expression**. Our objective in this section is to review the procedure for adding, subtracting, multiplying, and dividing rational expressions. We will then be able to simplify a complicated fraction such as

$$\frac{1 - \dfrac{1}{x}}{\dfrac{1}{x^2} + \dfrac{1}{x}}$$

into a form that will ease evaluation of the fraction or facilitate other operations with it.

MULTIPLICATION AND DIVISION OF RATIONAL EXPRESSIONS

The symbols appearing in rational expressions represent real numbers. We may, therefore, apply the rules of arithmetic to rational expressions.

$$\frac{a}{b} \cdot \frac{c}{d} = \frac{ac}{bd} \qquad \text{Multiplication of rational expressions}$$

$$\frac{\dfrac{a}{b}}{\dfrac{c}{d}} = \frac{a}{b} \cdot \frac{d}{c} = \frac{ad}{bc} \qquad \text{Division of rational expressions}$$

EXAMPLE 1

Divide $\dfrac{2x}{y}$ by $\dfrac{3y^3}{x-3}$.

Solution

$$\frac{\dfrac{2x}{y}}{\dfrac{3y^3}{x-3}} = \frac{2x}{y} \cdot \frac{x-3}{3y^3} = \frac{2x(x-3)}{3y^4}$$

The basic rule which allows us to simplify rational expressions is the cancellation principle.

Cancellation Principle

$$\frac{ab}{ac} = \frac{b}{c}, \qquad a \neq 0$$

This rule results from the fact that $\dfrac{a}{a} = 1$. Thus,

$$\frac{ab}{ac} = \frac{a}{a} \cdot \frac{b}{c} = 1 \cdot \frac{b}{c} = \frac{b}{c}$$

Once again we find that a rule for arithmetic of fractions carries over to rational expressions.

EXAMPLE 2
Simplify.

(a) $\dfrac{x^2-4}{x^2+5x+6} = \dfrac{(x+2)(x-2)}{(x+3)(x+2)} = \dfrac{x-2}{x+3}$, $\qquad x \neq -2$

(b) $\dfrac{\dfrac{3x^2(y-1)}{y+1}}{\dfrac{6x(y-1)^2}{(y+1)^3}} = \dfrac{3x^2(y-1)}{y+1} \cdot \dfrac{(y+1)^3}{6x(y-1)^2}$

$$= \frac{3x^2(y-1)(y+1)^3}{6x(y-1)^2(y+1)}$$

$$= \frac{x(y+1)^2}{2(y-1)}, \qquad y \neq 1, -1$$

(c) $\dfrac{x^2-x-6}{3x-x^2} = \dfrac{(x-3)(x+2)}{x(3-x)} = \dfrac{(x-3)(x+2)}{-x(x-3)}$

$$= \frac{x+2}{-x} = -\frac{x+2}{x}, \qquad x \neq 3$$

Note that we wrote $(3 - x)$ as $-(x - 3)$. This technique is often used to recognize factors that may be canceled.

PROGRESS CHECK

Simplify.

(a) $\dfrac{4 - x^2}{x^2 - x - 6}$ (b) $\dfrac{8 - 2x}{y} \div \dfrac{x^2 - 16}{y}$

Answers

(a) $\dfrac{2 - x}{x - 3}$ (b) $-\dfrac{2}{x + 4}$

WARNING

(a) Only multiplicative factors can be cancelled. Thus,

$$\frac{2x - 4}{x} \neq 2 - 4$$

Since x is *not a multiplicative factor* in the numerator, we may *not* perform cancellation.

(b) Note that

$$\frac{y^2 - x^2}{y - x} \neq y - x$$

To simplify correctly, write

$$\frac{y^2 - x^2}{y - x} = \frac{(y + x)(y - x)}{y - x} = y + x$$

ADDITION AND SUBTRACTION OF RATIONAL EXPRESSIONS

Since the variables in rational expressions represent real numbers, the rules of arithmetic for addition and subtraction of fractions apply to rational expressions. When rational expressions have the same denominator, the addition and subtraction rules are

$$\frac{a}{c} + \frac{b}{c} = \frac{a + b}{c}$$

$$\frac{a}{c} - \frac{b}{c} = \frac{a - b}{c}$$

For example,

$$\frac{2}{x - 1} - \frac{4}{x - 1} + \frac{5}{x - 1} = \frac{2 - 4 + 5}{x - 1} = \frac{3}{x - 1}$$

To add or subtract rational expressions with *different* denominators, we must first rewrite each rational expression as an equivalent one with the same denominator. Although any common denominator will do, we will concentrate on finding the **least common denominator** or **L.C.D.** of two or more rational expressions. We now outline the procedure and provide an example.

EXAMPLE 3

Find the L.C.D. of

$$\frac{1}{x^3 - x^2} \qquad \frac{-2}{x^3 - x} \qquad \frac{3x}{x^2 + 2x + 1}$$

Solution

Least Common Denominator	
Step 1. Factor the denominator of each rational expression.	*Step 1.* $\dfrac{1}{x^2(x - 1)} \quad \dfrac{-2}{x(x - 1)(x + 1)} \quad \dfrac{3x}{(x + 1)^2}$
Step 2. Determine the different factors in the denominators of the rational expressions, and the highest power to which each factor occurs in any denominator.	*Step 2.* factor $\quad$ highest exponent $\quad$ final factors $x \qquad\qquad 2 \qquad\qquad x^2$ $x - 1 \qquad\quad 1 \qquad\qquad x - 1$ $x + 1 \qquad\quad 2 \qquad\quad (x + 1)^2$
Step 3. The product of the factors determined in *Step 2* is the L.C.D.	*Step 3.* The L.C.D. is $x^2(x - 1)(x + 1)^2$

PROGRESS CHECK

Find the L.C.D. of the fractions

$$\frac{2a}{(3a^2 + 12a + 12)b} \qquad \frac{-7b}{a(4b^2 - 8b + 4)} \qquad \frac{3}{ab^3 + 2b^3}$$

Answer
$12ab^3(a + 2)^2(b - 1)^2$

The fractions 2/5 and 6/15 are said to be **equivalent** since we obtain 6/15 by multiplying 2/5 by $1 = 3/3$. We also say that algebraic fractions are **equivalent fractions** if we can obtain one from the other by multiplying both the numerator and denominator by the same expression.

To add rational expressions, we must first determine the L.C.D. and then convert each rational expression into an equivalent fraction with the L.C.D. as its denominator. By writing 1 in a properly chosen way, we can *force* each fraction to have the L.C.D. as its denominator. We now outline the procedure and provide an example.

EXAMPLE 4

Simplify

$$\frac{x + 1}{2x^2} - \frac{2}{3x(x + 2)}$$

Solution

Addition of Rational Expressions	
Step 1. Find the L.C.D.	*Step 1.* $$\text{L.C.D.} = 6x^2(x + 2)$$
Step 2. Multiply each rational expression by a fraction whose numerator and denominator are the same and consist of all factors of the L.C.D. which are missing in the denominator of the expression.	*Step 2.* $$\frac{x + 1}{2x^2} \cdot \frac{3(x + 2)}{3(x + 2)} = \frac{3x^2 + 9x + 6}{6x^2(x + 2)}$$ $$\frac{2}{3x(x + 2)} \cdot \frac{2x}{2x} = \frac{4x}{6x^2(x + 2)}$$
Step 3. Add the rational expressions. Do not multiply out the denominators since it may be possible to cancel.	*Step 3.* $$\frac{x + 1}{2x^2} - \frac{2}{3x(x + 2)}$$ $$= \frac{3x^2 + 9x + 6}{6x^2(x + 2)} - \frac{4x}{6x^2(x + 2)}$$ $$= \frac{3x^2 + 5x + 6}{6x^2(x + 2)}$$

PROGRESS CHECK

Find the sum.

(a) $\dfrac{2x}{x^2 - 4} + \dfrac{1}{x(x - 2)} - \dfrac{1}{x - 2}$

(b) $\dfrac{4r - 3}{9r^3} - \dfrac{2r + 1}{4r^2} + \dfrac{2}{3r}$

Answers

(a) $\dfrac{x + 1}{x(x + 2)}$ (b) $\dfrac{6r^2 + 7r - 12}{36r^3}$

COMPLEX FRACTIONS

At the beginning of this section we said that we wanted to simplify fractions such as

$$\frac{1 - \dfrac{1}{x}}{\dfrac{1}{x^2} + \dfrac{1}{x}}$$

This is an example of a **complex fraction**, which is a fractional form with fractions in the numerator or denominator or both.

There are two methods commonly used to simplify complex fractions. Fortunately, we already have all the tools needed and will apply both methods to the problem.

EXAMPLE 5
Simplify

$$\frac{1 - \dfrac{1}{x}}{\dfrac{1}{x^2} + \dfrac{1}{x}}$$

Solution

Simplifying Complex Fractions	
Method 1	**Example**
Step 1. Find the L.C.D. of all fractions appearing in the numerator and denominator.	*Step 1.* The L.C.D. of $\dfrac{1}{1}, \dfrac{1}{x}$, and $\dfrac{1}{x^2}$ is x^2.
Step 2. Multiply the numerator and denominator by the L.C.D. Since this is multiplication by 1, the result is an equivalent fraction.	*Step 2.* $$\frac{x^2\left(1 - \dfrac{1}{x}\right)}{x^2\left(\dfrac{1}{x^2} + \dfrac{1}{x}\right)} = \frac{x^2 - x}{1 + x} = \frac{x(x - 1)}{x + 1}$$
Method 2	**Example**
Step 1. Combine the terms in the numerator into a single rational expression.	*Step 1.* $$1 - \frac{1}{x} = \frac{x}{x} - \frac{1}{x} = \frac{x - 1}{x}$$
Step 2. Combine the terms in the denominator into a single rational expression.	*Step 2.* $$\frac{1}{x^2} + \frac{1}{x} = \frac{1}{x^2} + \frac{x}{x^2} = \frac{1 + x}{x^2}$$
Step 3. Apply the rules for division of rational expressions, that is, multiply the numerator by the reciprocal of the denominator.	*Step 3.* $$\frac{\dfrac{x - 1}{x}}{\dfrac{1 + x}{x^2}} = \frac{x - 1}{x} \cdot \frac{x^2}{1 + x} = \frac{x(x - 1)}{x + 1}$$

PROGRESS CHECK
Simplify.

(a) $\dfrac{2 + \dfrac{1}{x}}{1 - \dfrac{2}{x}}$ (b) $\dfrac{\dfrac{a}{b} + \dfrac{b}{a}}{\dfrac{1}{a} - \dfrac{1}{b}}$

Answers

(a) $\dfrac{2x + 1}{x - 2}$ (b) $-\dfrac{a^2 + b^2}{a - b}$

EXERCISE SET 1.5

Simplify each of the following.

1. $\dfrac{x + 4}{x^2 - 16}$

2. $\dfrac{y^2 - 25}{y + 5}$

3. $\dfrac{x^2 - 8x + 16}{x - 4}$

4. $\dfrac{5x^2 - 45}{2x - 6}$

5. $\dfrac{6x^2 - x - 1}{2x^2 + 3x - 2}$

6. $\dfrac{2x^3 + x^2 - 3x}{3x^2 - 5x + 2}$

7. $\dfrac{2}{3x - 6} \div \dfrac{3}{2x - 4}$

8. $\dfrac{5x + 15}{8} \div \dfrac{3x + 9}{4}$

9. $\dfrac{25 - a^2}{b + 3} \cdot \dfrac{2b^2 + 6b}{a - 5}$

10. $\dfrac{2xy^2}{x + y} \cdot \dfrac{x + y}{4xy}$

11. $\dfrac{x + 2}{3y} \div \dfrac{x^2 - 2x - 8}{15y^2}$

12. $\dfrac{3x}{x + 2} \div \dfrac{6x^2}{x^2 - x - 6}$

13. $\dfrac{6x^2 - x - 2}{2x^2 - 5x + 3} \cdot \dfrac{2x^2 - 7x + 6}{3x^2 + x - 2}$

14. $\dfrac{6x^2 + 11x - 2}{4x^2 - 3x - 1} \cdot \dfrac{5x^2 - 3x - 2}{3x^2 + 7x + 2}$

15. $(x^2 - 4) \cdot \dfrac{2x + 3}{x^2 + 2x - 8}$

16. $(a^2 - 2a) \cdot \dfrac{a + 1}{6 - a - a^2}$

17. $(x^2 - 2x - 15) \div \dfrac{x^2 - 7x + 10}{x^2 + 1}$

18. $\dfrac{2y^2 - 5y - 3}{y - 4} \div (y^2 + y - 12)$

19. $\dfrac{x^2 - 4}{x^2 + 2x - 3} \cdot \dfrac{x^2 + 3x - 4}{x^2 - 7x + 10} \cdot \dfrac{x + 3}{x^2 + 3x + 2}$

20. $\dfrac{x^2 - 9}{6x^2 + x - 1} \cdot \dfrac{2x^2 + 5x + 2}{x^2 + 4x + 3} \cdot \dfrac{x^2 - x - 2}{x^2 - 3x}$

Find the L.C.D. in each of the following.

21. $\dfrac{4}{x}$, $\dfrac{x - 2}{y}$

22. $\dfrac{x}{x - 1}$, $\dfrac{x + 4}{x + 2}$

23. $\dfrac{5 - a}{a}$, $\dfrac{7}{2a}$

24. $\dfrac{x + 2}{x}$, $\dfrac{x - 2}{x^2}$

25. $\dfrac{2b}{b - 1}$, $\dfrac{3}{(b - 1)^2}$

26. $\dfrac{2 + x}{x^2 - 4}$, $\dfrac{3}{x - 2}$

27. $\dfrac{4x}{x - 2}$, $\dfrac{5}{x^2 + x - 6}$

28. $\dfrac{3}{y^2 - 3y - 4}$, $\dfrac{2y}{y + 1}$

29. $\dfrac{3}{x + 1}$, $\dfrac{2}{x}$, $\dfrac{x}{x - 1}$

30. $\dfrac{4}{x}$, $\dfrac{3}{x - 1}$, $\dfrac{x}{x^2 - 2x + 1}$

Perform the indicated operations and simplify in each of the following.

31. $\dfrac{8}{a - 2} + \dfrac{4}{2 - a}$

32. $\dfrac{x}{x^2 - 4} + \dfrac{2}{4 - x^2}$

33. $\dfrac{x - 1}{3} + 2$

34. $\dfrac{1}{x - 1} + \dfrac{2}{x - 2}$

35. $\dfrac{1}{a + 2} + \dfrac{3}{a - 2}$

36. $\dfrac{a}{8b} - \dfrac{b}{12a}$

37. $\dfrac{4}{3x} - \dfrac{5}{xy}$

38. $\dfrac{4x - 1}{6x^3} + \dfrac{2}{3x^2}$

39. $\dfrac{5}{2x + 6} - \dfrac{x}{x + 3}$

40. $\dfrac{x}{x - y} - \dfrac{y}{x + y}$

41. $\dfrac{5x}{2x^2 - 18} + \dfrac{4}{3x - 9}$

42. $\dfrac{4}{r} - \dfrac{3}{r + 2}$

43. $\dfrac{1}{x - 1} + \dfrac{2x - 1}{(x - 2)(x + 1)}$

44. $\dfrac{2x}{2x + 1} - \dfrac{x - 1}{(2x + 1)(x - 2)}$

45. $\dfrac{2x}{x^2 + x - 2} + \dfrac{3}{x + 2}$

46. $\dfrac{2}{x - 2} + \dfrac{x}{x^2 - x - 6}$

47. $\dfrac{2x - 1}{x^2 + 5x + 6} - \dfrac{x - 2}{x^2 + 4x + 3}$

48. $\dfrac{2x - 1}{x^3 - 4x} - \dfrac{x}{x^2 + x - 2}$

49. $\dfrac{2x}{x^2 - 1} + \dfrac{x + 1}{x^2 + 3x - 4}$

50. $\dfrac{2x}{x + 2} + \dfrac{x}{x - 2} - \dfrac{1}{x^2 - 4}$

Simplify in each of the following exercises.

51. $\dfrac{1 + \dfrac{2}{x}}{1 - \dfrac{3}{x}}$

52. $\dfrac{x - \dfrac{1}{x}}{2 + \dfrac{1}{x}}$

53. $\dfrac{x + 1}{1 - \dfrac{1}{x}}$

54. $\dfrac{1 - \dfrac{r^2}{s^2}}{1 + \dfrac{r}{s}}$

55. $\dfrac{x^2 - 16}{\dfrac{1}{4} - \dfrac{1}{x}}$

56. $\dfrac{\dfrac{a}{a - b} - \dfrac{b}{a + b}}{a^2 - b^2}$

57. $2 - \dfrac{1}{1 + \dfrac{1}{a}}$

58. $\dfrac{\dfrac{4}{x^2 - 4} + 1}{\dfrac{x}{x^2 + x - 6}}$

59. $\dfrac{\dfrac{a}{b} - \dfrac{b}{a}}{\dfrac{1}{a} + \dfrac{1}{b}}$

60. $\dfrac{\dfrac{x}{x - 2} - \dfrac{x}{x + 2}}{\dfrac{2x}{x - 2} + \dfrac{x^2}{x - 2}}$

61. $3 - \dfrac{2}{1 - \dfrac{1}{1 + x}}$

62. $2 + \dfrac{3}{1 + \dfrac{2}{1 - x}}$

63. $\dfrac{y - \dfrac{1}{1 - \dfrac{1}{y}}}{y + \dfrac{1}{1 + \dfrac{1}{y}}}$

64. $1 - \dfrac{1 - \dfrac{1}{y}}{y - \dfrac{1}{y}}$

65. $1 - \dfrac{1}{1 + \dfrac{1}{1 - \dfrac{1}{1 + x}}}$

66. $1 + \dfrac{1}{1 - \dfrac{1}{1 + \dfrac{1}{1 + x}}}$

1.6
INTEGER EXPONENTS
POSITIVE INTEGER EXPONENTS

In this subsection, all exponents represent positive integers. We saw in Section 1.3 that

$$a^m a^n = a^{m+n}$$

and we will now develop additional rules for positive integer exponents. To simplify $(a^m)^n$, we note that

$$(a^m)^n = \underbrace{a^m \cdot a^m \cdot \ldots \cdot a^m}_{n \text{ factors}} = a^{m+m+\ldots+m} = a^{mn}$$

or

$$(a^m)^n = a^{mn}$$

In the same way, the definition of exponents leads to the rules

$$(ab)^m = a^m b^m$$
$$\left(\frac{a}{b}\right)^m = \frac{a^m}{b^m}, \quad b \neq 0$$

Finally, we turn to a^m/a^n, $a \neq 0$. If $m > n$, then $m - n$ is a positive integer. By writing m in the form $m = m - n + n$, we have

$$\frac{a^m}{a^n} = \frac{a^{m-n+n}}{a^n} = \frac{a^{m-n}a^n}{a^n} = a^{m-n}$$

In the last step we used the cancellation principle. If $n > m$, we have

$$\frac{a^m}{a^n} = \frac{1}{a^{n-m}}$$

and if $m = n$, then

$$\frac{a^m}{a^n} = \frac{a^m}{a^m} = 1$$

Table 2 summarizes the rules for positive integer exponents and provides illustrations of their use.

TABLE 2   **Positive Integer Exponents**

Rule	Examples
$a^m a^n = a^{m+n}$	$4^5 4^2 = 4^{5+2} = 4^7$ $x^3 x^2 = x^{3+2} = x^5$ $(2y)^3 (2y)^5 = (2y)^{3+5} = (2y)^8$
$(a^m)^n = a^{mn}$	$(2^2)^3 = 2^{2\cdot 3} = 2^6$ $(x^4)^3 = x^{4\cdot 3} = x^{12}$ $(a^2)^n = a^{2n}$ $[(x + 2)^4]^3 = (x + 2)^{4\cdot 3} = (x + 2)^{12}$
$(ab)^m = a^m b^m$	$(ab)^4 = a^4 b^4$ $(2x^2 y)^4 = 2^4 (x^2)^4 y^4 = 16x^8 y^4$
$\left(\dfrac{a}{b}\right)^m = \dfrac{a^m}{b^m}, \quad b \neq 0$	$-\left(\dfrac{2}{x}\right)^3 = -\dfrac{2^3}{x^3} = -\dfrac{8}{x^3}$ $\left(\dfrac{-x}{2y}\right)^4 = \dfrac{(-x)^4}{(2y)^4} = \dfrac{x^4}{2^4 y^4} = \dfrac{x^4}{16y^4}$
If $a \neq 0$, $\dfrac{a^m}{a^n} = a^{m-n}$ if $m > n$ $\dfrac{a^m}{a^n} = \dfrac{1}{a^{n-m}}$ if $n > m$ $\dfrac{a^m}{a^n} = 1 \quad$ if $m = n$	$\dfrac{(-3)^2}{(-3)^3} = \dfrac{1}{(-3)^{3-2}} = \dfrac{1}{-3} = -\dfrac{1}{3}$ $\dfrac{x^{2n+1}}{x^n} = x^{2n+1-n} = x^{n+1}$ $\dfrac{y^3}{y^3} = 1$

PROGRESS CHECK

Simplify, using only positive exponents.

(a) $(x^3)^4$ (b) $x^4 (x^2)^3$ (c) $\dfrac{a^{14}}{a^8}$

(d) $\dfrac{-2(x + 1)^n}{(x + 1)^{2n}}$ (e) $(3ab^2)^3$ (f) $\left(\dfrac{-ab^2}{c^3}\right)^3$

Answers

(a) x^{12} (b) x^{10} (c) a^6 (d) $-\dfrac{2}{(x + 1)^n}$ (e) $27a^3 b^6$

(f) $-\dfrac{a^3 b^6}{c^9}$

ZERO AND NEGATIVE EXPONENTS

We next expand our rules to include zero and negative exponents when the base is nonzero. We will assume that the previous rules for exponents apply to a^0 and see if this leads us to a definition of a^0. For example, applying the rule $a^m a^n = a^{m+n}$ yields

$$a^m a^0 = a^{m+0} = a^m$$

Dividing both sides by a^m, we obtain $a^0 = 1$. We therefore *define* a^0 for any nonzero real number by

$$a^0 = 1$$

The same approach will lead us to a definition for negative exponents. For consistency, we must have

$$a^m a^{-m} = a^{m-m} = a^0 = 1 \text{ or } a^m a^{-m} = 1 \qquad (1)$$

Dividing both sides of Equation (1) by a^m suggests that we *define* a^{-m} by

$$a^{-m} = \frac{1}{a^m}$$

Dividing Equation (1) by a^{-m}, we have

$$a^m = \frac{1}{a^{-m}}$$

Thus, a^{-m} is the reciprocal of a^m, and a^m is the reciprocal of a^{-m}. The rule for handling negative exponents can be expressed as follows.

A factor moves from numerator to denominator (or from denominator to numerator) by changing the sign of the exponent.

Table 3 summarizes and illustrates these results.

TABLE 3   **Zero and Integer Exponents**

Definition	Example
$a^0 = 1, \quad a \neq 0$	$3^0 = 1 \qquad \left(\frac{2}{5}\right)^0 = 1 \qquad 4(xy)^0 = 4$
	$\frac{2}{(x-1)^0} = \frac{2}{1} = 2 \qquad -3(y^2+1)^0 = -3$
$a^m = \frac{1}{a^{-m}}, \quad a \neq 0$	$\frac{-2}{(a-1)^{-2}} = -2(a-1)^2 \qquad (2x)^{-3} = \frac{1}{(2x)^3} = \frac{1}{8x^3}$
$a^{-m} = \frac{1}{a^m}, \quad a \neq 0$	$(x^2y^{-3})^{-5} = (x^2)^{-5}(y^{-3})^{-5} = x^{-10}y^{15} = \frac{y^{15}}{x^{10}}$

PROGRESS CHECK

Simplify, using only positive exponents.

(a) $x^{-2}y^{-3}$ (b) $\dfrac{-3x^4y^{-2}}{9x^{-8}y^6}$ (c) $\left(\dfrac{x^{-3}}{x^{-4}}\right)^{-1}$

Answers

(a) $\dfrac{1}{x^2y^3}$ (b) $-\dfrac{x^{12}}{3\,y^8}$ (c) $\dfrac{1}{x}$

 WARNING ─────────────────────────

Don't confuse negative numbers and negative exponents.

(a) $2^{-4} = \dfrac{1}{2^4}$

Note that $2^{-4} \neq -2^4$

(b) $(-2)^{-3} = \dfrac{1}{(-2)^3} = \dfrac{1}{-8} = -\dfrac{1}{8}$

Note that $(-2)^{-3} \neq \dfrac{1}{2^3} = \dfrac{1}{8}$

EXERCISE SET 1.6

The right-hand side of each of the following is incorrect. Find the correct term.

1. $x^2 \cdot x^4 = x^8$

2. $(y^2)^5 = y^7$

3. $\dfrac{b^6}{b^2} = b^3$

4. $\dfrac{x^2}{x^6} = x^4$

5. $(2x)^4 = 2x^4$

6. $\left(\dfrac{4}{3}\right)^4 = \dfrac{4}{3^4}$

Simplify, using the rules for exponents. Write the answers using only positive exponents.

7. $\left(-\dfrac{1}{2}\right)^4 \left(-\dfrac{1}{2}\right)^3$

8. $(x^m)^{3m}$

9. $(y^4)^{2n}$

10. $\dfrac{(-4)^6}{(-4)^{10}}$

11. $-\left(\dfrac{x}{y}\right)^3$

12. $-3r^3 r^3$

13. $(x^3)^5 \cdot x^4$

14. $\dfrac{x^{12}}{x^8}$

15. $(-2x^2)^5$

16. $-(2x^2)^5$

17. $x^{3n} \cdot x^n$

18. $(-2)^m (-2)^n$

19. $\dfrac{x^n}{x^{n+2}}$

20. $\left(\dfrac{3x^3}{y^2}\right)^5$

21. $(-5x^3)(-6x^5)$

22. $(x^2)^3 (y^2)^4 (x^3)^7$

23. $\dfrac{(r^2)^4}{(r^4)^2}$

24. $[(3b + 1)^5]^5$

25. $\left(\dfrac{3}{2}x^2 y^3\right)^n$

26. $\dfrac{(-2a^2 b)^4}{(-3ab^2)^3}$

27. $(2x + 1)^3 (2x + 1)^7$

28. $\dfrac{y^3 (y^3)^4}{(y^4)^6}$

29. $(-2a^2 b^3)^{2n}$

30. $\left(-\dfrac{2}{3}a^2 b^3 c^2\right)^3$

31. $2^0 + 3^{-1}$

32. $(xy)^0 - 2^{-1}$

33. $\dfrac{3}{(2x^2 + 1)^0}$

34. $(-3)^{-3}$

35. $\dfrac{1}{3^{-4}}$

36. x^{-5}

37. $(-x)^3$

38. $-x^{-5}$

39. $\dfrac{1}{y^{-6}}$

40. $(2a)^{-6}$

41. $5^{-3} 5^5$

42. $4y^5 y^{-2}$

43. $(3^2)^{-3}$

44. $(x^{-2})^4$

45. $(x^{-3})^{-3}$

46. $[(x + y)^{-2}]^2$

47. $\dfrac{2^2}{2^{-3}}$

48. $\dfrac{x^8}{x^{-10}}$

49. $\dfrac{2x^4y^{-2}}{x^2y^{-3}}$

50. $(x^4y^{-2})^{-1}$

51. $(3a^{-2}b^{-3})^{-2}$

52. $\dfrac{1}{(2xy)^{-2}}$

53. $\left(-\dfrac{1}{2}x^3y^{-4}\right)^{-3}$

54. $\dfrac{(x^{-2})^2}{(3y^{-2})^3}$

55. $\dfrac{3a^5b^{-2}}{9a^{-4}b^2}$

56. $\left(\dfrac{x^3}{x^{-2}}\right)^2$

57. $\left(\dfrac{2a^2b^{-4}}{a^{-3}c^{-3}}\right)^2$

58. $\dfrac{2x^{-3}y^2}{x^{-3}y^{-3}}$

59. $(a - 2b^2)^{-1}$

60. $\left(\dfrac{y^{-2}}{y^{-3}}\right)^{-1}$

61. $\dfrac{(a + b)^{-1}}{(a - b)^{-2}}$

62. $(a^{-1} + b^{-1})^{-1}$

63. $\dfrac{a^{-1} + b^{-1}}{a^{-1} - b^{-1}}$

64. $\left(\dfrac{a}{b}\right)^{-1} + \left(\dfrac{b}{a}\right)^{-1}$

65. Show that $\left(\dfrac{a}{b}\right)^{-n} = \left(\dfrac{b}{a}\right)^{n}$

Evaluate each expression in Exercises 66–69.

66. $(1.20^2)^{-1}$

67. $[(-3.67)^2]^{-1}$

68. $\left(\dfrac{7.65^{-1}}{7.65^2}\right)^2$

69. $\left(\dfrac{4.46^2}{4.46^{-1}}\right)^{-1}$

1.7
RATIONAL EXPONENTS
AND RADICALS

Consider a square whose area is 25 square centimeters, and whose sides are of length a. We can then write

$$a^2 = 25$$

so that a is a number whose square is 25. We say that a is the **square root** of b if $a^2 = b$. Similarly, we say that a is a **cube root** of b if $a^3 = b$, and, in general, if n is a natural number, we say that

> a is an **nth root** of b if $a^n = b$

Thus, 5 is a square root of 25 since $5^2 = 25$, and -2 is a cube root of -8 since $(-2)^3 = -8$.

Since $(-5)^2 = 25$, we conclude that -5 is also a square root of 25. More generally, if $b > 0$ and a is a square root of b, then $-a$ is also a square root of b. If $b < 0$, there is no real number a such that $a^2 = b$, since the square of a real number is always nonnegative. The cases are summarized in Table 4.

TABLE 4

		Number of nth roots of b such that $b = a^n$	Form of nth roots	Examples	
b	n			b	
> 0	Even	2	$a, -a$	4	Square roots are 2, -2
< 0	Even	None	None	-1	No square roots
> 0	Odd	1	$a > 0$	8	Cube root is 2
< 0	Odd	1	$a < 0$	-8	Cube root is -2
0	All	1	0	0	Square root is 0

We would like to define rational exponents in a manner that will be consistent with the rules developed in the previous section for integer exponents. If the rule $(a^m)^n = a^{mn}$ is to hold, then we must have

$$(b^{1/n})^n = b^{n/n} = b$$

But a is an nth root of b if $a^n = b$. Then for every natural number n, we say that

$$b^{1/n} \text{ is an } n\text{th root of } b$$

If n is even and b is positive, Table 4 indicates that there are two numbers a such that $a^n = b$. For example,

$$4^2 = 16 \quad \text{and} \quad (-4)^2 = 16$$

Thus, there are two candidates for $16^{1/2}$, namely 4 and -4. To avoid ambiguity we shall say that $16^{1/2} = 4$. That is, if n is even and b is positive, we always *choose* the positive number a such that $a^n = b$ to be the nth root and call a the **principal nth root** of b. Thus, $b^{1/n}$ denotes the principal nth root of b.

EXAMPLE 1
Evaluate.

(a) $144^{1/2} = 12$

(b) $(-8)^{1/3} = -2$

(c) $(-25)^{1/2}$ is not a real number

(d) $-\left(\dfrac{1}{16}\right)^{1/4} = -\dfrac{1}{2}$

Now we are prepared to define $b^{m/n}$, where m is an integer (positive or negative), n is a natural number, and $b \geq 0$ when n is even. We want the rules for exponents to also hold for rational exponents. That is, we want to have

$$4^{3/2} = 4^{(1/2)\,(3)} = (4^{1/2})^3 = 2^3 = 8$$

and

$$4^{3/2} = 4^{(3)(1/2)} = (4^3)^{1/2} = (64)^{1/2} = 8$$

To achieve this, we define $b^{m/n}$, for an integer m, a natural number n, and a real number b, by

$$b^{m/n} = (b^{1/n})^m = (b^m)^{1/n}$$

where b must be nonnegative when n is even. With this definition, all the rules of exponents continue to hold when the exponents are rational numbers.

EXAMPLE 2
Simplify.

(a) $(-8)^{4/3} = [(-8)^{1/3}]^4 = (-2)^4 = 16$

(b) $x^{1/2} \cdot x^{3/4} = x^{1/2\,+\,3/4} = x^{5/4}$

(c) $(x^{3/4})^2 = x^{(3/4)(2)} = x^{3/2}$

(d) $\left(\dfrac{x^8}{y^{-6}}\right)^{-1/2} = \dfrac{(x^8)^{-1/2}}{(y^{-6})^{-1/2}} = \dfrac{x^{(8)\,(-1/2)}}{y^{(-6)\,(-1/2)}} = \dfrac{x^{-4}}{y^3} = \dfrac{1}{x^4y^3}$

(e) $(3x^{2/3}\,y^{-5/3})^3 = 3^3 \cdot x^{(2/3)(3)}\,y^{(-5/3)(3)} = 27x^2y^{-5} = \dfrac{27x^2}{y^5}$

PROGRESS CHECK
Simplify.

(a) $27^{4/3}$

(b) $(-64)^{2/3}$

(c) $\left(\dfrac{1}{4}\right)^{5/2}$

(d) $\dfrac{2x^{1/5}}{x^{4/5}}$

(e) $(a^{1/2}b^{-2})^{-2}$

(f) $\left(\dfrac{x^{1/3}y^{2/3}}{z^{5/6}}\right)^{12}$

Answers

(a) *81*

(b) *16*

(c) $\dfrac{1}{32}$

(d) $\dfrac{2}{x^{3/5}}$

(e) $\dfrac{b^4}{a}$

(f) $\dfrac{x^4y^8}{z^{10}}$

The symbol $\sqrt{b}$ is an alternate way of writing $b^{1/2}$, that is, $\sqrt{b}$ denotes the nonnegative square root of b. The symbol "$\sqrt{}$" is called a **radical sign** and $\sqrt{b}$ is called the **principal square root** of b. Thus,

$$\sqrt{25} = 5 \qquad \sqrt{0} = 0 \qquad \sqrt{-25} \;\text{ is undefined}$$

In general, the symbol $\sqrt[n]{b}$ is an alternate way of writing $b^{1/n}$, the principal nth root of b, and is defined as follows.

$$\sqrt[n]{b} = b^{1/n} = a \text{ where } a^n = b$$

Since $b^{1/n}$ is undefined if b is negative and n is even, $\sqrt[n]{b}$ is also undefined under these conditions. As a matter of convenience, in this section we shall henceforth assume that any *variable* under a radical sign is nonnegative when an *even* root is indicated.

In short, $\sqrt[n]{b}$ is the **radical form** of $b^{1/n}$. We can switch back and forth from rational exponent form to radical form and vice versa. For instance,

$$\sqrt[3]{7} = 7^{1/3} \quad \text{and} \quad (11)^{1/5} = \sqrt[5]{11}$$

Finally, we define $b^{m/n}$, where m is an integer and n is a natural number, as follows.

and	$b^{m/n} = (b^m)^{1/n} = \sqrt[n]{b^m}$
	$b^{m/n} = (b^{1/n})^m = (\sqrt[n]{b})^m$

Thus,

$$7^{2/3} = (7^2)^{1/3} = \sqrt[3]{7^2}$$
$$7^{2/3} = (7^{1/3})^2 = \left(\sqrt[3]{7}\right)^2$$

EXAMPLE 3
Write in radical form.

(a) $(3y)^{2/5} = \sqrt[5]{3^2 y^2}$

(b) $x^{-3/2} = \dfrac{1}{x^{3/2}} = \dfrac{1}{\sqrt{x^3}}$

(c) $(-8a)^{3/7} = \sqrt[7]{(-8a)^3}$

(d) $(x^2 + y^2)^{1/2} = \sqrt{x^2 + y^2}$

EXAMPLE 4
Write in rational exponent form.

(a) $\sqrt{(5xy)^3} = (5xy)^{3/2}$

(b) $\dfrac{1}{\sqrt[7]{x^4}} = \dfrac{1}{x^{4/7}} = x^{-4/7}$

(c) $\sqrt[3]{(-2x)^5} = (-2x)^{5/3}$

(d) $\sqrt{2x + 3y} = (2x + 3y)^{1/2}$

PROGRESS CHECK
Change from radical form to rational exponent or vice versa.

(a) $\sqrt[4]{2rs^3}$

(b) $(x + y)^{5/2}$

(c) $y^{-5/4}$

(d) $\dfrac{1}{\sqrt[4]{m^5}}$

Answers

(a) $(2rs)^{3/4}$

(b) $\sqrt{(x + y)^5}$

(c) $\dfrac{1}{\sqrt[4]{y^5}}$

(d) $m^{-5/4}$

WARNING

Note that

$$\sqrt{x^2} = |x|$$

It is a common error to write $\sqrt{x^2} = x$, but this leads to the conclusion that $\sqrt{(-6)^2} = -6$, which contradicts our earlier definition stating that the symbol $\sqrt{}$ represents the principal or *nonnegative* square root of a number. It is therefore essential to write $\sqrt{x^2} = |x|$ unless we know that $x \geq 0$.

SIMPLIFYING RADICALS

A radical is said to be in **simplified form** when the following conditions are satisfied.

(1) $\sqrt[n]{x^m}$ has $m < n$.

(2) $\sqrt[n]{x^m}$ has no common factors between m and n.

(3)   A fraction is free of radicals in the denominator.

The first two conditions can always be met by using the properties of radicals and by writing radicals in exponent form. For example,

$$\sqrt[3]{x^4} = \sqrt[3]{x^3 \cdot x} = \sqrt[3]{x^3}\,\sqrt[3]{x} = x\,\sqrt[3]{x}$$

and

$$\sqrt[6]{x^4} = x^{4/6} = x^{2/3} = \sqrt[3]{x^2}$$

The third condition can always be satisfied by multiplying the fraction by a properly chosen form of unity, a process called **rationalizing the denominator.** For example, to rationalize $1/\sqrt{3}$, we proceed as follows.

$$\frac{1}{\sqrt{3}} = \frac{1}{\sqrt{3}} \cdot \frac{\sqrt{3}}{\sqrt{3}} = \frac{\sqrt{3}}{\sqrt{3^2}} = \frac{\sqrt{3}}{3}$$

In this connection, a useful formula is

$$(\sqrt{m} + \sqrt{n})(\sqrt{m} - \sqrt{n}) = m - n$$

which we will apply in the following examples.

EXAMPLE 5
Rationalize the denominator.

(a) $\sqrt{\dfrac{x}{y}}$ (b) $\dfrac{x}{\sqrt[3]{x^2 y}}$ (c) $\dfrac{4}{\sqrt{5} - \sqrt{2}}$ (d) $\dfrac{5}{\sqrt{x} + 2}$

Solution

(a) $\sqrt{\dfrac{x}{y}} = \dfrac{\sqrt{x}}{\sqrt{y}} = \dfrac{\sqrt{x}}{\sqrt{y}} \cdot \dfrac{\sqrt{y}}{\sqrt{y}} = \dfrac{\sqrt{xy}}{\sqrt{y^2}} = \dfrac{\sqrt{xy}}{y}$

(b) $\dfrac{x}{\sqrt[3]{x^2 y}} = \dfrac{x}{\sqrt[3]{x^2 y}} \cdot \dfrac{\sqrt[3]{xy^2}}{\sqrt[3]{xy^2}}$ This multiplier will produce $\sqrt[3]{x^3 y^3}$

$\qquad = \dfrac{x\sqrt[3]{xy^2}}{\sqrt[3]{x^3 y^3}} = \dfrac{x\sqrt[3]{xy^2}}{xy} = \dfrac{\sqrt[3]{xy^2}}{y}$

(c) $\dfrac{4}{\sqrt{5} - \sqrt{2}} = \dfrac{4}{\sqrt{5} - \sqrt{2}} \cdot \dfrac{\sqrt{5} + \sqrt{2}}{\sqrt{5} + \sqrt{2}}$

$\qquad = \dfrac{4(\sqrt{5} + \sqrt{2})}{5 - 2} = \dfrac{4}{3}(\sqrt{5} + \sqrt{2})$

(d) $\dfrac{5}{\sqrt{x} + 2} = \dfrac{5}{\sqrt{x} + 2} \cdot \dfrac{\sqrt{x} - 2}{\sqrt{x} - 2} = \dfrac{5(\sqrt{x} - 2)}{x - 4}$

PROGRESS CHECK
Rationalize the denominator.

(a) $\dfrac{-9xy^3}{\sqrt{3xy}}$

(b) $\dfrac{-6}{\sqrt{2} + \sqrt{6}}$

(c) $\dfrac{4}{\sqrt{x} - \sqrt{y}}$

Answers

(a) $-3y^2 \sqrt{3xy}$

(b) $\dfrac{3}{2} (\sqrt{2} - \sqrt{6})$

(c) $\dfrac{4(\sqrt{x} + \sqrt{y})}{x - y}$

EXAMPLE 6
Write in simplified form.

(a) $\sqrt[4]{x^2 y^5} = \sqrt[4]{x^2}\, \sqrt[4]{y^4 \cdot y} = \sqrt{x}\, \sqrt[4]{y^4}\, \sqrt[4]{y}$ Since $\sqrt[4]{x^2} = x^{2/4} = x^{1/2} = \sqrt{x}$

$\qquad\qquad = y \sqrt{x}\, \sqrt[4]{y}$

(b) $\sqrt{\dfrac{8x^3}{y}} = \dfrac{\sqrt{(4x^2)(2x)}}{\sqrt{y}} = \dfrac{\sqrt{4x^2}\, \sqrt{2x}}{\sqrt{y}} = \dfrac{2x\sqrt{2x}}{\sqrt{y}}$

$\qquad\qquad = \dfrac{2x \sqrt{2x}}{\sqrt{y}} \cdot \dfrac{\sqrt{y}}{\sqrt{y}} = \dfrac{2x\sqrt{2xy}}{y}$

(c) $\sqrt[6]{\dfrac{x^3}{y^2}} = \dfrac{\sqrt[6]{x^3}}{\sqrt[6]{y^2}} = \dfrac{\sqrt{x}}{\sqrt[3]{y}} = \dfrac{\sqrt{x}}{\sqrt[3]{y}} \cdot \dfrac{\sqrt[3]{y^2}}{\sqrt[3]{y^2}} = \dfrac{\sqrt{x}\, \sqrt[3]{y^2}}{y}$

PROGRESS CHECK
Write in simplified form.

(a) $\sqrt{\dfrac{18x^6}{y}}$

(b) $\sqrt[3]{ab^4 c^7}$

(c) $\dfrac{-2xy^3}{\sqrt[4]{32x^3 y^5}}$

Answers

(a) $\dfrac{3x^3 \sqrt{2y}}{y}$

(b) $bc^2 \sqrt[3]{abc}$

(c) $-\dfrac{y}{2} \sqrt[4]{8xy^3}$

OPERATIONS WITH RADICALS

We can add or subtract expressions involving exactly the same radical forms. For example,

$$2\sqrt{2} + 3\sqrt{2} = 5\sqrt{2} \quad \text{since} \quad 2\sqrt{2} + 3\sqrt{2} = (2+3)\sqrt{2} = 5\sqrt{2}$$

$$3\sqrt[3]{x^2 y} - 7\sqrt[3]{x^2 y} = -4\sqrt[3]{x^2 y}$$

EXAMPLE 7

(a) $7\sqrt{5} + 4\sqrt{3} - 9\sqrt{5} = -2\sqrt{5} + 4\sqrt{3}$

(b) $\sqrt[3]{x^2 y} - \dfrac{1}{2}\sqrt{xy} - 3\sqrt[3]{x^2 y} + 4\sqrt{xy} = -2\sqrt[3]{x^2 y} - \dfrac{7}{2}\sqrt{xy}$

WARNING

$$\sqrt{9} + \sqrt{16} \neq \sqrt{25}$$

You can perform addition only with identical radical forms. *Adding unlike radicals is one of the most common mistakes made by students in algebra!* You can easily verify that

$$\sqrt{9} + \sqrt{16} = 3 + 4 = 7$$

The product of $\sqrt[n]{a}$ and $\sqrt[m]{b}$ can be simplified only when $m = n$. Thus,

$$\sqrt[5]{x^2y} \cdot \sqrt[5]{xy} = \sqrt[5]{x^3y^2}$$

but

$$\sqrt[3]{x^2y} \cdot \sqrt[5]{xy}$$

cannot be simplified.

EXAMPLE 8
Multiply and simplify.

(a) $2\sqrt[3]{xy^2} \cdot \sqrt[3]{x^2y^2} = 2\sqrt[3]{x^3y^4} = 2xy\sqrt[3]{y}$

(b) $\sqrt[5]{a^2b}\sqrt{ab}\sqrt[5]{ab^2} = \sqrt[5]{a^3b^3}\sqrt{ab}$

EXERCISE SET 1.7
Simplify and write answers using only positive exponents.

1. $16^{3/4}$

2. $(-125)^{-1/3}$

3. $(-64)^{-2/3}$

4. $c^{1/4}\,c^{-2/3}$

5. $\dfrac{2x^{1/3}}{x^{-3/4}}$

6. $\dfrac{y^{-2/3}}{y^{1/5}}$

7. $\left(\dfrac{x^{3/2}}{x^{2/3}}\right)^{1/6}$

8. $\dfrac{125^{4/3}}{125^{2/3}}$

9. $(x^{1/3}y^2)^6$

10. $(x^6y^4)^{-1/2}$

11. $\left(\dfrac{x^{15}}{y^{10}}\right)^{3/5}$

12. $\left(\dfrac{x^{18}}{y^{-6}}\right)^{2/3}$

Write each of the following in radical form.

13. $\left(\dfrac{1}{4}\right)^{2/5}$

14. $x^{2/3}$

15. $a^{3/4}$

16. $(-8x^2)^{2/5}$

17. $(12x^3y^{-2})^{2/3}$

18. $\left(\dfrac{8}{3}x^{-2}y^{-4}\right)^{-3/2}$

Write each of the following in exponent form.

19. $\sqrt[4]{8^3}$

20. $\sqrt[5]{3^2}$

21. $\dfrac{1}{\sqrt[5]{(-8)^2}}$

22. $\dfrac{1}{\sqrt[3]{x^7}}$

23. $\dfrac{1}{\sqrt[4]{\dfrac{4}{9}a^3}}$

24. $\sqrt[5]{(2a^2b^3)^4}$

Evaluate.

25. $\sqrt{\dfrac{4}{9}}$

26. $\sqrt{\dfrac{25}{4}}$

27. $\sqrt[4]{-81}$

28. $\sqrt[3]{\dfrac{1}{27}}$

29. $\sqrt{(-5)^2}$

30. $\sqrt{\left(-\dfrac{1}{3}\right)^2}$

31. $\sqrt{\left(\dfrac{5}{4}\right)^2}$ **32.** $\sqrt{\left(-\dfrac{7}{2}\right)^2}$ ▦ **33.** $(14.43)^{3/2}$

In Exercises 34–35, provide a real value for x and for y to demonstrate the result.

34. $\sqrt{x^2} \neq x$ **35.** $\sqrt{x^2 + y^2} \neq x + y$

36. Find the step in the following "proof" that is incorrect. Explain.

$1 = 1^{1/2} = [(-1)^2]^{1/2} = (-1)^{2/2} = (-1)^1 = -1$

Simplify each of the following and write the answer in simplified form. Every variable represents a positive real number.

37. $\sqrt{48}$ **38.** $\sqrt{100}$ **39.** $\sqrt[3]{54}$

40. $\sqrt{x^8}$ **41.** $\sqrt[3]{y^7}$ **42.** $\sqrt[4]{b^{14}}$

43. $\sqrt[4]{96x^{10}}$ **44.** $\sqrt{x^5 y^4}$ **45.** $\sqrt{x^5 y^3}$

46. $\sqrt[3]{24b^{10}c^{14}}$ **47.** $\sqrt[4]{16x^8 y^5}$ **48.** $\sqrt{20x^5 y^7 z^4}$

49. $\sqrt{\dfrac{1}{5}}$ **50.** $\dfrac{4}{3\sqrt{11}}$ **51.** $\dfrac{1}{\sqrt{3y}}$

52. $\sqrt{\dfrac{2}{y}}$ **53.** $\dfrac{4x^2}{\sqrt{2x}}$ **54.** $\dfrac{8a^2 b^2}{2\sqrt{2b}}$

55. $\sqrt[3]{x^2 y^7}$ **56.** $\sqrt[4]{48x^8 y^6 z^2}$

Simplify and combine terms.

57. $2\sqrt{3} + 5\sqrt{3}$ **58.** $4\sqrt[3]{11} - 6\sqrt[3]{11}$

59. $3\sqrt{x} + 4\sqrt{x}$ **60.** $3\sqrt{2} + 5\sqrt{2} - 2\sqrt{2}$

61. $2\sqrt{27} + \sqrt{12} - \sqrt{48}$ **62.** $\sqrt{20} - 4\sqrt{45} + \sqrt{80}$

63. $\sqrt[3]{40} + \sqrt{45} - \sqrt[3]{135} + 2\sqrt{80}$ **64.** $\sqrt{2abc} - 3\sqrt{8abc} + \sqrt{\dfrac{abc}{2}}$

65. $2\sqrt{5} - (3\sqrt{5} + 4\sqrt{5})$ **66.** $2\sqrt{18} - (3\sqrt{12} - 2\sqrt{75})$

Multiply and simplify in each of the following.

67. $\sqrt{3}(\sqrt{3} + 4)$ **68.** $\sqrt{8}(\sqrt{2} - \sqrt{3})$

69. $3\sqrt[3]{x^2 y}\,\sqrt[3]{xy^2}$ **70.** $-4\sqrt[5]{x^2 y^3}\,\sqrt[5]{x^4 y^2}$

71. $(\sqrt{2} - \sqrt{3})^2$ **72.** $(\sqrt{8} - 2\sqrt{2})(\sqrt{2} + 2\sqrt{8})$

73. $(\sqrt{3x} + \sqrt{2y})(\sqrt{3x} - 2\sqrt{2y})$ **74.** $(\sqrt[3]{2x} + 3)(\sqrt[3]{2x} - 3)$

Rationalize the denominator.

75. $\dfrac{3}{\sqrt{2} + 3}$ **76.** $\dfrac{-3}{\sqrt{7} - 9}$ **77.** $\dfrac{-2}{\sqrt{3} - 4}$

78. $\dfrac{3}{\sqrt{x} - 5}$ **79.** $\dfrac{-3}{3\sqrt{a} + 1}$ **80.** $\dfrac{4}{2 - \sqrt{2y}}$

81. $\dfrac{-3}{5 + \sqrt{5y}}$ **82.** $\dfrac{\sqrt{3}}{\sqrt{3} - 5}$ **83.** $\dfrac{\sqrt{2} + 1}{\sqrt{2} - 1}$

84. $\dfrac{\sqrt{5} + \sqrt{3}}{\sqrt{5} - \sqrt{3}}$ **85.** $\dfrac{\sqrt{6} + \sqrt{2}}{\sqrt{3} - \sqrt{2}}$ **86.** $\dfrac{2\sqrt{a}}{\sqrt{2x} + \sqrt{y}}$

In Exercises 87–88 provide a real value for x and for y and a positive integer value for n that demonstrate the result.

87. $\sqrt{x} + \sqrt{y} \neq \sqrt{x + y}$ **88.** $\sqrt[n]{x^n + y^n} \neq x + y$

Find the step in the following "proof" that is incorrect. Explain.

89. $1 = \sqrt{1} = \sqrt{(-1)(-1)} = \sqrt{-1}\sqrt{-1} = -1$

90. Prove that $|ab| = |a|\,|b|$. $\left(\textit{Hint: } \text{Begin with } |ab| = \sqrt{(ab)^2}\right)$

1.8
COMPLEX NUMBERS

One of the central problems in algebra is that of finding solutions to a given polynomial equation. This problem will be discussed in later chapters of this book. However, observe at this point that there is no real number which satisfies a simple polynomial equation such as

$$x^2 = -4$$

since the square of a real number is always nonnegative.

To resolve this problem, mathematicians created a new number system built upon an "imaginary unit" i defined by $i = \sqrt{-1}$. This number i has the property that when we square both sides of the equation we have $i^2 = -1$, a result that cannot be obtained with real numbers. In summary,

$$i = \sqrt{-1}$$
$$i^2 = -1$$

We also assume that i behaves according to all the algebraic laws we have already developed (with the exception of the rules for inequalities). This allows us to simplify higher powers of i. Thus,

$$i^3 = i^2 \cdot i = (-1)i = -i$$
$$i^4 = i^2 \cdot i^2 = (-1)(-1) = 1$$

Now it's easy to simplify i^n when n is any natural number. Since $i^4 = 1$, we simply seek the highest multiple of 4 which is less than or equal to n. For example,

$$i^5 = i^4 \cdot i = (1) \cdot i = i$$
$$i^{27} = i^{24} \cdot i^3 = (i^4)^6 \cdot i^3 = (1)^6 \cdot i^3 = i^3 = -i$$

EXAMPLE 1
Simplify.

(a) $i^{51} = i^{48} \cdot i^3 = (i^4)^{12} \cdot i^3 = (1)^{12} \cdot i^3 = i^3 = -i$

(b) $-i^{74} = -i^{72} \cdot i^2 = -(i^4)^{18} \cdot i^2 = -(1)^{18} \cdot i^2 = (-1)(-1) = 1$

It is easy also to write square roots of negative numbers in terms of i. For example,

$$\sqrt{-25} = i\sqrt{25} = 5i$$

and, in general,

$$\sqrt{-a} = i\sqrt{a} \quad \text{for } a > 0$$

Any number of the form bi, where b is a real number, is called an **imaginary number**.

WARNING ————————————————————————————————

$$\sqrt{-4}\,\sqrt{-9} \neq \sqrt{36}$$

The rule $\sqrt{a} \cdot \sqrt{b} = \sqrt{ab}$ holds only when $a \geq 0$ and $b \geq 0$. Instead, write

$$\sqrt{-4}\,\sqrt{-9} = 2i \cdot 3i = 6i^2 = -6$$

————————————————————————————————

Having created imaginary numbers, we next combine real and imaginary numbers. We say that $a + bi$, where a and b are real numbers, is a **complex number.** The number a is called the **real part** of $a + bi$, and bi is called the **imaginary part.** The following are examples of complex numbers.

$$3 + 2i \qquad 2 - i \qquad -2i \qquad \frac{4}{5} + \frac{1}{5}i$$

Note that every real number a can be written as a complex number by choosing $b = 0$. Thus,

$$a = a + 0i$$

We see therefore that the real number system is a subset of the complex number system. Once again, we have established a number system which incorporates all of the previous number systems and is itself more complicated than the earlier systems.

EXAMPLE 2
Write as a complex number.

(a) $\quad -\dfrac{1}{2} = -\dfrac{1}{2} + 0i$

(b) $\quad \sqrt{-9} = i\sqrt{9} = 3i = 0 + 3i$

(c) $\quad -1 - \sqrt{-4} = -1 - i\sqrt{4} = -1 - 2i$

We next seek to define operations with complex numbers in such a way that the rules for real numbers and the imaginary unit i continue to hold. We begin with equality and say that two complex numbers are **equal** if their real parts are equal and their imaginary parts are equal. Thus,

$$a + bi = c + di \qquad \text{if and only if}$$
$$a = c \text{ and } b = d$$

EXAMPLE 3
Solve the equation $x + 3i = 6 - yi$ for x and for y.

Solution
Equating the real parts, we have $x = 6$; equating the imaginary parts, $3i = -yi$ or $y = -3$.

Complex numbers are added and subtracted by adding or subtracting the real parts and by adding or subtracting the imaginary parts. That is,

$$(a + bi) + (c + di) = (a + c) + (b + d)i$$
$$(a + bi) - (c + di) = (a - c) + (b - d)i$$

Note that the sum or difference of two complex numbers is again a complex number. Thus, the complex numbers are closed under addition. It is easy to verify that the associative and commutative laws of addition also hold under this definition.

EXAMPLE 4
Perform the indicated operations.

(a) $(7 - 2i) + (4 - 3i) = (7 + 4) + (-2 - 3)i = 11 - 5i$

(b) $14 - (3 - 8i) = (14 - 3) + 8i = 11 + 8i$

PROGRESS CHECK
Perform the indicated operations.

(a) $(-9 + 3i) + (6 - 2i)$ (b) $7i - (3 + 9i)$

Answers

(a) $-3 + i$ (b) $-3 - 2i$

We now define multiplication of complex numbers in a manner which permits the commutative, associative, and distributive laws to hold along with the definition $i^2 = -1$. We must have

$$(a + bi)(c + di) = a(c + di) + bi(c + di)$$
$$= ac + adi + bci + bdi^2$$
$$= ac + (ad + bc)i + bd(-1)$$
$$= (ac - bd) + (ad + bc)i$$

Thus,

$$(a + bi)(c + di) = (ac - bd) + (ad + bc)i$$

This result is significant since it demonstrates that the product of two complex numbers is again a complex number. It need not be memorized; simply use the distributive law to form all the products and the substitution $i^2 = -1$ to simplify.

EXAMPLE 5
Find the product of $(2 - 3i)$ and $(7 + 5i)$.

Solution

$$(2 - 3i)(7 + 5i) = 2(7 + 5i) - 3i(7 + 5i)$$
$$= 14 + 10i - 21i - 15i^2$$
$$= 14 - 11i - 15(-1)$$
$$= 29 - 11i$$

PROGRESS CHECK

Find the product.

(a) $(-3 - i)(4 - 2i)$ (b) $(-4 - 2i)(2 - 3i)$

Answers
(a) $-14 + 2i$ *(b)* $-14 + 8i$

EXERCISE SET 1.8

Simplify.

1. i^{60} 2. i^{27} 3. i^{83}
4. $-i^{54}$ 5. $-i^{33}$ 6. i^{-15}
7. i^{-84} 8. $-i^{39}$ 9. $-i^{-25}$

Write as a complex number in the form $a + bi$.

10. 2 11. $-\dfrac{1}{2}$ 12. -0.3

13. $\sqrt{-25}$ 14. $-\sqrt{-5}$ 15. $-\sqrt{-36}$

16. $-\sqrt{-18}$ 17. $3 - \sqrt{-49}$ 18. $-\dfrac{3}{2} - \sqrt{-72}$

19. $0.3 - \sqrt{-98}$ 20. $-0.5 + \sqrt{-32}$ 21. $-2 - \sqrt{-16}$

Solve for x and for y.

22. $(x + 2) + (2y - 1)i = -1 + 5i$ 23. $(3x - 1) + (y + 5)i = 1 - 3i$

24. $\left(\dfrac{1}{2}x + 2\right) + (3y - 2)i = 4 - 7i$ 25. $(2y + 1) - (2x - 1)i = -8 + 3i$

26. $(y - 2) + (5x - 3)i = 5$

Compute and write the answer in the form $a + bi$.

27. $2i + (3 - i)$ 28. $-3i + (2 - 5i)$

29. $2 + 3i + (3 - 2i)$ 30. $(3 - 2i) - \left(2 + \dfrac{1}{2}i\right)$

31. $-3 - 5i - (2 - i)$ 32. $\left(\dfrac{1}{2} - i\right) + \left(1 - \dfrac{2}{3}i\right)$

33. $-2i(3 + i)$ 34. $3i(2 - i)$

35. $i\left(-\dfrac{1}{2} + i\right)$ 36. $\dfrac{i}{2}\left(\dfrac{4 - i}{2}\right)$

37. $(2 - i)(2 + i)$ 38. $(5 + i)(2 - 3i)$
39. $(-2 - 2i)(-4 - 3i)$ 40. $(2 + 5i)(1 - 3i)$
41. $(3 - 2i)(2 - i)$ 42. $(4 - 3i)(2 + 3i)$

In Exercises 43–46, evaluate the polynomial $x^2 - 2x + 5$ for the given complex value of x.

43. $1 + 2i$ 44. $2 - i$
45. $1 - i$ 46. $1 - 2i$

47. Prove that the commutative law of addition holds for the set of complex numbers.

48. Prove that the commutative law of multiplication holds for the set of complex numbers.

49. Prove that $0 + 0i$ is the additive identity and $1 + 0i$ is the multiplicative identity for the set of complex numbers.

50. Prove that $-a - bi$ is the additive inverse of the complex number $a + bi$.
51. Prove the distributive property for the set of complex numbers.
52. For what values of x is $\sqrt{x - 3}$ a real number?
53. For what values of y is $\sqrt{2y - 10}$ a real number?

TERMS AND SYMBOLS

set (p. 1)	algebraic expression (p. 15)	cancellation principle (p. 29)
element, member (p. 2)	constant (p. 15)	least common denominator (p. 30)
{ } (p. 2)	algebraic operations (p. 15)	L.C.D. (p. 30)
ϵ (p. 2)	evaluate (p. 15)	equivalent fraction (p. 31)
$\notin$ (p. 2)	base (p. 16)	complex fraction (p. 32)
subset (p. 2)	exponent (p. 16)	nth root (p. 40)
natural numbers (p. 2)	power (p. 16)	principal nth root (p. 41)
integers (p. 2)	polynomial (p. 17)	radical sign (p. 42)
rational numbers (p. 2)	monomial (p. 17)	principal square root (p. 42)
irrational numbers (p. 3)	coefficient (p. 17)	radical form (p. 42)
real number system (p. 3)	degree of a monomial (p. 17)	simplified form of a radical (p. 44)
field axioms (p. 5)	degree of a polynomial (p. 17)	rationalizing the denominator (p. 44)
equal (p. 5)	constant term (p. 17)	imaginary unit i (p. 48)
factor (p. 6)	leading coefficient (p. 17)	imaginary number (p. 48)
real number line (p. 10)	zero polynomial (p. 17)	complex number (p. 49)
origin (p. 9)	like terms (p. 18)	real part (p. 49)
nonnegative (p. 10)	factoring (p. 24)	imaginary part (p. 49)
nonpositive (p. 10)	prime polynomial (p. 27)	
$<, >, \leq, \geq$ (p. 10)	irreducible polynomial (p. 27)	
inequality symbols (p. 10)	algebraic fraction (p. 28)	
inequalities (p. 10)	rational expression (p. 28)	
absolute value (p. 11)		
| | (p. 12)		
$\overline{AB}$ (p. 13)		
variable (p. 15)		

KEY IDEAS FOR REVIEW

■ A set is simply a collection of objects or numbers.

■ The real number system is composed of the rational and irrational numbers. The rational numbers are those that can be written as the ratio of two integers, p/q, with $q \neq 0$; the irrational numbers cannot be written as a ratio of integers.

■ The real number system satisfies a number of important properties. These are
    closure
    commutativity
    associativity
    identities
    inverses
    distributivity

■ If two numbers are identical, we say that they are equal. Equality satisfies these basic properties
    reflexive property
    symmetric property
    transitive property
    substitution property

■ The field properties may be used to *prove* many other properties of the real numbers.

■ There is a one-to-one correspondence between the set of all real numbers and the set of all points on the real number line. That is, for every point on the line there is a real number and for every real number there is a point on the line.

■ Algebraic statements using inequality symbols have straightforward geometric interpretations using the real number line. Thus, $a < b$ says that a lies to the left of b on the real number line.

■ Inequalities can be operated upon in the same manner as statements involving an equal sign, with one important exception. When an inequality is multiplied by a negative number, the sense of the inequality is reversed.

■ Absolute value specifies distance independent of the direction. Three important properties of absolute value are

$$|a| \geq 0 \qquad |a| = |-a| \qquad |a - b| = |b - a|$$

■ The distance between points A and B whose coordinates are a and b, respectively, is given by

$$\overline{AB} = |b - a|$$

■ Algebraic expressions of the form

$$P = c_n x^n + c_{n-1} x^{n-1} + \ldots + c_1 x + c_0$$

are called polynomials.

■ To add (subtract) polynomials, simply add (subtract) like terms. To multiply polynomials, form all possible products using the rule for exponents: $a^m a^n = a^{m+n}$.

■ A polynomial is said to be factored when it is written as a product of polynomials of lower degree.

■ Most of the rules of arithmetic for handling fractions carry over to rational expressions. For example, the L.C.D. has the same meaning except that we deal with polynomials in factored form rather than with integers.

■ The rules for positive integer exponents also apply to zero and negative integer exponents and to rational exponents.

■ Radical notation is simply another way of writing a rational exponent. That is, $\sqrt[n]{b} = b^{1/n}$.

■ If n is even and b is positive, there are two real numbers a such that $b^{1/n} = a$. Under these circumstances, we insist that the nth root be positive. That is, $\sqrt[n]{b}$ is a positive number if n is even and b is positive. Thus, $\sqrt{16} = 4$.

■ We must write $\sqrt{x^2} = |x|$ to insure that the result is a positive number.

■ To be in simplified form, a radical must satisfy the following conditions.

　　$\sqrt[n]{x^m}$ has $m < n$.

　　$\sqrt[n]{x^m}$ has no common factors between m and n.

　　The denominator must be rationalized.

■ Complex numbers were created because there are no real numbers which satisfy simple polynomial equations such as $x^2 + 5 = 0$.

■ Using the imaginary unit $i = \sqrt{-1}$, a complex number is of the form $a + bi$, where a and b are real numbers.

■ The real number system is a subset of the complex number system.

■ The complex numbers satisfy the field properties described in Section 1.1.

REVIEW EXERCISES

In Exercises 1–3, write each set by listing its elements within braces.

1. The set of natural numbers between -5 and 4, inclusive.
2. The set of integers between -3 and -1, inclusive.
3. The subset of $x \in S$, $S = \{0.5, 1, 1.5, 2\}$ such that x is an even integer.

For Exercises 4–7, determine whether the statement is true (T) or false (F).

4. $\sqrt{7}$ is a real number.
5. -35 is a natural number.
6. -14 is not an integer.
7. 0 is an irrational number.

In Exercises 8–11, identify the property of the real number system that justifies the statement. All variables represent real numbers.

8. $(3a) + (-3a) = 0$
9. $(3 + 4)x = 3x + 4x$
10. $2x + 2y + z = 2x + z + 2y$
11. $9x \cdot 1 = 9x$

In Exercises 12–14, sketch the given set of numbers on a real number line.

12. The negative real numbers.
13. The real numbers x such that $x > 4$.
14. The real numbers x such that $-1 \le x < 1$.
15. Find the value of $|-3| - |1 - 5|$.
16. Find $\overline{PQ}$ if the coordinates of P and Q are 6 and 9/2, respectively.
17. A salesperson receives $3.25x + 0.15y$ dollars, where x is the number of hours worked and y is the number of miles of automobile usage. Find the amount due the salesperson if $x = 12$ hours and $y = 80$ miles.
18. Which of the following expressions are not polynomials.

　　(a) $-2xy^2 + x^2y$

　　(b) $3b^2 + 2b - 6$

　　(c) $x^{-1/2} + 5x^2 - x$

　　(d) $7.5x^2 + 3x - \dfrac{1}{2}x^0$

In Exercises 19–20 indicate the leading coefficient and the degree of each polynomial.

19. $-0.5x^7 + 6x^3 - 5$
20. $2x^2 + 3x^4 - 7x^5$

In Exercises 21–23, perform the indicated operations.

21. $(3a^2b^2 - a^2b + 2b - a) - (2a^2b^2 + 2a^2b - 2b - a)$
22. $x(2x - 1)(x + 2)$
23. $3x(2x + 1)^2$

In Exercises 24–29, factor each expression.

24. $2x^2 - 2$
25. $x^2 - 25y^2$
26. $2a^2 + 3ab + 6a + 9b$
27. $4x^2 + 19x - 5$
28. $x^8 - 1$
29. $6x^4 + 7x^2 - 3$

In Exercises 30–33, perform the indicated operations and simplify.

30. $\dfrac{14(y-1)}{3(x^2-y^2)} \cdot \dfrac{9(x+y)}{-7xy^2}$

31. $\dfrac{4-x^2}{2y^2} \div \dfrac{x-2}{3y}$

32. $\dfrac{a+b}{a+2b} \cdot \dfrac{a^2-4b^2}{a^2-b^2}$

33. $\dfrac{x^2-2x-3}{2x^2-x} \div \dfrac{x^2-4x+3}{3x^3-3x^2}$

In Exercises 34–37, find the L.C.D.

34. $\dfrac{-1}{2x^2}, \dfrac{2}{x^2-4}, \dfrac{3}{x-2}$

35. $\dfrac{4}{x}, \dfrac{5}{x^2-x}, \dfrac{-3}{(x-1)^2}$

36. $\dfrac{2}{(x-1)y}, \dfrac{-4}{y^2}, \dfrac{x+2}{5(x-1)^2}$

37. $\dfrac{y-1}{x^2(y+1)}, \dfrac{x-2}{2xy-2x}, \dfrac{3x}{4y^2+8y+4}$

In Exercises 38–41, perform the indicated operations and simplify.

38. $2 + \dfrac{4}{a^2-4}$

39. $\dfrac{3}{x^2-16} - \dfrac{2}{x-4}$

40. $\dfrac{\dfrac{3}{x+2} - \dfrac{2}{x-1}}{x-1}$

41. $x^2 + \dfrac{\dfrac{1}{x}+1}{x-\dfrac{1}{x}}$

In Exercises 42–50, simplify and express the answers using only positive exponents.

42. $(2a^2b^{-3})^{-3}$

43. $2(a^2-1)^0$

44. $\left(\dfrac{x^3}{y^{-6}}\right)^{-4/3}$

45. $\dfrac{x^{3+n}}{x^n}$

46. $\sqrt{80}$

47. $\dfrac{2}{\sqrt{12}}$

48. $\sqrt{x^7y^5}$

49. $\sqrt[4]{32x^8y^6}$

50. $\dfrac{\sqrt{x}}{\sqrt{x}+\sqrt{y}}$

In Exercises 51–52, perform the indicated operations. Simplify the answer.

51. $\sqrt[4]{x^2y^2} + 2\sqrt[4]{x^2y^2}$

52. $(\sqrt{3}+\sqrt{5})^2$

53. Solve for x and for y: $(x-2) + (2y-1)i = -4 + 7i$

54. Simplify i^{47}.

In Exercises 55–57, perform the indicated operations and write all answers in the form $a + bi$.

55. $2 + (6-i)$

56. $(2+i)^2$

57. $(4-3i)(2+3i)$

PROGRESS TEST 1A

In Problems 1–2, write each set by listing its elements within braces.

1. The set of positive, even integers less than 13.
2. The subset of $x \in S$, $S = \{-1, 2, 3, 5, 7\}$, such that x is a multiple of 3.

In Problems 3–4, determine whether the statement is true (T) or false (F).

3. -1.36 is an irrational number.

4. π is equal to $\dfrac{22}{7}$.

In Problems 5–6, identify the property of the real number system that justifies the statement. All variables represent real numbers.

5. $xy(z+1) = (z+1)xy$

6. $(-6)\left(-\dfrac{1}{6}\right) = 1$

In Problems 7–8, sketch the given set of numbers on a real number line.

7. The integers that are greater than -3 and less than or equal to 3.

8. The real numbers x such that $-2 \le x < 1/2$.

9. Find the value of $|2 - 3| - |4 - 2|$.

10. Find $\overline{AB}$ if the coordinates of A and B are -6 and -4, respectively.

11. The area of a region is given by the expression $3x^2 - xy$. Find the area when $x = 5$ meters and $y = 10$ meters.

12. Evaluate the expression $\dfrac{-|y - 2x|}{|xy|}$ when $x = 3$ and $y = -1$.

13. Which of the following expressions are not polynomials?

(a) x^5 (b) $5x^{-4}y + 3x^2 - y$

(c) $4x^3 + x$ (d) $2x^2 + 3x^0$

In Problems 14–15, indicate the leading coefficients and the degree of each polynomial.

14. $-2.2x^5 + 3x^3 - 2x$ 15. $14x^6 - 2x + 1$

In Problems 16–17, perform the indicated operations.

16. $3xy + 2x + 3y + 2 - (1 - y - x + xy)$ 17. $(a + 2)(3a^2 - a + 5)$

In Problems 18–19, factor each expression.

18. $8a^3b^5 - 12a^5b^2 + 16a^2b$ 19. $4 - 9x^2$

In Problems 20–21, perform the indicated operations and simplify.

20. $\dfrac{m^4}{3n^2} \div \left(\dfrac{m^2}{9n} \cdot \dfrac{n}{2m^3} \right)$ 21. $\dfrac{16 - x^2}{x^2 - 3x - 4} \cdot \dfrac{x - 1}{x + 4}$

22. Find the L.C.D. of $\dfrac{-1}{2x^2}, \dfrac{2}{4x^2 - 4}, \dfrac{3}{x - 2}$

In Problems 23–24, perform the indicated operations and simplify.

23. $\dfrac{2x}{x^2 - 9} + \dfrac{5}{3x + 9}$ 24. $\dfrac{2 - \dfrac{4}{x + 1}}{x - 1}$

In Problems 25–28, simplify and express the answers only using positive exponents.

25. $\left(\dfrac{x^{7/2}}{x^{2/3}} \right)^{-6}$ 26. $\dfrac{y^{2n}}{y^{n-1}}$ 27. $\dfrac{-1}{(x - 1)^0}$ 28. $(2a^2b^{-1})^2$

In Problems 29–31, perform the indicated operations.

29. $3\sqrt[3]{24} - 2\sqrt[3]{81}$ 30. $(\sqrt{7} - 5)^2$ 31. $\dfrac{1}{2}\sqrt{\dfrac{xy}{4}} - \sqrt{9xy}$

32. For what values of x is $\sqrt{2 - x}$ a real number?

In Problems 33–34, perform the indicated operations and write all answers in the form $a + bi$.

33. $(2 - i) + (-3 + i)$ 34. $(5 + 2i)(2 - 3i)$

PROGRESS TEST 1B

In Problems 1–2, write each set by listing its elements within braces.

1. The set of positive, odd integers less than 10.

2. The subset of $x \in S$, $S = \{0, 15, 12, 24\}$, such that x is divisible by 3.

In Problems 3–4, determine whether the statement is true (T) or false (F).

3. 19.6 is a real number.   4. π is equal to 3.14.

In Problems 5–6, identify the property of the real number system that justifies the statement. All variables represent real numbers.

5. $a + b + c = c + a + b$ 6. $2(3 + x) = 6 + 2x$

In Problems 7–8, sketch the given set of numbers on a real number line.

7.   The natural numbers that are less than 5.

8.   The real numbers x such that $\frac{3}{2} < x < 3$.

9.   Find the value of $\dfrac{|2 - 5| + |1 - 5|}{|-7|}$.

10.   Find $\overline{AB}$ if the coordinates of A and B are -2 and 5, respectively.

11.   The area of a trapezoid is given by the formula $A = \frac{1}{2} h (b + b')$. Find the area if $h = 4$ meters, $b = 3$ meters, and $b' = 4$ meters.

12.   Evaluate the expression $|x|/|x - y|$ when $x = -2$ and $y = -3$.

13.   Which of the following expressions are not polynomials.

(a) $3x^2 + x^{-1} - 2$ (b) $2x^3 - xy^2 + x$

(c) $2x^2y^2 + xy - 4$ (d) $x^2y + x^{1/2}y + 2$

In Problems 14–15, indicate the leading coefficient and the degree of each polynomial.

14. $-3x^3 + 4x^5$ 15. $1.5x^{10} - x^9 + 17x^8$

In Problems 16–17, perform the indicated operations.

16. $(2s^2t^3 - st^2 + st - s + t) - (3s^2t^2 - 2s^2t - 4st^2 - t + 3)$

17. $(b + 3) (-3b^2 + 2b + 4)$

In Problems 18–19, factor each expression.

18. $5r^3s^4 - 40r^4s^3t$ 19. $2x^2 + 7x - 4$

In Problems 20–21, perform the indicated operations and simplify.

20. $\dfrac{3x^2(y - 1)}{6u^2v^3} \div \dfrac{(y - 1)^2}{2uv^2}$ 21. $\dfrac{x^2 + 7x - 8}{x - x^2} \cdot \dfrac{x}{x^2 + 8x}$

22.   Find the L.C.D. of $\dfrac{y - 1}{x^2(y + 1)}, \dfrac{x - 2}{2xy - 2x}, \dfrac{3x}{4y^2 + 8y + 4}$

In Problems 23–24, perform the indicated operations and simplify.

23. $\dfrac{-4}{x - 1} - \dfrac{3}{1 - x} + \dfrac{x}{x - 1}$ 24. $\dfrac{x + 2}{x^2 - x} - \dfrac{2x}{x + 1}$

In Problems 25–28, simplify and express the answers using only positive exponents.

25. $\dfrac{4x^{-3}}{x^{-2}}$ 26. $(b^2)^5(b^3)^6$ 27. $\left(\dfrac{x^8}{y^{12}}\right)^{3/4}$ 28. $\dfrac{2(x + 2)^0}{-2}$

In Problems 29–31, simplify the given expression.

29. $\sqrt{x^{14}y^{17}}$ 30. $\dfrac{-4}{2\sqrt{x} - 2}$ 31. $\sqrt[3]{a^3b^5}$

32.   For what values of x is $\dfrac{1}{\sqrt{x - 2}}$ a real number?

In Problems 33–34, perform the indicated operations and write all answers in the form $a + bi$.

33. $(4 - 2i) - \left(2 - \frac{1}{2} i\right)$ 34. $(3 - 2i) (2 - i)$

CHAPTER TWO
EQUATIONS AND INEQUALITIES

A major concern of algebra is the solution of equations. It is reasonable to ask questions such as these: Does a given equation have a solution? Is it possible for an equation to have more than one solution? Is there a procedure for solving an equation? In this chapter we will explore the answers to these questions for polynomial equations of the first and second degree. We will also see that the ability to solve equations enables us to tackle a wide variety of applications and word problems.

Linear inequalities also play an important role in solving word problems. For example, if we are required to combine food products in such a way that a specified minimum or maximum of protein is provided, we need to use inequalities. Many important industries, including steel and petroleum refineries, use computers daily to solve problems which involve thousands of inequalities. The solutions enable these concerns to optimize their "product mix" and their profitability.

2.1
LINEAR EQUATIONS IN ONE UNKNOWN

Expressions of the form

$$x - 2 = 0 \qquad x^2 - 9 = 0$$

$$3(2x - 5) = 3 \qquad 2x + 5 = \sqrt{x - 7}$$

$$\frac{1}{2x + 3} = 5 \qquad x^3 - 3x^2 = 32$$

are examples of equations in the unknown x. An **equation** states that two algebraic expressions are equal. We refer to these expressions as the **left-hand** and **right-hand sides** of the equation.

Our task is to find values of the unknown for which the equation holds true. These values are called **solutions** or **roots** of the equation, and the set of all solutions is called the **solution set**. For example, 2 is a solution of the equation $3x - 1 = 5$ since $3(2) - 1 = 5$. However, -2 is *not* a solution since $3(-2) - 1 \neq 5$.

The solutions of an equation depend upon the number system we are using. For example, the equation $2x = 5$ has no integer solutions but does have a solution among the rational numbers, namely $\frac{5}{2}$. Similarly, the equation $x^2 = -4$ has no solutions among the real numbers but does have solutions if we consider complex numbers, namely $2i$ and $-2i$. The solution sets of these two equations are $\{\frac{5}{2}\}$ and $\{2i, -2i\}$, respectively.

We shall say that an equation is an **identity** if it is true for every real number for which both sides of the equation are defined. For example,

$$x^2 - 1 = (x + 1)(x - 1)$$

is an identity since it is true for all real numbers, that is, every real number is a solution of the equation. The equation

$$x - 5 = 3$$

is a false statement for all values of x except 8. If, as in the preceding equation, there are real numbers for which the sides of an equation are both defined but unequal, then the equation is called a **conditional equation.**

When we say that we want to ''solve an equation'' we mean that we want to find *all* the solutions or roots. If we can replace an equation by another, simpler equation that has the same solutions, we will have an approach to solving equations. Equations having the same solutions are called **equivalent equations**. For example, $3x - 1 = 5$ and $3x = 6$ are equivalent equations since it can be shown that $\{2\}$ is the solution set of both equations.

There are two important rules that allow us to replace an equation by an equivalent equation.

Equivalent Equations

The solutions of a given equation are not affected by the following operations.
(1) Addition or subtraction of the same number or expression to both sides of the equation.
(2) Multiplication or division of both sides of the equation by a number other than 0.

EXAMPLE 1
Solve $3x + 4 = 13$.

Solution
We apply the preceding rules to this equation. The strategy is to isolate x, so we *subtract $+ 4$ from both sides* of the equation.

$$3x + 4 - 4 = 13 - 4$$

$$3x = 9$$

Dividing both sides by 3 we obtain the solution

$$x = 3$$

It's generally a good idea to check by substitution, to make sure that 3 does indeed satisfy the original equation.

$$3x + 4 \overset{?}{=} 13$$

$$3(3) + 4 \overset{?}{=} 13$$

$$13 \overset{\checkmark}{=} 13$$

To be technically accurate, the *solution* of the equation in Example 1 is 3, while $x = 3$ is an equation which is *equivalent* to the original equation. Now that this distinction is understood, we will join in the common usage which says that the equation $3x + 4 = 13$ "has the solution $x = 3$."

When the given equation contains rational expressions, we eliminate fractions by first multiplying by the least common denominator of all of the fractions. This technique is illustrated in Examples 2, 3, and 4 which follow.

EXAMPLE 2
Solve the equation.

$$\frac{5}{6}x - \frac{4}{3} = \frac{3}{5} x + 1$$

Solution
We first eliminate fractions by multiplying both sides of the equation by the L.C.D. of all fractions, which is 30. Thus, we obtain

$$30\left(\frac{5}{6}x - \frac{4}{3}\right) = 30\left(\frac{3}{5} x + 1\right)$$

$$25x - 40 = 18x + 30$$

$$7x = 70$$

$$x = 10$$

The student should verify that $x = 10$ is a solution of the original equation.

PROGRESS CHECK
Solve and check.

(a) $-\frac{2}{3} (x - 5) = \frac{3}{2} (x + 1)$

(b) $\frac{1}{3}x + 2 - 3 \left(\frac{x}{2} + 4\right) = 2\left(\frac{x}{4} - 1\right)$

Answers

(a) $\frac{11}{13}$ (b) $-\frac{24}{5}$

The equations we have solved are all of the first degree and involve only one unknown. Such equations are called **first-degree equations in one**

unknown, or more simply, **linear equations**. The general form of such equations is

$$ax + b = 0$$

where a and b are any real numbers and $a \neq 0$. Let's see how we would solve this equation.

$$ax + b = 0$$
$$ax + b - b = 0 - b \qquad \text{Subtract } b \text{ from both sides.}$$
$$ax = -b$$
$$\frac{ax}{a} = \frac{-b}{a} \qquad \text{Divide both sides by } a \neq 0.$$
$$x = -\frac{b}{a}$$

We have thus obtained the following result.

Roots of a Linear Equation

The linear equation $ax + b = 0$, $a \neq 0$, has exactly one solution: $-\dfrac{b}{a}$

Sometimes we are led to linear equations in the course of solving other equations. The following example illustrates this situation.

EXAMPLE 3
Solve $\dfrac{5x}{x + 3} - 3 = \dfrac{1}{x + 3}$.

Solution
The L.C.D. of all fractions is $x + 3$. Multiplying both sides of the equation by $x + 3$ to eliminate fractions, we obtain

$$5x - 3(x + 3) = 1$$
$$5x - 3x - 9 = 1$$
$$2x = 10$$
$$x = 5$$

Checking our solution we have

$$\frac{5(5)}{5 + 3} - 3 \overset{?}{=} \frac{1}{5 + 3}$$
$$\frac{25}{8} - 3 \overset{?}{=} \frac{1}{8}$$
$$\frac{1}{8} \overset{\checkmark}{=} \frac{1}{8}$$

We said earlier that multiplication of both sides of an equation by any nonzero number results in an equivalent equation. What happens if we multiply an equation by an expression that contains an unknown? In

Example 3 this procedure worked just fine and gave us a solution. But this will not always be so, because the answer we obtain may produce a zero denominator when substituted in the original equation. The following rule must therefore be carefully observed.

> When multiplying or dividing both sides of an equation by an expression that contains the unknown, the resulting equation may not be equivalent to the original equation. The answer obtained must be substituted in the original equation to verify that it is a solution.

EXAMPLE 4
Solve and check.

$$\frac{8x + 1}{x - 2} + 4 = \frac{7x + 3}{x - 2}$$

Solution
The L.C.D. of all fractions is $x - 2$. Multiplying both sides of the equation by $x - 2$ we eliminate fractions and obtain

$$8x + 1 + 4(x - 2) = 7x + 3$$
$$8x + 1 + 4x - 8 = 7x + 3$$
$$5x = 10$$
$$x = 2$$

Checking our answer we find that $x = 2$ is not a solution since substituting $x = 2$ in the original equation yields a denominator of zero. Thus, the given equation has no solution.

PROGRESS CHECK
Solve and check.

(a) $\dfrac{3}{x} - 1 = \dfrac{1}{2} - \dfrac{6}{x}$ (b) $-\dfrac{2x}{x + 1} = 1 + \dfrac{2}{x + 1}$

Answers
(a) $x = 6$ (b) *no solution*

EXERCISE SET 2.1
In the following exercises determine whether the given statement is true (T) or false (F).

1. $x = -5$ is a solution of $2x + 3 = -7$.

2. $x = \dfrac{5}{2}$ is a solution of $3x - 4 = \dfrac{5}{2}$.

3. $x = \dfrac{6}{4 - k}$, $k \neq 4$, is a solution of $kx + 6 = 4x$.

4. $x = \dfrac{7}{3k}$, $k \neq 0$, is a solution of $2kx + 7 = 5x$.

Solve the given linear equation and check your answer in the following exercises.

5. $3x + 5 = -1$ 6. $5r + 10 = 0$

7. $2 = 3x + 4$

8. $\frac{1}{2}s + 2 = 4$

9. $\frac{3}{2}t - 2 = 7$

10. $-1 = -\frac{2}{3}x + 1$

11. $0 = -\frac{1}{2}a - \frac{2}{3}$

12. $4r + 4 = 3r - 2$

13. $-5x + 8 = 3x - 4$

14. $2x - 1 = 3x + 2$

15. $-2x + 6 = -5x - 4$

16. $6x + 4 = -3x - 5$

17. $2(3b + 1) = 3b - 4$

18. $-3(2x + 1) = -8x + 1$

19. $4(x - 1) = 2(x + 3)$

20. $-3(x - 2) = 2(x + 4)$

21. $2(x + 4) - 1 = 0$

22. $3a + 2 - 2(a - 1) = 3(2a + 3)$

23. $-4(2x + 1) - (x - 2) = -11$

24. $3(a + 2) - 2(a - 3) = 0$

Solve for x.

25. $kx + 8 = 5x$

26. $8 - 2kx = -3x$

27. $2 - k + 5(x - 1) = 3$

28. $3(2 + 3k) + 4(x - 2) = 5$

Solve and check.

29. $\frac{x}{2} = \frac{5}{3}$

30. $\frac{3x}{4} - 5 = \frac{1}{4}$

31. $\frac{2}{x} + 1 = \frac{3}{x}$

32. $\frac{5}{a} - \frac{3}{2} = \frac{1}{4}$

33. $\frac{2y - 3}{y + 3} = \frac{5}{7}$

34. $\frac{1 - 4x}{1 - 2x} = \frac{9}{8}$

35. $\frac{1}{x - 2} + \frac{1}{2} = \frac{2}{x - 2}$

36. $\frac{4}{x - 4} - 2 = \frac{1}{x - 4}$

37. $\frac{2}{x - 2} + \frac{2}{x^2 - 4} = \frac{3}{x + 2}$

38. $\frac{3}{x - 1} + \frac{2}{x + 1} = \frac{5}{x^2 - 1}$

39. $\frac{x}{x - 1} - 1 = \frac{3}{x + 1}$

40. $\frac{2}{x - 2} + 1 = \frac{x + 2}{x - 2}$

41. $\frac{4}{b} - \frac{1}{b + 3} = \frac{3b + 2}{b^2 + 2b - 3}$

42. $\frac{3}{x^2 - 2x} + \frac{2x - 1}{x^2 + 2x - 8} = \frac{2}{x + 4}$

43. $\frac{3r + 1}{r + 3} + 2 = \frac{5r - 2}{r + 3}$

44. $\frac{2x - 1}{x - 5} + 3 = \frac{3x - 2}{5 - x}$

Indicate whether the equation is an identity (I) or a conditional equation (C).

45. $x^2 + x - 2 = (x + 2)(x - 1)$

46. $(x - 2)^2 = x^2 - 4x + 2$

47. $2x + 1 = 3x - 1$

48. $3x - 5 = 4x - x - 2 - 3$

Write (T) if the equations in each exercise are all equivalent equations, and (F) if they are not equivalent.

49. $2x - 3 = 5$ $2x = 8$ $x = 4$

50. $5(x - 1) = 10$ $x - 1 = 2$ $x = 3$

51. $x(x - 1) = 5x$ $x - 1 = 5$ $x = 6$

52. $x = 5$ $x^2 = 25$

53. $3(x^2 + 2x + 1) = -6$ $x^2 + 2x + 1 = -2$ $(x + 1)^2 = -2$

54. $(x + 3)(x - 1) = x^2 - 2x + 1$ $(x + 3)(x - 1) = (x - 1)^2$
    $x + 3 = x - 1$

2.2
APPLICATIONS

Many applied problems lead to linear equations that must be solved. The solution procedure described in Section 2.1 was already familiar to you and probably presents no difficulties. The challenging aspect of applied problems is that of translating from words to appropriate algebraic forms. This translation process requires a kind of ability that you can acquire only with practice.

The steps listed here can serve to guide you in solving word problems.

Step 1. Read the problem carefully to understand what is required.

Step 2. Separate what is known and what is to be found.

Step 3. In many problems, the unknown quantity answers questions such as "how much" or "how many." Let an algebraic symbol, say x, represent the unknown.

Step 4. Represent other quantities in the problem, if possible, in terms of x.

Step 5. Find the relationship in the problem which lets you write an equation (or an inequality).

Step 6. Solve and check.

The words and phrases in Table 1 should prove helpful in translating a word problem into an algebraic expression that can be solved.

TABLE 1

Word or phrase	Algebraic symbol	Example	Algebraic expression
Sum	+	Sum of two numbers	$a + b$
Difference	−	Difference of two numbers	$a - b$
		Difference of a number and 3	$x - 3$
Product	× or ·	Product of two numbers	$a \cdot b$
Quotient	÷ or /	Quotient of two numbers	$\frac{a}{b}$ or a/b
Exceeds		a exceeds b by 3	$a = b + 3$
More than		a is 3 more than b	or
More of		There are 3 more of a than of b	$a - 3 = b$
Twice		Twice a number	$2x$
		Twice the difference of x and 3	$2(x - 3)$
		3 more than twice a number	$2x + 3$
		3 less than twice a number	$2x - 3$
Is or equals	=	The sum of a number and 3 is 15	$x + 3 = 15$

EXAMPLE 1
If you pay $66 for a car radio after receiving a 25% discount, what was the price of the radio before the discount?

Solution
Let x = the price of the radio (in dollars) before the discount.

Then

$$0.25x = \text{the amount discounted}$$

and the price of the radio after the discount is given by

$$x - 0.25x$$

Hence

$$x - 0.25x = 66$$
$$0.75x = 66$$
$$x = \frac{66}{0.75} = 88$$

The price of the radio was $88 before the discount.

COIN PROBLEMS

Coin problems are easy to interpret if you keep this in mind—always distinguish between the *number* of coins and the *value* of the coins. Having done that, you will also find it helpful to use a chart, as in the following example.

EXAMPLE 2
A purse contains $3.20 in quarters and dimes. If there are 3 more quarters than dimes, how many coins of each type are there?

Solution
In this problem, we may let the unknown represent the number of either quarters or dimes. We make a choice. Let

$$n = \text{the number of quarters}$$

Then $\qquad\qquad n - 3 = \text{the number of dimes}$

since there are 3 more quarters than dimes.

The following table is useful in further analysis of the problem.

	Number of coins $\times$	Number of cents in each coin	$=$ Value in cents
Quarters	n	25	$25n$
Dimes	$n - 3$	10	$10(n - 3)$

We know that

$$\text{total value} = (\text{value of quarters}) + (\text{value of dimes})$$
$$320 = 25n + 10(n - 3)$$
$$320 = 25n + 10n - 30$$
$$350 = 35n$$
$$10 = n$$

Then

$$n = \text{number of quarters} = 10$$
$$n - 3 = \text{number of dimes} = 7$$

Now verify that the total value of all the coins is $3.20.

SIMPLE INTEREST

Interest is the fee charged for borrowing money. In this section we will deal only with simple interest, which assumes the fee to be a fixed percentage r of the amount borrowed. We call the amount borrowed the **principal** and denote it by P.

If the principal P is borrowed at a simple interest rate r, then the interest due at the end of each year is Pr and the total interest I due at the end of t years is

$$I = Prt$$

Consequently, if S is the total amount owed at the end of t years, then

$$S = P + Prt$$

since both the principal and interest are to be repaid.

The basic formulas for simple interest calculations that we have derived are

$$I = Prt$$
$$S = P + Prt$$

EXAMPLE 3

A part of $7000 is borrowed at 6% simple annual interest and the remainder at 8%. If the total amount of interest due after 3 years is $1380, how much was borrowed at each rate?

Solution
Let

$$n = \text{the amount borrowed at 6\%}$$

Then

$$7000 - n = \text{the amount borrowed at 8\%}$$

since the total amount is $7000. We can display the information in table form using the equation $I = Prt$.

	P	$\times$	r	$\times$	t	$=$	Interest
6% portion	n		0.06		3		$0.18n$
8% portion	$7000 - n$		0.08		3		$0.24(7000 - n)$

Note that we write the rate r in its decimal form, so that 6% = 0.06 and 8% = 0.08.

Since the total interest of $1380 is the sum of the interest from the two portions, we have

$$1380 = 0.18n + 0.24(7000 - n)$$

$$1380 = 0.18n + 1680 - 0.24n$$

$$0.06n = 300$$

$$n = 5000$$

We conclude that $5000 was borrowed at 6% and $2000 was borrowed at 8%.

DISTANCE (UNIFORM MOTION) PROBLEMS

Here is the key to the solution of distance problems.

$$\text{Distance} = \text{rate} \times \text{time}$$

$$\text{or}$$

$$d = r \cdot t$$

The relationships that permit you to write an equation are sometimes obscured by the words. Here are some questions to ask as you set up a distance problem.

(a) Are there two distances that are equal? Will two objects have traveled the same distance? Is the distance on a return trip the same as the distance going?

(b) Is the sum (or difference) of two distances equal to a constant? When two objects are traveling toward each other, they meet when the sum of the distances traveled by each equals the original distance between them.

EXAMPLE 4

Two trains leave New York for Chicago. The first train travels at an average speed of 60 miles per hour while the second train, which departs an hour later, travels at an average speed of 80 miles per hour. How long will it take the second train to overtake the first train?

Solution

Since we are interested in the time the second train travels, we choose to let

$$t = \text{the number of hours the second train travels}$$

Then

$$t + 1 = \text{the number of hours the first train travels}$$

since the first train departs one hour earlier.

	Rate	$\times$	Time	$=$	Distance
First train	60		$t + 1$		$60(t + 1)$
Second train	80		t		$80t$

At the moment the second train overtakes the first, they must both have traveled the *same* distance. Thus,

$$60(t + 1) = 80t$$

$$60t + 60 = 80t$$

$$60 = 20t$$

$$3 = t$$

It takes the second train 3 hours to catch up with the first train.

MIXTURE PROBLEMS

One type of mixture problem involves mixing commodities, say two or more types of nuts, to obtain a mix which has a desired value. If the commodities are measured in pounds, the relationships we need are

> Number of pounds × Price per pound = Value of the commodity
>
> Pounds in mix = Sum of the number of pounds in each commodity
>
> Value of mix = Sum of the values of each commodity

EXAMPLE 5

How many pounds of Brazilian coffee worth $5 per pound must be mixed with 20 pounds of Colombian coffee worth $4 per pound to produce a mixture worth $4.20 per pound?

Solution

Let n = number of pounds of Brazilian coffee. We display all the information, using cents in place of dollars.

Type of coffee	Number of pounds	×	Price per pound	=	Value (in cents)
Brazilian	n		500		$500n$
Colombian	20		400		8000
Mixture	$n + 20$		420		$420(n + 20)$

Note that the weight of the mixture equals the sum of the weights of the Brazilian and Colombian coffees that make up the mixture. Since the value of the mixture is the sum of the values of the two types of coffee,

$$\text{Value of mixture} = (\text{Value of Brazilian}) + (\text{Value of Colombian})$$
$$420(n + 20) = 500n + 8000$$
$$420n + 8400 = 500n + 8000$$
$$400 = 80n$$
$$5 = n$$

We must add 5 pounds of Brazilian coffee to make the required mixture.

WORK PROBLEMS

Work problems typically involve two or more people or machines working on the same task. The key to these problems is to express the *rate of work per unit of time*, whether it is per hour, day, week, or other. For example, if a machine can do a job in 5 days, then

$$\text{rate of machine} = \frac{1}{5} \text{ job per day}$$

If this machine were used for two days, it would perform $\frac{2}{5}$ of the job. In summary,

> If a machine (or person) can complete a job in n days, then
>
> $$\text{Rate of machine (or person)} = \frac{1}{n} \text{ job per day}$$
>
> Work done = Rate × Time

EXAMPLE 6

Using a small mower, at 12 noon a student begins to mow a lawn, a job which would take him 9 hours. At 1 P.M. another student, using a tractor, joins him and they complete the job together at 3 P.M. How many hours would it take to do the job by tractor only?

Solution

Let x = number of hours to do the job by tractor alone. The small mower works from 12 noon to 3 P.M., or 3 hours; the tractor is used from 1 P.M. to 3 P.M., or 2 hours.

All of the information can be displayed in table form.

	Rate	×	Time	=	Work done
Small mower	$\dfrac{1}{9}$		3		$\dfrac{3}{9}$
Tractor	$\dfrac{1}{x}$		2		$\dfrac{2}{x}$

Since

$$\begin{array}{c}\text{work done by} \\ \text{small mower}\end{array} + \begin{array}{c}\text{work done by} \\ \text{tractor}\end{array} = 1 \ \ \text{whole job}$$

$$\frac{3}{9} + \frac{2}{x} = 1$$

To solve, multiply both sides by the L.C.D., which is $9x$.

$$9x\left(\frac{3}{9} + \frac{2}{x}\right) = 9x \cdot 1$$

$$3x + 18 = 9x$$

$$x = 3$$

Thus, by tractor alone, the job can be done in 3 hours.

LITERAL EQUATIONS

The circumference C of a circle is given by the formula

$$C = 2\pi r$$

where r is the radius of the circle. For every value of r, the formula gives us a value of C. If $r = 20$, we have

$$C = 2\pi(20) = 40\pi$$

It is sometimes convenient to be able to turn a formula around, that is, to be able to solve for a different variable. For example, if we want to express the radius of a circle in terms of the circumference, we have

$$C = 2\pi r$$

$$\frac{C}{2\pi} = \frac{2\pi r}{2\pi} \qquad \text{Divide by } 2\pi$$

$$\frac{C}{2\pi} = r$$

Now, given a value of C, we can determine a value of r.

EXAMPLE 7

If an amount P is borrowed at the simple annual interest rate r, then the amount S due at the end of t years is

$$S = P + Prt$$

Solve for P.

Solution

$$S = P + Prt$$

$$S = P(1 + rt) \qquad \text{Common factor } P$$

$$\frac{S}{1 + rt} = P \qquad \text{Divide both sides by } (1 + rt)$$

EXERCISE SET 2.2

In Exercises 1–3, let n represent the unknown. Translate from words to an algebraic expression or equation.

1. The number of blue chips is 3 more than twice the number of red chips.
·2. The number of station wagons on a parking lot is 20 fewer than three times the number of sedans.
3. Five less than 6 times a number is 26.
4. Janis is 3 years older than her sister. Thirty years from now the sum of their ages will be 111. Find the current ages of the sisters.
5. John is presently 12 years older than Fred. Four years ago, John was twice as old as Fred. How old is each now?
6. The larger of two numbers is 3 more than twice the smaller. If their sum is 18, find the numbers.
7. Find three consecutive integers whose sum is 21.
8. A certain number is 5 less than another number. If their sum is 11, find the two numbers.
9. A resort guarantees that the average temperature over the period Friday, Saturday, and Sunday will be exactly 80°F, or else each guest pays only half price for the facilities. If the temperatures on Friday and Saturday were 90°F and 82°F, respectively, what must the temperature be on Sunday so that the resort does not lose half of its revenue?
10. A patient's temperature was taken at 6 A.M., 12 noon, 3 P.M., and 8 P.M. The first, third, and fourth readings were 102.5°, 101.5°, and 102°F, respectively. The nurse forgot to write down the second reading but recorded that the average of the four readings was 101.5°F. What was the second temperature reading?
11. A 12-meter-long steel beam is to be cut into two pieces so that one piece will be 4 meters longer than the other. How long will each piece be?
12. A rectangular field whose length is 10 meters longer than its width is to be enclosed with exactly 100 meters of fencing material. What are the dimensions of the field?
13. A vending machine contains $3.00 in nickels and dimes. If the number of dimes is five more than twice the number of nickels, how many coins of each type are there?
14. A wallet contains $460 in $5, $10, and $20 bills. The number of $5 bills exceeds twice the number of $10 bills by 4, while the number of $20 bills is 6 fewer than the number of $10 bills. How many bills of each type are there?
15. A movie theatre charges $3 admission for an adult and $1.50 for a child. If

700 tickets were sold and the total revenue received was $1650, how many tickets of each type were sold?

16. A student buys 5¢, 10¢, and 15¢ stamps, with a total value of $6.70. If the number of 5¢ stamps is 2 more than the number of 10¢ stamps, and the number of 15¢ stamps is 5 more than one half the number of 10¢ stamps, how many stamps of each denomination did the student obtain?

17. An amateur theatre group is converting a classroom to an auditorium for a forthcoming play. The group will sell $3, $5, and $6 tickets, and will receive exactly $503 from the sale of tickets. If the number of $5 tickets is twice the number of $6 tickets and the number of $3 tickets is one more than three times the number of $6 tickets, how many tickets of each type are there?

18. To pay for their child's college education, the parents invest $10,000 in two separate investments. Part is in a certificate of deposit, paying 8.5% annual interest, the rest is in a mutual fund paying 7% annual interest. The annual income from the certificate of deposit is $200 more than the annual income from the mutual fund. How much money was put into each type of investment?

19. A bicycle store selling 3-speed and 10-speed models has $16,000 in inventory. The profit on a 3-speed bicycle is 11% while the profit on a 10-speed model is 22%. If the profit on the entire stock is 19%, how much was invested in each type of bicycle?

20. A film shop carrying black-and-white film and color film has $4000 in inventory. The profit on black-and-white film is 12% while the profit on color film is 21%. If the annual profit on color film is $150 less than the annual profit on black-and-white film, how much was invested in each type of film?

21. A firm borrows $12,000 at a simple annual interest rate of 8% for a period of 3 years. At the end of the first year, the firm finds that its needs are reduced. The firm returns a portion of the original loan and retains the remainder until the end of the three year period. If the total interest paid was $1760, how much was returned at the end of the first year?

22. A finance company lends a certain amount of money to Firm A at 7% annual interest; an amount $100 less than that lent to Firm A is lent to Firm B at 8%; and an amount $200 more than that lent to Firm A is lent to Firm C at 8.5% for one year. If the total annual income is $126.50, how much was lent to each firm?

23. Two trucks leave Philadelphia for Miami. The first truck to leave travels at an average speed of 50 kilometers per hour. The second truck, which leaves two hours later, travels at an average speed of 55 kilometers per hour. How long will it take the second truck to overtake the first truck?

24. Jackie either drives or bicycles from home to school. Her average speed when driving is 36 miles per hour and her average speed when bicycling is 12 miles per hour. If it takes her ½ hour less to drive than to bicycle to school, how long does it take her to go to school and how far is the school from her home?

25. Professors Roberts and Jones, who live 676 miles apart, are exchanging houses and jobs for the summer. They start out at exactly the same time for their new locations and meet after 6.5 hours of driving. If their average speeds differ by 4 miles per hour, what are their average speeds?

26. Steve leaves school by moped for spring vacation. Forty minutes later, his roommate, Frank, notices that Steve forgot to take his camera, so he decides to try to catch up with him by car. If Steve's average speed is 25 miles per hour and Frank averages 45 miles per hour, how long does it take Frank to overtake Steve?

27. An express train and a local train start out from the same point at the same time and travel in opposite directions. The express train travels twice as fast as the local train. If after 4 hours they are 480 kilometers apart, what was the average speed of each train?

28. How many pounds of raisins worth $1.50 per pound must be mixed with 10 pounds of peanuts worth $1.20 per pound to produce a mixture worth $1.40 per pound?

29. How many ounces of Ceylon tea worth $1.50 per ounce and how many ounces of Formosa tea worth $2.00 per ounce must be mixed to obtain a mixture of 8 ounces which is worth $1.85 per ounce?

30. A copper alloy which is 40% copper is to be combined with a copper alloy which is 80% copper to produce 120 kilograms of an alloy which is 70% copper. How many kilograms of each alloy must be used?

31. A vat contains 27 gallons of water and 9 gallons of acetic acid. How many gallons of water must be evaporated if the resulting solution is to be 40% acetic acid?

32. A producer of packaged frozen vegetables wants to market the product at $1.20 per kilogram. How many kilograms of green beans worth $1 per kilogram must be mixed with 100 kilograms of corn worth $1.30 per kilogram and 90 kilograms of peas worth $1.40 per kilogram to produce a satisfactory mix?

33. A certain number is three times another. If the difference of their reciprocals is 8, find the numbers.

34. If $\frac{1}{3}$ is subtracted from 3 times the reciprocal of a certain number, the result is $\frac{25}{6}$. Find the number.

35. Computer A can carry out an engineering analysis in 4 hours, while computer B can do the same job in 6 hours. How long would it take to complete the job if both computers work together?

36. Jackie, Lisa, and Susan can paint a certain room in 3, 4, and 2 hours, respectively. How long would it take to paint the room if they all work together?

37. A senior copy editor together with a junior copy editor can edit a book in 3 days. The junior editor, working alone, takes twice as long to complete the job as the senior editor would require if working alone. How long would it take each editor to complete the job by herself?

38. Hose A can fill a certain vat in 3 hours. After 2 hours of pumping, hose A is turned off. Hose B is then turned on and completes filling the vat in 3 hours. How long would it take hose B alone to fill the vat?

39. A printing shop starts a job at 10 A.M. on press A. Using this press alone, it takes 8 hours to complete the job. At 2 P.M., press B is also turned on and both presses together finish the job at 4 P.M.. How long would it take press B alone to do the job?

40. A boat travels 20 kilometers upstream in the same time that it takes the same boat to travel 30 kilometers downstream. If the rate of the stream is 5 kilometers per hour, find the speed of the boat in still water.

41. An airplane flying against the wind travels 300 miles in the same time that it takes the same plane to travel 400 miles with the wind. If the wind speed is 20 miles per hour, find the speed of the airplane in still air.

In Exercises 42–51, solve for the indicated variable in terms of the remaining variables.

42. $A = Pr$ for r 43. $C = 2\pi r$ for r

44. $V = \frac{1}{3} \pi r^2 h$ for h

45. $F = \frac{5}{9} C + 32$ for C

46. $S = \frac{1}{2} gt^2 + vt$ for v

47. $A = \frac{1}{2} h (b + b')$ for b

48. $A = P(1 + rt)$ for r

49. $\frac{1}{f} = \frac{1}{f_1} + \frac{1}{f_2}$ for f_2

50. $a = \frac{v_1 - v_0}{t}$ for v_0

51. $S = \frac{a - rL}{L - r}$ for L

2.3
LINEAR INEQUALITIES

The terms *solution* and *solution set* as defined for equations have a parallel meaning when applied to inequalities. Thus, $x = 2$ is a solution of the inequality $2x + 3 > 5$ since

$$2(2) + 3 = 7 > 5$$

shows that 2 satisfies the inequality. We seek the solution set or set of all real values of the unknown that satisfy the inequality.

   The properties of inequalities listed in Section 1.2 enable us to use the same procedures in solving inequalities as those used in solving equations, with one exception. Multiplication or division of an inequality by a *negative* number reverses the sense of the inequality.

EXAMPLE 1
Solve the inequality.

$$2x + 11 \geq 5x - 1$$

Solution
We will perform addition and subtraction to collect terms in x just as we did for equations.

$$2x + 11 \geq 5x - 1$$
$$2x \geq 5x - 12$$
$$-3x \geq -12$$

We now divide both sides of the inequality by -3, a negative number, and therefore reverse the sense of the inequality.

$$\frac{-3x}{-3} \leq \frac{-12}{-3} \, .$$
$$x \leq 4$$

PROGRESS CHECK
Solve the inequality $3x - 2 \geq 5x + 4$.

Answer
$x \leq -3$

WARNING

Given the inequality

$$-2x \geq -6$$

it is a common error to conclude that dividing by -2 gives $x \leq -3$. Multiplication or division by a negative number changes the sense of the inequality but the *signs* obey the usual rules of algebra. Thus,

$$-2x \geq -6$$

$$\frac{-2x}{-2} \leq \frac{-6}{-2}$$

$$x \leq 3$$

There are three methods that are commonly used to designate certain subsets of the real number line: graphs on a real number line, interval notation, and set-builder notation. Since there are other occasions upon which we will use these notational schemes, this is a convenient time to introduce them and to apply them to inequalities.

Consider the inequality $a \leq x < b$. The graph of this inequality is shown in Figure 1. The portion of the real number line that is in color is the solution set of the inequality. The circle at point a has been filled in to indicate that a is also a solution of the inequality; the circle at point b has been left open to indicate that b is not a member of the solution set.

a $\qquad\qquad\qquad$ b

FIGURE 1

An **interval** is a set of numbers on the real number line that forms a line segment, a half-line, or the entire real number line. The subset of the real number line shown in Figure 1 would be written in **interval notation** as $[a, b)$ where a and b are the **endpoints** of the interval. The bracket [or] indicates that the endpoint is included, while the parenthesis (or) indicates that the endpoint is not included. The interval $[a, b]$ is called a **closed interval** since both endpoints are included. The interval (a,b) is called an **open interval** since neither endpoint is included. Finally, the intervals $[a, b)$ and $(a, b]$ are called **half-open intervals.**

The set of all real numbers satisfying a given property P is written as

$$\{x | x \text{ satisfies property } P\}$$

which is read as "the set of all x such that x satisfies property P." This form is called **set-builder notation** and provides a third means of designating subsets of the real number line. Thus, the interval $[a, b)$ shown in Figure 1 is written as

$$\{x | a \leq x < b\}$$

which indicates that x must satisfy the inequalities $x \geq a$ and $x < b$.

EXAMPLE 2

Graph each of the given intervals on a real number line and indicate the same subset of the real number line in set-builder notation.

(a) $(-3, 2]$ (b) $(1, 4)$ (c) $[-4, -1]$

Solution

(a)

$\{x \mid -3 < x \leqslant 2\}$

(b)

$\{x \mid 1 < x < 4\}$

(c)

$\{x \mid -4 \leqslant x \leqslant -1\}$

To describe the inequalities $x > 2$ and $x \leq 3$ in interval notation we need to introduce the symbols ∞ and $-\infty$ (read "infinity" and "minus infinity," respectively). The inequalities $x > 2$ and $x \leq 3$ are then written as $(2, \infty)$ and $(-\infty, 3]$, respectively, in interval notation and would be graphed on a real number line as shown in Figure 2. Note that ∞ and $-\infty$ are symbols (and not numbers) indicating that the intervals extend indefinitely. An interval using one of these symbols is called an **infinite interval**. In particular, the interval $(-\infty, \infty)$ designates the real number line.

FIGURE 2

EXAMPLE 3

Graph each inequality and write the solution set in interval notation.

(a) $x \leq -2$ (b) $x \geq -1$ (c) $x < 3$

Solution

(a)

$(-\infty, -2]$

(b)

$[-1, \infty)$

(c)

$(-\infty, 3)$

EXAMPLE 4

Solve the inequality.

$$\frac{x}{2} - 9 < \frac{1 - 2x}{3}$$

Graph the solution set and write the solution set in both interval notation and set-builder notation.

Solution
To clear the inequality of fractions, we multiply both sides by the L.C.D. of all fractions, which is 6.

$$3x - 54 < 2(1 - 2x)$$
$$3x - 54 < 2 - 4x$$
$$7x < 56$$
$$x < 8$$

We may write the solution set as $\{x \mid x < 8\}$ or as the infinite interval $(-\infty, 8)$. The graph of the solution set is shown in Figure 3.

FIGURE 3

EXAMPLE 5
Solve the inequality.

$$\frac{2(x + 1)}{3} < \frac{2x}{3} - \frac{1}{6}$$

Solution
The L.C.D. of all fractions is 6. Multiplying both sides of the inequality by 6 we obtain

$$4(x + 1) < 4x - 1$$
$$4x + 4 < 4x - 1$$
$$4 < -1$$

Our procedure has led to a contradiction, indicating that there is no solution to the inequality.

PROGRESS CHECK
Solve and write the answer in interval notation.

(a) $\frac{3x - 1}{4} + 1 > 2 + \frac{x}{3}$ (b) $\frac{2x - 3}{2} \geq x + \frac{2}{5}$

Answers

(a) $(3, \infty)$ (b)   *no solution*

We can solve double inequalities such as

$$1 < 3x - 2 \leq 7$$

by operating on both inequalities at the same time.

$$3 < 3x \leq 9 \qquad \text{Add } +2 \text{ to each member.}$$
$$1 < x \leq 3 \qquad \text{Divide each member by 3.}$$

The solution set is the half-open interval $(1,3]$.

EXAMPLE 6
Solve the inequality $-3 \leq 1 - 2x < 6$ and write the answer in interval notation.

Solution
Operating on both inequalities we have

$$-4 \le -2x < 5 \qquad \text{Add } (-1) \text{ to each member.}$$

$$2 \ge x > -\frac{5}{2} \qquad \text{Divide each member by } -2.$$

The solution set is the half-open interval $\left(-\frac{5}{2}, 2\right]$.

PROGRESS CHECK
Solve the inequality $-5 < 2 - 3x < -1$ and write the answer in interval notation.

Answer

$\left(1, \frac{7}{3}\right)$

EXAMPLE 7
A taxpayer may choose to pay a 20% tax on the gross income or a 25% tax on the gross income less $4000. Above what income level should the taxpayer elect to pay at the 20% rate?

Solution
If we let x = gross income, then the choice available to the taxpayer is
(a) pay at the 20% rate on the gross income, that is, pay $0.20x$, or
(b) pay at the 25% rate on the gross income less $4000, that is, pay $0.25(x - 4000)$.
To determine when (a) produces a lower tax than (b), we must solve

$$0.20x \le 0.25(x - 4000)$$

$$0.20x \le 0.25x - 1000$$

$$-0.05x \le -1000$$

$$x \ge \frac{1000}{0.05} = 20,000$$

The taxpayer should choose to pay at the 20% rate if the gross income is $20,000 or more.

PROGRESS CHECK
A customer is offered the following choice of telephone services: unlimited local calls at a fixed $20 monthly charge, or a base rate of $8 per month plus 6¢ per message unit. At what level of usage does it cost less to choose the unlimited service?

Answer
When the anticipated use exceeds 200 message units.

EXERCISE SET 2.3
Express the given inequality in interval notation.

1. $-5 \le x < 1$ 2. $-4 < x \le 1$ 3. $x > 9$

4. $x \le -2$ 5. $-12 \le x \le -3$ 6. $x \ge -5$

7. $3 < x < 7$ 8. $x < 17$ 9. $-6 < x \le -4$

Express the given interval as an inequality.

10. $(-4, 3]$ 11. $[5, 8]$ 12. $(-\infty, -2]$

13. $(3, \infty)$ 14. $[-3, 10)$ 15. $(-\infty, 5]$

16. $(-2, -1)$ 17. $[0, \infty)$ 18. $(-5, 7)$

Solve the inequality and graph the result.

19. $x + 4 < 8$ 20. $x + 5 < 4$ 21. $x + 3 < -3$

22. $x - 2 \le 5$ 23. $x - 3 \ge 2$ 24. $x + 5 \ge -1$

25. $2 < a + 3$ 26. $-5 > b - 3$ 27. $2y < -1$

28. $3x < 6$ 29. $2x \ge 0$ 30. $-\frac{1}{2} y \ge 4$

31. $2r + 5 < 9$ 32. $3x - 2 > 4$ 33. $3x - 1 \ge 2$

34. $\dfrac{1}{2x + 3} > 0$ 35. $\dfrac{4}{5 - 3x} < 0$ 36. $\dfrac{3}{3x - 1} > 0$

Solve the given inequality in Exercises 37–60 and write the solution set in interval notation.

37. $4x + 3 \le 11$ 38. $\frac{1}{2}y - 2 \le 2$ 39. $\frac{3}{2}x + 1 \ge 4$

40. $-5x + 2 > -8$ 41. $4(2x + 1) < 16$ 42. $3(3r - 4) \ge 15$

43. $2(x - 3) < 3(x + 2)$ 44. $4(x - 3) \ge 3(x - 2)$

45. $3(2a - 1) > 4(2a - 3)$ 46. $2(3x - 1) + 4 < 3(x + 2) - 8$

47. $\frac{2}{3}(x + 1) + \frac{5}{6} \ge \frac{1}{2}(2x - 1) + 4$ 48. $\frac{1}{4}(3x + 2) - 1 \le -\frac{1}{2}(x - 3) + \frac{3}{4}$

49. $\frac{x - 1}{3} + \frac{1}{5} < \frac{x + 2}{5} - \frac{1}{3}$ 50. $\frac{x}{5} - \frac{1 - x}{2} > \frac{x}{2} - 3$

51. $3(x + 1) + 6 \ge 2(2x - 1) + 4$ 52. $4(3x + 2) - 1 \le -2(x - 3) + 15$

53. $-2 < 4x \le 5$ 54. $3 \le 6x < 12$

55. $-4 \le 2x + 2 \le -2$ 56. $5 \le 3x - 1 \le 11$

57. $3 \le 1 - 2x < 7$ 58. $5 < 2 - 3x \le 11$

59. $-8 < 2 - 5x \le 7$ 60. $-10 < 5 - 2x < -5$

61. A student has grades of 42 and 70 in the first two tests of the semester. If an average of 70 is required to obtain a C grade, what is the minimum score the student must achieve on the third exam to obtain a C?

62. A compact car can be rented from firm A for $160 per week with no charge for mileage, or from firm B for $100 per week plus 20 cents for each mile driven. If the car is driven m miles, for what values of m does it cost less to rent from firm A?

63. An appliance salesperson is paid $30 per day plus $25 for each appliance sold. How many appliances must be sold for the salesperson's income to exceed $130 per day?

64. A pension trust invests $6000 in a bond which pays 5% simple interest per year. Additional funds are to be invested in a more speculative bond paying 9% simple interest per year, so that the return on the total investment will be at least 6%. What is the minimum amount that must be invested in the more speculative bond?

65. A book publisher spends $19,000 on editorial expenses and $6 per book for manufacturing and sales expenses in the course of publishing a psychology textbook. If the book sells for $12.50, how many copies must be sold to show a profit?

66. If the area of a right triangle is not to exceed 80 square inches and the base is 10 inches, what values may be assigned to the altitude h?
67. A total of 70 meters of fencing material is available with which to enclose a rectangular area. If the width of the rectangle is 15 meters, what values can be assigned to the length L?

2.4
ABSOLUTE VALUE IN EQUATIONS AND INEQUALITIES

In Section 1.2 we discussed the use of absolute value notation to indicate distance and provided this formal definition.

$$|x| = \begin{cases} x & \text{if } x \geq 0 \\ -x & \text{if } x < 0 \end{cases}$$

The following example illustrates the application of this definition to the solution of equations involving absolute value.

EXAMPLE 1
Solve the equation $|2x - 7| = 11$.

Solution
We apply the definition of absolute value to the two cases.

Case 1. $2x - 7 \geq 0$
With the first part of the definition,

$$|2x - 7| = 2x - 7 = 11$$
$$2x = 18$$
$$x = 9$$

Case 2. $2x - 7 < 0$
With the second part of the definition,

$$|2x - 7| = -(2x - 7) = 11$$
$$-2x + 7 = 11$$
$$x = -2$$

It's a good idea to check the answers by substituting in the original equation.

$$|2(9) - 7| \overset{?}{=} 11$$
$$|18 - 7| \overset{?}{=} 11$$
$$11 = 11$$

$$|2(-2) - 7| \overset{?}{=} 11$$
$$|-4 - 7| \overset{?}{=} 11$$
$$|-11| \overset{?}{=} 11$$
$$11 = 11$$

PROGRESS CHECK
Solve each equation and check the solution(s).

(a) $|x + 8| = 9$

(b) $|3x - 4| = 7$

Answers

(a) *1, −17*

(b) $\frac{11}{3}, -1$

When used in inequalities, absolute value notation plays an important and frequently used role in higher mathematics. To solve inequalities involving absolute value, we recall that $|x|$ is the distance between the origin and the point on the real number line corresponding to x. The solution set of the inequality $|x| < a$ is then seen to consist of the real numbers in the open interval $(-a, a)$ shown in Figure 4. Similarly, if

FIGURE 4

$|x| > a$, then we must have $x > a$ or $x < -a$. The solution set of the inequality $|x| > a$ is seen to consist of the real numbers in the infinite intervals $(-\infty, -a)$ and (a, ∞) as shown in Figure 5. Of course, $|x| \le a$ and $|x| \ge a$ would include the endpoints a and $-a$ and the circles would be filled in.

FIGURE 5

EXAMPLE 2
Solve $|2x - 5| \le 7$, graph the solution set, and write the solution set in interval notation.

Solution
We must solve the equivalent double inequality

$$-7 \le 2x - 5 \le 7$$
$$-2 \le 2x \le 12 \qquad \text{Add } +5 \text{ to each member.}$$
$$-1 \le x \le 6 \qquad \text{Divide each member by 2.}$$

The graph of the solution set is then

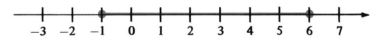

Thus, the solution set is the closed interval $[-1, 6]$.

PROGRESS CHECK
Solve each inequality, graph the solution set, and write the solution set in interval notation.

(a) $|x| < 3$ (b) $|3x - 1| \le 8$ (c) $|x| < -2$

Answers
(a) $(-3, 3)$

(b) $\left[-\dfrac{7}{3}, 3\right]$

(c)   *No solution. Since $|x|$ is always nonnegative, $|x|$ cannot be less than -2.*

EXAMPLE 3
Solve the inequality $|2x - 6| > 4$, write the solution set in interval notation, and graph the solution.

Solution

We must solve the equivalent inequalities

$$2x - 6 > 4 \quad \text{and} \quad 2x - 6 < -4$$
$$2x > 10 \qquad\qquad 2x < 2$$
$$x > 5 \qquad\qquad x < 1$$

The solution set consists of the real numbers in the infinite intervals $(-\infty, 1)$ and $(5, \infty)$. The graph of the solution set is then

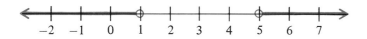

PROGRESS CHECK

Solve each inequality, write the solution set in interval notation, and graph the solution.

(a) $|5x - 6| > 9$ (b) $|2x - 2| \geq 8$

Answers

(a) $\left(-\infty, -\dfrac{3}{5}\right), (3, \infty)$

(b) $(-\infty, -3], [5, \infty)$

WARNING

Students sometimes write

$$1 > x > 5$$

This is a misuse of the inequality notation since it states that x is simultaneously less than 1 *and* greater than 5, which is impossible. What is usually intended is the pair of infinite intervals $(-\infty, 1)$ and $(5, \infty)$, and the inequalities must be written

$$x < 1 \; or \; x > 5$$

EXERCISE SET 2.4

Solve and check.

1. $|x + 2| = 3$ 2. $|r - 5| = \dfrac{1}{2}$ 3. $|2x - 4| = 2$

4. $|5y + 1| = 11$ 5. $|-3x + 1| = 5$ 6. $|2t + 2| = 0$

7. $3|-4x - 3| = 27$ 8. $\dfrac{1}{|x|} = 5$ 9. $\dfrac{1}{|s - 1|} = \dfrac{1}{3}$

Solve and graph the solution set.

10. $|x + 3| < 5$ 11. $|x + 1| > 3$ 12. $|3x + 6| \leq 12$

13. $|4x - 1| > 3$ 14. $|3x + 2| \geq -1$ 15. $\left|\dfrac{1}{3} - x\right| < \dfrac{2}{3}$

Solve and write the solution set using interval notation.

16. $|x - 2| \leq 4$ 17. $|x - 3| \geq 4$ 18. $|2x + 1| < 5$

19. $\dfrac{|2x - 1|}{4} < 2$ 20. $\dfrac{|3x + 2|}{2} \leq 4$ 21. $\dfrac{|2x + 1|}{3} < 0$

22. $\left|\dfrac{4}{3x - 2}\right| < 1$ 23. $\left|\dfrac{5 - x}{3}\right| > 4$ 24. $\left|\dfrac{2x + 1}{3}\right| \leq 5$

In Exercises 25–26, x and y are real numbers.

25. Prove that $\left|\dfrac{x}{y}\right| = \dfrac{|x|}{|y|}$ (*Hint:* Treat as four cases.)

26. Prove that $|x|^2 = x^2$

27. A machine that packages 100 vitamin pills per bottle can make an error of 2 pills per bottle. If x is the number of pills in a bottle, write an inequality, using absolute value, that indicates a maximum error of 2 pills per bottle. Solve the inequality.

28. The weekly income of a worker in a manufacturing plant differs from $300 by no more than $50. If x is the weekly income, write an inequality, using absolute value, which expresses this relationship. Solve the inequality.

2.5
THE QUADRATIC EQUATION

We now turn our attention to equations involving second-degree polynomials. A **quadratic equation** is an equation of the form

$$ax^2 + bx + c = 0, \qquad a \neq 0$$

where a, b, and c are real numbers. In this section we will explore techniques for solving this important class of equations. We will also show that there are several kinds of equations that can be transformed to quadratic equations and then solved.

THE FORM $ax^2 + c = 0$

When the quadratic equation $ax^2 + bx + c = 0$ has the coefficient $b = 0$, we have an equation of the form

$$ax^2 + c = 0$$

Solving for x we have

$$x^2 = -\frac{c}{a}$$

or

$$x = \pm \sqrt{-\frac{c}{a}}$$

That is,

$$x = \sqrt{-\frac{c}{a}} \quad \text{and} \quad x = -\sqrt{-\frac{c}{a}}$$

are solutions of the original equation.

EXAMPLE 1

Solve the equation $3x^2 - 8 = 0$.

Solution

$$3x^2 - 8 = 0$$

$$x^2 = \frac{8}{3}$$

$$x = \pm\sqrt{\frac{8}{3}} = \pm\frac{\sqrt{24}}{3} = \pm\frac{2\sqrt{6}}{3}$$

The solutions are $\frac{2}{3}\sqrt{6}$ and $-\frac{2}{3}\sqrt{6}$.

EXAMPLE 2

Solve the equation $(x - 5)^2 = -9$.

Solution

Although this equation is not strictly of the form $ax^2 + c = 0$, we can use the same approach. From

$$(x - 5)^2 = -9$$

we conclude that

$$x - 5 = \pm\sqrt{-9}$$

$$x - 5 = \pm 3i$$

$$x = 5 \pm 3i$$

The solutions of the given equation are the complex numbers $5 + 3i$ and $5 - 3i$. We should have anticipated complex solutions since the original equation requires the square of a number to be negative.

PROGRESS CHECK

Solve the given equation.

(a) $5x^2 + 13 = 0$ (b) $(2x - 7)^2 - 5 = 0$

Answers

(*a*) $\pm\dfrac{i\sqrt{65}}{5}$ (*b*) $\dfrac{7 \pm \sqrt{5}}{2}$

We have seen that the solutions of a quadratic equation may be complex numbers whereas the solution of a linear equation is a real number. In addition, quadratic equations appear to have two solutions. We will have more to say about these observations when we study the roots of polynomial equations in a later chapter.

SOLVING BY FACTORING

If a and b are real numbers and $ab = 0$, what must be true about a and b? We will show that at least one of the numbers, a or b, must be zero. Suppose that $a \neq 0$. Then we can divide both sides of the equation by a.

$$\frac{ab}{a} = \frac{0}{a}$$

$$b = 0$$

Similarly, if we assume $b \neq 0$ we would conclude that $a = 0$. We have established the following result.

> If a and b are real numbers and $ab = 0$, then $a = 0$ or $b = 0$.

If we can factor a quadratic equation into linear factors, we can then find the roots of the equation by setting each factor equal to 0. That is, suppose

$$ax^2 + bx + c = (rx + u)(sx + v) = 0$$

Then

$$rx + u = 0 \qquad \text{or} \qquad sx + v = 0$$

$$x = -\frac{u}{r} \qquad\qquad\qquad x = -\frac{v}{s}$$

Thus, $-u/r$ and $-v/s$ are the real roots of the quadratic. Note that each root corresponds to a linear factor and it is therefore possible to have the same root occur more than once.

EXAMPLE 3
Solve the equation $2x^2 - 3x - 2 = 0$ by factoring.

Solution
Factoring, we have

$$2x^2 - 3x - 2 = 0$$

$$(2x + 1)(x - 2) = 0$$

Since the product of the factors is 0, at least one factor must be 0. Setting each factor equal to 0,

$$2x + 1 = 0 \qquad \text{or} \qquad x - 2 = 0$$

$$x = -\frac{1}{2} \qquad\qquad\qquad x = 2$$

EXAMPLE 4
Solve the equation $3x^2 - 4x = 0$ by factoring.

Solution

$$3x^2 - 4x = 0$$

Factoring, we have $\qquad x(3x - 4) = 0$

Setting each factor equal to zero,

$$x = 0 \qquad \text{or} \qquad x = \frac{4}{3}$$

EXAMPLE 5
Solve the equation $x^2 + x + 1 = 0$ by factoring.

Solution
We saw in Section 1.4 that the polynomial $x^2 + x + 1$ is irreducible. It follows that factoring cannot be used to solve the given equation. This illustrates the need for developing more complete solution techniques in the following sections.

PROGRESS CHECK

Solve each of the given equations by factoring.

(a) $4x^2 - x = 0$ (b) $3x^2 - 11x - 4 = 0$ (c) $2x^2 + 4x + 1 = 0$

Answers

(a) $0, \dfrac{1}{4}$ (b) $-\dfrac{1}{3}, 4$ (c) *cannot be factored*

COMPLETING THE SQUARE

We now present a method which yields the solutions to *any* quadratic equation. First, we observe that the equation

$$ax^2 + bx + c = 0, \quad a \neq 0 \tag{1}$$

can always be written in the form

$$x^2 + dx + e = 0 \tag{2}$$

by dividing each term of Equation (1) by a and by letting $d = b/a$ and $e = c/a$. We then rewrite Equation (2) in the form

$$x^2 + dx = -e \tag{3}$$

We next focus on the left-hand side of Equation (3) and seek a constant k such that the addition of k^2 will ''complete'' a perfect square, that is,

$$x^2 + dx + k^2 = (x + k)^2 \tag{4}$$

Expanding and solving Equation (4),

$$x^2 + dx + k^2 = x^2 + 2kx + k^2$$

$$dx = 2kx$$

$$k = \frac{d}{2}$$

The constant k we seek is exactly half the coefficient of x appearing in Equation (2). We now add $k^2 = d^2/4$ to both sides of Equation (3) to obtain

$$x^2 + dx + \frac{d^2}{4} = -e + \frac{d^2}{4}$$

or

$$\left(x + \frac{d}{2}\right)^2 = c \tag{5}$$

where $c = -e + d^2/4$ is a constant. Equation (5) is then in a form which is easily solved. The following example illustrates the method of **completing the square.**

EXAMPLE 6

Solve $2x^2 - 5x + 4 = 0$ by completing the square.

Solution

We now outline and explain each step of the process.

Completing the Square	
Step 1. Divide the equation by the coefficient of x^2.	Step 1. $x^2 - \frac{5}{2}x + 2 = 0$
Step 2. Rewrite the equation with the constant on the right-hand side.	Step 2. $x^2 - \frac{5}{2}x = -2$
Step 3. Complete the square, $(x + k)^2$, where k is half the coefficient of x. Balance the equation by adding k^2 to the right-hand side. Simplify.	Step 3. $x^2 - \frac{5}{2}x + \left(-\frac{5}{4}\right)^2 = -2 + \left(-\frac{5}{4}\right)^2$ $\left(x - \frac{5}{4}\right)^2 = -\frac{32}{16} + \frac{25}{16}$ $\left(x - \frac{5}{4}\right)^2 = -\frac{7}{16}$
Step 4. Solve for x.	Step 4. $x - \frac{5}{4} = \pm\sqrt{-\frac{7}{16}} = \frac{\pm i\sqrt{7}}{4}$ $x = \frac{5}{4} \pm \frac{i\sqrt{7}}{4}$
	or $x = \frac{5 \pm i\sqrt{7}}{4}$

PROGRESS CHECK

Solve by completing the square.

(a) $x^2 - 3x + 2 = 0$ (b) $3x^2 - 4x + 2 = 0$

Answers

(a)   1, 2   (b) $\dfrac{2 \pm i\sqrt{2}}{3}$

THE QUADRATIC FORMULA

Let's apply the method of completing the square to the general quadratic equation

$$ax^2 + bx + c = 0, \qquad a \neq 0$$

Following the steps we illustrated in the last section, we have

$ax^2 + bx + c = 0$

$x^2 + \dfrac{b}{a}x + \dfrac{c}{a} = 0$ Divide the equation by the coefficient of x^2.

$x^2 + \dfrac{b}{a}x = -\dfrac{c}{a}$ Rewrite with the constant on the right-hand side.

$x^2 + \dfrac{b}{a}x + \left(\dfrac{b}{2a}\right)^2 = \left(\dfrac{b}{2a}\right)^2 - \dfrac{c}{a}$ Complete the square and balance the equation.

$\left(x + \dfrac{b}{2a}\right)^2 = \dfrac{b^2}{4a^2} - \dfrac{c}{a} = \dfrac{b^2 - 4ac}{4a^2}$ Simplify.

$x + \dfrac{b}{2a} = \pm\sqrt{\dfrac{b^2 - 4ac}{4a^2}}$ Solve for x.

$$x = -\frac{b}{2a} \pm \frac{\sqrt{b^2 - 4ac}}{2a}$$

$$x = \frac{-b \pm \sqrt{b^2 - 4ac}}{2a}$$

By applying the method of completing the square to the standard form of the quadratic equation, we have derived a *formula* that gives us the roots or solutions for *any* quadratic equation in one variable.

Quadratic Formula

$$x = \frac{-b \pm \sqrt{b^2 - 4ac}}{2a}, \qquad a \neq 0$$

That is, the roots of the quadratic equation $ax^2 + bx + c = 0$

are $x = \dfrac{-b + \sqrt{b^2 - 4ac}}{2a}$ and $x = \dfrac{-b - \sqrt{b^2 - 4ac}}{2a}$

EXAMPLE 7
Solve $2x^2 - 3x - 3 = 0$ by use of the quadratic formula.

Solution
Since $a = 2$, $b = -3$, and $c = -3$, we have

$$x = \frac{-b \pm \sqrt{b^2 - 4ac}}{2a}$$

$$= \frac{-(-3) \pm \sqrt{(-3)^2 - 4(2)(-3)}}{2(2)}$$

$$= \frac{3 \pm \sqrt{33}}{4}$$

EXAMPLE 8
Solve $5x^2 - 3x = -2$ by the quadratic formula.

Solution
We first rewrite the given equation as $5x^2 - 3x + 2 = 0$. Then $a = 5$, $b = -3$, and $c = 2$. Substituting in the quadratic formula, we have

$$x = \frac{-b \pm \sqrt{b^2 - 4ac}}{2a}$$

$$= \frac{-(-3) \pm \sqrt{(-3)^2 - 4(5)(2)}}{2(5)}$$

$$= \frac{3 \pm \sqrt{-31}}{10} = \frac{3 \pm i\sqrt{31}}{10}$$

PROGRESS CHECK
Solve by use of the quadratic formula.

(a) $x^2 - 8x = -10$ (b) $4x^2 - 2x + 1 = 0$

Answers

(a) $4 \pm \sqrt{6}$ (b) $\dfrac{1 \pm i\sqrt{3}}{4}$

 WARNING

There are a number of errors that students make in using the quadratic formula.

(a) To solve $x^2 - 3x = -4$, you must write the equation in the form $x^2 - 3x + 4 = 0$ to properly identify a, b, and c.

(b) The quadratic formula is

$$x = \frac{-b \pm \sqrt{b^2 - 4ac}}{2a}$$

Note that

$$x \neq -b \pm \frac{\sqrt{b^2 - 4ac}}{2a}$$

since the term $-b$ must also be divided by $2a$.

Now that you have a formula that works for *any* quadratic equation, you may be tempted to use it all the time. However, if you see an equation of the form

$$x^2 = 15$$

it is certainly easier to immediately supply the answer: $x = \pm \sqrt{15}$. Similarly, if you are faced with

$$x^2 + 3x + 2 = 0$$

it is faster to solve if you see that

$$x^2 + 3x + 2 = (x + 1)(x + 2)$$

The method of completing the square is generally not used for solving quadratic equations once you have learned the quadratic formula. The *technique* of completing the square is helpful in a variety of applications and we will use it in a later chapter when we graph second-degree equations.

THE DISCRIMINANT

By analyzing the quadratic formula

$$x = \frac{-b \pm \sqrt{b^2 - 4ac}}{2a}$$

we can learn a great deal about the roots of the quadratic equation $ax^2 + bx + c = 0$. The key to the analysis is the **discriminant** $b^2 - 4ac$ found under the radical.

(a) If $b^2 - 4ac$ is negative, we have the square root of a negative number and the values of x will be complex.

(b) If $b^2 - 4ac$ is positive, we have the square root of a positive number and the values of x will be real.

(c) If $b^2 - 4ac = 0$, then $x = -b/2a$, which we call a **double root** or **repeated root** of the quadratic equation. For example, if $x^2 - 10x + 25 = 0$, then $b^2 - 4ac = 0$ and $x = 5$. But

$$x^2 - 10x + 25 = (x - 5)(x - 5) = 0$$

We call $x = 5$ a double root because the factor $(x - 5)$ is a double factor of $x^2 - 10x + 25 = 0$. This hints at the importance of the relationship between roots and factors, a relationship which we will explore in a later chapter on roots of polynomial equations.

If the roots of the quadratic equation are real and a, b, and c are rational numbers, the discriminant enables us to determine whether the roots are rational or irrational. Since $\sqrt{k}$ is a rational number only if k is a perfect square, we see that the quadratic formula produces a rational result only if $b^2 - 4ac$ is a perfect square. We summarize as follows.

The quadratic equation $ax^2 + bx + c = 0$ has exactly two roots, the nature of which are determined by the discriminant $b^2 - 4ac$.

Discriminant $b^2 - 4ac$	*Roots*
Positive	Two real roots
Negative	Two complex roots
0	A double root
a, b, c rational $\begin{cases} \text{A perfect square} \\ \text{Not a perfect square} \end{cases}$	Rational Irrational

---

EXAMPLE 9
Without solving, determine the nature of the roots of the quadratic equation $3x^2 - 4x + 6 = 0$.

Solution
We evaluate $b^2 - 4ac$ using $a = 3$, $b = -4$, and $c - 6$.

$$b^2 - 4ac = (-4)^2 - 4(3)(6) = 16 - 72 = -56$$

The discriminant is negative and the equation has two complex roots.

EXAMPLE 10
Without solving, determine the nature of the roots of the equation $2x^2 - 7x = -1$.

Solution
We rewrite the equation in the standard form

$$2x^2 - 7x + 1 = 0$$

and then substitute $a = 2$, $b = -7$, and $c = 1$ in the discriminant. Thus,

$$b^2 - 4ac = (-7)^2 - 4(2)(1) = 49 - 8 = 41$$

The discriminant is positive and is not a perfect square; thus, the roots are real, unequal, and irrational.

PROGRESS CHECK

Without solving, determine the nature of the roots of the quadratic equation by using the discriminant.

(a) $4x^2 - 20x + 25 = 0$ (b) $5x^2 - 6x = -2$
(c) $10x^2 = x + 2$ (d) $x^2 + x - 1 = 0$

Answers
(a)   a real, double root   *(b)   2 complex roots*
(c)   2 real, rational roots   *(d)   2 real, irrational roots*

FORMS LEADING TO QUADRATICS

Certain types of equations can be transformed into quadratic equations that can be solved by the methods discussed in this chapter. One form that leads to a quadratic equation is the radical equation, such as $x - \sqrt{x - 2} = 4$, which is solved in Example 11, which follows. To solve the equation we isolate the radical and raise both sides to suitable powers. The following is the key to the solution of such equations.

If P and Q are algebraic expressions, then the solution set of the equation

$$P = Q$$

is a subset of the solution set of the equation

$$P^n = Q^n$$

where n is a natural number.

This suggests that we can solve radical equations if we observe a precaution.

If both sides of an equation are raised to the same power, the solutions of the resulting equation must be checked to see that they satisfy the original equation.

EXAMPLE 11

Solve $x - \sqrt{x - 2} = 4$.

Solution
Isolate the radical on one side of the equation before solving.

$$x - 4 = \sqrt{x - 2}$$

$$x^2 - 8x + 16 = x - 2 \qquad \text{Square both sides.}$$

$$x^2 - 9x + 18 = 0 \qquad \text{Simplify.}$$

$$(x - 3)(x - 6) = 0 \qquad \text{Factor.}$$

$$x = 3 \quad \text{or} \quad x = 6$$

We check by substituting in the original equation.

$$\text{checking } x = 3 \qquad\qquad \text{checking } x = 6$$

$$3 - \sqrt{3 - 2} \overset{?}{=} 4 \qquad\qquad 6 - \sqrt{6 - 2} \overset{?}{=} 4$$

$$3 - 1 \overset{?}{=} 4 \qquad\qquad 6 - \sqrt{4} \overset{?}{=} 4$$

$$2 \neq 4 \qquad\qquad 4 \overset{\checkmark}{=} 4$$

We conclude that 6 is a solution of the original equation and 3 is not a solution of the original equation. We say that 3 is an **extraneous solution** that was introduced when we raised each side of the original equation to the second power.

PROGRESS CHECK
Solve $x - \sqrt{1 - x} = -5$.

Answer
−3

EXAMPLE 12
Solve $\sqrt{2x - 4} - \sqrt{3x + 4} = -2$.

Solution
Before squaring, rewrite the equation so that a radical is on each side of the equation.

$$\sqrt{2x - 4} = \sqrt{3x + 4} - 2$$

$$2x - 4 = (3x + 4) - 4\sqrt{3x + 4} + 4 \qquad \text{Square both sides.}$$

$$-x - 12 = -4\sqrt{3x + 4} \qquad \text{Isolate the radical.}$$

$$x^2 + 24x + 144 = 16(3x + 4) \qquad \text{Square both sides.}$$

$$x^2 - 24x + 80 = 0$$

$$(x - 20)(x - 4) = 0$$

$$x = 20, x = 4$$

Verify that both 20 and 4 are solutions of the original equation.

PROGRESS CHECK
Solve $\sqrt{5x - 1} - \sqrt{x + 2} = 1$.

Answer
2

Although the equation

$$x^4 - x^2 - 2 = 0$$

is not a quadratic in the unknown x, it is a quadratic in the unknown x^2, that is,

$$(x^2)^2 - (x^2) - 2 = 0$$

This may be seen more clearly by replacing x^2 by a new unknown u such that $u = x^2$. Substituting, we have

$$u^2 - u - 2 = 0$$

which is a quadratic equation in the unknown u. Solving,

$$(u + 1)(u - 2) = 0$$

$$u = -1 \quad \text{or} \quad u = 2$$

Since $x^2 = u$, we must next solve the equations

$$x^2 = -1 \quad \text{and} \quad x^2 = 2$$

$$x = \pm i \qquad x = \pm \sqrt{2}$$

The original equation has four solutions: i, $-i$, $\sqrt{2}$, and $-\sqrt{2}$.

The technique we have used is called a **substitution of variable.** Although simple in concept, this is a powerful method that is commonly used in calculus.

PROGRESS CHECK

Indicate an appropriate substitution of variable and solve each of the following equations.

(a) $3x^4 - 10x^2 - 8 = 0$

(b) $4x^{2/3} + 7x^{1/3} - 2 = 0$

(c) $\dfrac{2}{x^2} + \dfrac{1}{x} - 10 = 0$

(d) $\left(1 + \dfrac{2}{x}\right)^2 - 8\left(1 + \dfrac{2}{x}\right) + 15 = 0$

Answers

(a) $u = x^2; \quad \pm 2, \pm \dfrac{i\sqrt{6}}{3}$

(b) $u = x^{1/3}; \dfrac{1}{64}, -8$

(c) $u = \dfrac{1}{x}; \quad -\dfrac{2}{5}, \dfrac{1}{2}$

(d) $u = 1 + \dfrac{2}{x}; \quad 1, \dfrac{1}{2}$

EXERCISE SET 2.5

Solve the given equation.

1. $3x^2 - 27 = 0$ 2. $4x^2 - 64 = 0$ 3. $5y^2 - 25 = 0$

4. $6x^2 - 12 = 0$ 5. $(2r + 5)^2 = 8$ 6. $(3x - 4)^2 = -6$

7. $(3x - 5)^2 - 8 = 0$ 8. $(4t + 1)^2 - 3 = 0$ 9. $9x^2 + 64 = 0$

10. $81x^2 + 25 = 0$

Solve by factoring.

11. $x^2 - 3x + 2 = 0$ 12. $x^2 - 6x + 8 = 0$ 13. $x^2 + x - 2 = 0$

14. $3r^2 - 4r + 1 = 0$ 15. $x^2 + 6x = -8$ 16. $x^2 + 6x + 5 = 0$

17. $y^2 - 4y = 0$ 18. $2x^2 - x = 0$ 19. $2x^2 - 5x = -2$

20. $2s^2 - 5s - 3 = 0$ 21. $t^2 - 4 = 0$ 22. $4x^2 - 9 = 0$

23. $6x^2 - 5x + 1 = 0$ 24. $6x^2 - x = 2$

Solve by completing the square.

25. $x^2 - 2x = 8$ 26. $t^2 - 2t = 15$ 27. $2r^2 - 7r = 4$

28. $9x^2 + 3x = 2$ 29. $3x^2 + 8x = 3$ 30. $2y^2 + 4y = 5$

31. $2y^2 + 2y = -1$ 32. $3x^2 - 4x = -3$ 33. $4x^2 - x = 3$

34. $2x^2 + x = 2$ 35. $3x^2 + 2x = -1$ 36. $3u^2 - 3u = -1$

Solve by the quadratic formula.

37. $2x^2 + 3x = 0$ 38. $2x^2 + 3x + 3 = 0$ 39. $5x^2 - 4x + 3 = 0$

40. $2x^2 - 3x - 2 = 0$ 41. $5y^2 - 4y + 5 = 0$ 42. $x^2 - 5x = 0$

43. $3x^2 + x - 2 = 0$ 44. $2x^2 + 4x - 3 = 0$ 45. $3y^2 - 4 = 0$

46. $2x^2 + 2x + 5 = 0$ 47. $4u^2 + 3u = 0$ 48. $4x^2 - 1 = 0$

Solve by any method.

49. $2x^2 + 2x - 5 = 0$ 50. $2t^2 + 2t + 3 = 0$ 51. $3x^2 + 4x - 4 = 0$

52. $x^2 + 2x = 0$ 53. $2x^2 + 5x + 4 = 0$ 54. $2r^2 - 3r + 2 = 0$

55. $4u^2 - 1 = 0$ 56. $x^2 + 2 = 0$ 57. $4x^3 + 2x^2 + 3x = 0$

58. $4s^3 + 4s^2 - 15s = 0$

Solve for the indicated variable in terms of the remaining variables.

59. $a^2 + b^2 = c^2$, for b 60. $s = \frac{1}{2} gt^2$, for t

61. $V = \frac{1}{3} \pi r^2 h$, for r 62. $A = \pi r^2$, for r

63. $s = \frac{1}{2} gt^2 + vt$, for t 64. $F = g \dfrac{m_1 m_2}{d^2}$, for d

Without solving, determine the nature of the roots of each quadratic equation.

65. $x^2 - 2x + 3 = 0$ 66. $3x^2 + 2x - 5 = 0$

67. $4x^2 - 12x + 9 = 0$ 68. $2x^2 + x + 5 = 0$

69. $-3x^2 + 2x + 5 = 0$ 70. $-3y^2 + 2y - 5 = 0$

71. $3x^2 + 2x = 0$ 72. $4x^2 + 20x + 25 = 0$

73. $2r^2 = r - 4$ 74. $3x^2 = 5 - x$

75. $3x^2 + 6 = 0$ 76. $4x^2 - 25 = 0$

77. $6r = 3r^2 + 1$ 78. $4x = 2x^2 + 3$

79. $12x = 9x^2 + 4$ 80. $4s^2 = -4s - 1$

Determine a value or values of k that will result in the quadratic having a double root.

81. $kx^2 - 4x + 1 = 0$ 82. $2x^2 + 3x + k = 0$

83. $x^2 - kx - 2k = 0$ 84. $kx^2 - 4x + k = 0$

Find the solution set.

85. $x + \sqrt{x + 5} = 7$ 86. $x - \sqrt{13 - x} = 1$

87. $2x + \sqrt{x + 1} = 8$ 88. $3x - \sqrt{1 + 3x} = 1$

89. $\sqrt{3x + 4} - \sqrt{2x + 1} = 1$ 90. $\sqrt{4 - 4x} - \sqrt{x + 4} = 3$

91. $\sqrt{2x - 1} + \sqrt{x - 4} = 4$ 92. $\sqrt{5x + 1} + \sqrt{4x - 3} = 7$

Indicate an appropriate substitution of variable and solve each of the equations.

93. $3x^4 + 5x^2 - 2 = 0$ 94. $2x^6 + 15x^3 - 8 = 0$

95. $\dfrac{6}{x^2} + \dfrac{1}{x} - 2 = 0$ 96. $\dfrac{2}{x^4} - \dfrac{3}{x^2} - 9 = 0$

97. $2x^{2/5} + 5x^{1/5} + 2 = 0$ 98. $3x^{4/3} - 4x^{2/3} - 4 = 0$

99. $2\left(\dfrac{1}{x} + 1\right)^2 - 3\left(\dfrac{1}{x} + 1\right) - 20 = 0$

100. $3\left(\dfrac{1}{x} - 2\right)^2 + 2\left(\dfrac{1}{x} - 2\right) - 1 = 0$

Provide a proof of the stated theorem.

101. If r_1 and r_2 are the roots of the equation $ax^2 + bx + c = 0$, then (a) $r_1 r_2 = c/a$ and (b) $r_1 + r_2 = -b/a$.

102. If a, b, and c are rational numbers, and the discriminant of the equation $ax^2 + bx + c = 0$ is positive, then the quadratic has either two rational roots or two irrational roots.

Use the theorems of Exercise 101 to find a value or values of k so that the indicated condition is satisfied.

103. $kx^2 + 3x + 5 = 0$; sum of the roots is 6.
104. $2x^2 - 3kx - 2 = 0$; sum of the roots is -3.
105. $3x^2 - 10x + 2k = 0$; product of the roots is -4.
106. $2kx^2 + 5x - 1 = 0$; product of the roots is $\frac{1}{2}$.
107. $2x^2 - kx + 9 = 0$; one root is double the other.
108. $3x^2 - 4x + k = 0$; one root is triple the other.
109. $6x^2 - 13x + k = 0$; one root is the reciprocal of the other.

2.6
APPLICATIONS—
QUADRATIC EQUATIONS

As your knowledge of mathematical techniques and ideas grows you will be capable of tackling an ever wider variety of applied problems. In Section 2.2 we explored many types of word problems that led to linear equations. We can now attack a similar group of applied problems that lead to quadratic equations.

One word of caution: It is possible to arrive at a solution which is meaningless. For example, a negative solution which represents hours worked or the age of an individual is meaningless and must be rejected.

EXAMPLE 1
The larger of two positive numbers exceeds the smaller by 2. If the sum of the squares of the two numbers is 74, find the two numbers.

Solution
If we let

$$x = \text{the larger number}$$

then

$$x - 2 = \text{the smaller number}$$

The sum of the squares is then

$$x^2 + (x - 2)^2 = 74$$
$$x^2 + x^2 - 4x + 4 = 74$$
$$2x^2 - 4x - 70 = 0$$
$$x^2 - 2x - 35 = 0$$
$$(x + 5)(x - 7) = 0$$
$$x = 7 \qquad \text{Reject } x = -5$$

The numbers are then 7 and $(7 - 2) = 5$. Verify that the sum of the squares is indeed 74.

EXAMPLE 2

The length of a pool is 3 times its width and the pool is surrounded by a grass walk 4 feet wide. If the total area covered and enclosed by the walk is 684 square feet, find the dimensions of the pool.

Solution

A diagram such as Figure 6 is useful in solving geometric problems. If we let $x = $ width of the pool, then $3x = $ length of the pool, and the region enclosed by the walk has length $3x + 8$ and width $x + 8$. The total area is the product of the length and width, so that

$$(3x + 8)(x + 8) = 684$$
$$3x^2 + 32x + 64 = 684$$
$$3x^2 + 32 x - 620 = 0$$
$$(3x + 62)(x - 10) = 0$$
$$x = 10 \qquad \text{Reject } x = -\frac{62}{3}.$$

The dimensions of the pool are 10 feet by 30 feet.

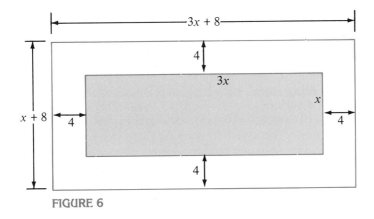

FIGURE 6

EXAMPLE 3

Working together, two cranes can unload a ship in 4 hours. The slower crane, working alone, requires 6 hours more than the faster crane to do the job. How long does it take each crane to do the job by itself?

Solution

Let $x = $ number of hours for the faster crane to do the job. Then $x + 6 = $ number of hours for the slower crane to do the job. The rate of the faster crane is $1/x$ since this is the portion of the whole job completed in 1 hour; similarly, the rate of the slower crane is $1/(x + 6)$. We display this information in a table.

	Rate	×	Time	=	Work
Faster crane	$\dfrac{1}{x}$		4		$\dfrac{4}{x}$
Slower crane	$\dfrac{1}{x + 6}$		4		$\dfrac{4}{x + 6}$

When the two cranes work together, we must have

$$\frac{\text{work done by}}{\text{fast crane}} + \frac{\text{work done by}}{\text{slow crane}} = 1 \text{ whole job}$$

or

$$\frac{4}{x} + \frac{4}{x+6} = 1$$

To solve, we multiply by the L.C.D. $x(x + 6)$, obtaining

$$4(x + 6) + 4x = x^2 + 6x$$
$$0 = x^2 - 2x - 24$$
$$0 = (x + 4)(x - 6)$$
$$x = -4 \quad \text{or} \quad x = 6$$

The solution $x = -4$ is rejected since it makes no sense to speak of negative hours of work. Thus,

$$x = 6 \text{ is the number of hours in which the} \atop \text{fast crane can do the job alone.}$$

$$x + 6 = 12 \text{ is the number of hours in which the} \atop \text{slow crane can do the job alone.}$$

EXERCISE SET 2.6

1. Working together, computers A and B can complete a data-processing job in 2 hours. Computer A working alone can do the job in 3 hours less than computer B working alone. How long does it take each computer to do the job by itself?

2. A graphics designer and her assistant working together complete an advertising layout in 6 days. The assistant working alone can complete the job in 16 more days than the designer working alone. How long does it take each person to do the job by herself?

3. A roofer and his assistant working together can finish a roofing job in 4 hours. The roofer working alone can finish the job in 6 hours less than the assistant working alone. How long does it take each person to do the job by himself?

4. A 16-inch by 20-inch mounting board is used to mount a photograph. How wide a uniform border is there if the photograph occupies $\frac{2}{3}$ of the area of the mounting board?

5. The length of a rectangle exceeds twice its width by 4 feet. If the area of the rectangle is 48 square feet, find the dimensions.

6. The length of a rectangle is 4 centimeters less than twice its width. Find the dimensions if the area of the rectangle is 96 square centimeters.

7. The area of a rectangle is 48 square centimeters. If the length and width are each increased by 4 centimeters, the area of the newly formed rectangle is 120 square centimeters. Find the dimensions of the original rectangle.

8. The base of a triangle is 2 feet more than twice its altitude. If the area is 12 square feet, find the dimensions.

9. Find the width of a strip that has been mowed around a rectangular field which is 60 feet by 80 feet if one half of the lawn has not yet been mowed.

10. The sum of the reciprocals of two consecutive numbers is $\frac{7}{12}$. Find the numbers.

11. The sum of a number and its reciprocal is $\frac{26}{5}$. Find the number.

12. The difference of a number and its reciprocal is $\frac{35}{6}$. Find the number.

13. The smaller of two numbers is 4 less than the larger. If the sum of their squares is 58, find the numbers.

14. The sum of the reciprocals of two consecutive odd numbers is $\frac{8}{15}$. Find the numbers.

15. The sum of the reciprocals of two consecutive even numbers is $\frac{7}{24}$. Find the numbers.

16. A number of students rented a car for $160 for a one-week camping trip. If another student had joined the original group, each person's share of expenses would have been reduced by $8. How many students were there in the original group?

17. An investor placed an order totaling $1200 for a certain number of shares of a stock. If the price of each share of stock were $2 more, the investor would get 30 shares less for the same amount of money. How many shares did the investor buy?

18. A fraternity charters a bus for a ski trip at a cost of $360. If 6 more students join the trip, each person's cost decreases by $2. How many students were in the original group of travelers?

19. A salesman worked a certain number of days to earn $192. If he had been paid $8 more per day he would have earned the same amount of money in two fewer days. How many days did he work?

20. A freelance photographer works a certain number of days for a newspaper to earn $480. If she had been paid $8 less per day, she would have earned the same amount in two more days. What was her daily rate of pay?

2.7

SECOND-DEGREE INEQUALITIES

Unfortunately, the quadratic formula cannot be used to find those real numbers that satisfy a second-degree inequality such as

$$x^2 - 2x > 15$$

However, the given inequality is equivalent to

$$x^2 - 2x - 15 > 0$$

or, after factoring,

$$(x + 3)(x - 5) > 0$$

Observe that a product of two real numbers is positive only if both factors have the same sign. Thus, we must analyze the *signs* of $(x + 3)$ *and* $(x - 5)$.

In the general situation we are interested in knowing all values of x for which the expression $(ax + b)$ will be positive and those values for which it will be negative. Since $ax + b = 0$ when $x = -b/a$, we see that

> The linear factor $ax + b$ equals 0 at $x = -\dfrac{b}{a}$ and has opposite signs to the left and right of $-\dfrac{b}{a}$ on a number line.

Returning to the preceding example, Figure 7 gives an analysis of the signs of the factors $(x + 3)$ and $(x - 5)$ for all real values of x.

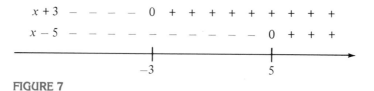

FIGURE 7

From Figure 7 we see that $x + 3$ and $x - 5$ have the same sign when $x < -3$ or when $x > 5$. The solution set of the given inequality is

$$\{x \mid x < -3 \text{ or } x > 5\}$$

which consists of the real numbers in the open intervals $(-\infty, -3)$ and $(5, \infty)$. The solution set is shown in Figure 8.

FIGURE 8

EXAMPLE 1
Solve the inequality $x^2 \leq -3x + 4$ and graph the solution set on a real number line.

Solution
We rewrite the inequality and factor.

$$x^2 \leq -3x + 4$$
$$x^2 + 3x - 4 \leq 0$$
$$(x - 1)(x + 4) \leq 0$$

We seek values of x for which the factors $(x - 1)$ and $(x + 4)$ have opposite signs or are zero. Figure 9 gives an analysis of the signs of the factors $x - 1$ and $x + 4$ and we see that the solution set consists of the set

$$\{x \mid -4 \leq x \leq 1\}$$

FIGURE 9

which is the closed interval $[-4, 1]$, shown in Figure 10.

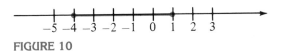

FIGURE 10

PROGRESS CHECK

Solve the inequality $2x^2 \geq 5x + 3$ and graph the solution set on a real number line.

Answer

$\{x \mid x \leq -\frac{1}{2} \text{ or } x \geq 3\}.$

Although $\dfrac{ax + b}{cx + d} < 0$ is not a second-degree inequality, the solution of this inequality is the same as the solution of the inequality $(ax + b)(cx + d) < 0$ since both expressions always have the same sign.

EXAMPLE 2

Solve the inequality $\dfrac{y + 1}{2 - y} \leq 0$.

Solution

Figure 11 gives an analysis of the signs of $y + 1$ and $2 - y$.

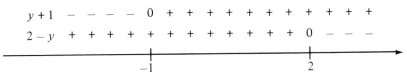

FIGURE 11

Since $\dfrac{y + 1}{2 - y}$ can be negative only if the factors $y + 1$ and $2 - y$ have opposite signs, the solution set is $\{y \mid y \leq -1 \text{ or } y > 2\}$. Note that $y = 2$ would result in division by 0.

PROGRESS CHECK

Solve the inequality $\dfrac{2x - 3}{1 - 2x} \geq 0$.

Answer

$\left\{x \mid \frac{1}{2} < x \leq \frac{3}{2}\right\} \text{ or } \left(\frac{1}{2}, \frac{3}{2}\right]$

EXAMPLE 3

Solve the inequality $(x - 2)(2x + 5)(3 - x) < 0$.

Solution

Although this is a third-degree inequality, the same approach will work. Figure 12 gives an analysis of the signs of $x - 2$, $2x + 5$, and $3 - x$. The product of three

FIGURE 12

factors is negative when there are an odd number of negative factors. The solution set is then

$$\left\{x \mid -\frac{5}{2} < x < 2 \text{ or } x > 3\right\} \quad \text{or} \quad \left(-\frac{5}{2}, 2\right), (3, \infty)$$

PROGRESS CHECK
Solve the inequality $(2y - 9)(6 - y)(y + 5) \ge 0$.

Answer

$$\left\{y \mid y \le -5 \text{ or } \frac{9}{2} \le y \le 6\right\} \quad \text{or} \quad (-\infty, -5], \left[-\frac{9}{2}, 6\right]$$

EXERCISE SET 2.7
Determine the solution set of each inequality.

1. $x^2 + 5x + 6 > 0$ 2. $x^2 + 3x - 4 \le 0$

3. $2x^2 - x - 1 < 0$ 4. $3x^2 - 4x - 4 \ge 0$

5. $4x - 2x^2 < 0$ 6. $r^2 + 4r \ge 0$

7. $\dfrac{x + 5}{x + 3} \le 0$ 8. $\dfrac{x - 6}{x + 4} \ge 0$

9. $\dfrac{2r + 1}{r - 3} \le 0$ 10. $\dfrac{x - 1}{2x - 3} \ge 0$

11. $\dfrac{3s + 2}{2s - 1} \ge 0$ 12. $\dfrac{4x + 5}{x^2} \le 0$

13. $(x + 2)(3x - 2)(x - 1) > 0$ 14. $(x - 4)(2x + 5)(2 - x) \le 0$

Indicate the solution set of each inequality on a real number line.

15. $x^2 + x - 6 > 0$ 16. $x^2 - 3x - 10 \ge 0$

17. $2x^2 - 3x - 5 < 0$ 18. $3x^2 - 4x - 4 \le 0$

19. $\dfrac{2r + 3}{2r - 1} < 0$ 20. $\dfrac{3x + 2}{2x - 3} \ge 0$

21. $\dfrac{x - 1}{x + 1} \ge 0$ 22. $\dfrac{2x - 1}{x + 2} \le 0$

23. $6x^2 + 8x + 2 \ge 0$ 24. $2x^2 + 5x + 2 \le 0$

25. $(y - 3)(2 - y)(2y + 4) \ge 0$ 26. $(2x + 5)(3x - 2)(x + 1) < 0$

27. $(x - 3)(1 + 2x)(3x + 5) > 0$ 28. $(1 - 2x)(2x + 1)(x - 3) \le 0$

In Exercises 29–32, find the values of x for which the given expression has real values.

29. $\sqrt{(x - 2)(x + 1)}$ 30. $\sqrt{(2x + 1)(x - 3)}$

31. $\sqrt{2x^2 + 7x + 6}$ 32. $\sqrt{2x^2 + 3x + 1}$

33. A manufacturer of solar heaters finds that when x units are made and sold, the profit (in thousands of dollars) is given by $x^2 - 50x - 5000$. For what values of x will the firm show a loss?

34. A ball thrown directly upward from level ground at an initial velocity of 40 feet per second attains a height d given by $d = 40t - 16t^2$ after t seconds. During what time interval is the ball at a height of at least 16 feet?

TERMS AND SYMBOLS

equation (p. 58)	**graph of the solution set** (p. 74)	**completing the square** (p. 85)
left-hand side (p. 58)	**interval** (p. 74)	**quadratic formula** (p. 87)
right-hand side (p. 58)	**interval notation** (p. 74)	**discriminant** (p. 88)
solution (p. 59)	**open interval** (p. 74)	**double root** (p. 89)
root (p. 59)	**closed interval** (p. 74)	**repeated root** (p. 89)
solution set (p. 59)	**half-open interval** (p. 74)	**extraneous root** (p. 91)
equivalent equation (p. 59)	$\infty, -\infty$ (p. 75)	**substitution of variable** (p. 92)
first-degree equation in one unknown (p. 60)	**infinite interval** (p. 75)	
linear equation (p. 61)	**quadratic equation** (p. 82)	

KEY IDEAS FOR REVIEW

◼ To solve an equation we generally form a succession of simpler, equivalent equations.

◼ In the process of solving an equation, we may add or subtract any number or expression to both sides of the equation. We may also multiply both sides by any nonzero number. If we multiply the equation by an expression containing a variable, then the answers must be substituted in the original equation to verify that they are solutions.

◼ The linear equation $ax + b = 0$, $a \neq 0$, has precisely one solution: $-b/a$.

◼ In the process of solving linear inequalities, remember that multiplication or division by a negative number reverses the sense of the inequality.

◼ The solution set of a linear inequality can be indicated by graphing on a real number line, by set-builder notation, or by interval notation.

◼ The quadratic equation $ax^2 + bx + c = 0$, $a \neq 0$, always has two solutions, which may be found by using the quadratic formula. If $b = 0$ or if the quadratic can be factored, then faster solution methods are available.

◼ The solutions or roots of a quadratic equation may be complex numbers. The expression $b^2 - 4ac$, called the discriminant, which appears under the radical of the quadratic formula, permits the nature of the roots to be analyzed without solving the equation.

◼ Radical equations often can be transformed into quadratic equations. Since the process involves raising both sides of an equation to a power, the answers must be checked to see that they satisfy the original equation.

◼ The method called "substitution of a variable" can be used to transform certain equations into quadratics. This technique is a handy tool and will be used in other chapters of this book.

■ If a second-degree inequality can be written in the factored form

$$(ax + b)(cx + d) < 0$$

or

$$(ax + b)(cx + d) > 0$$

then the solution set is easily found. First, determine the intervals in which each factor is positive and the intervals in which they are negative. If the product of the factors is negative (< 0), then the solution set consists of the intervals in which the factors are opposite in sign; if the product is positive (> 0), the solution set consists of the intervals in which the factors are of like sign.

REVIEW EXERCISES

In Exercises 1–4 solve for x.

1. $3x - 5 = 3$

2. $2(2x - 3) - 3(x + 1) = -9$

3. $\dfrac{2 - x}{3 - x} = 4$

4. $k - 2x = 4kx$

5. The width of a rectangle is 4 less than twice its length. If the perimeter is 12 centimeters, find the dimensions of each side.

6. A donation box contains coins consisting of dimes and quarters. The number of dimes is 4 more than twice the number of quarters. If the total value of the coins is $2.65, how many coins of each type are there?

7. It takes 4 hours for a bush pilot in Australia to pick up mail at a remote village and return to home base. If the average speed going is 150 miles per hour and the average speed returning is 100 miles per hour, how far from the home base is the village?

8. Copying machines A and B, working together, can prepare enough copies of the annual report for the Board of Directors in 2 hours. Machine A, working alone, requires 3 hours to do the job. How long would it take machine B to do the job by itself?

9. Indicate whether the statement is true (T) or false (F): The equation $3x^2 = 9$ is an identity.

10. Indicate whether the statement is true (T) or false (F): $x = 3$ is a solution of the equation $3x - 1 = 10$.

11. Solve and graph $3 \le 2x + 1$.

12. Solve and graph $-4 < -2x + 1 \le 10$.

In Exercises 13–15 solve and express the solution set in interval notation.

13. $2(a + 5) > 3a + 2$

14. $\dfrac{-1}{2x - 5} < 0$

15. $\dfrac{2x}{3} + \dfrac{1}{2} \ge \dfrac{x}{2} - 1$

16. Solve $|3x + 2| = 7$ for x.

17. Solve and graph $|4x - 1| = 5$.

18. Solve and graph $|2x + 1| > 7$.

19. Solve $|2 - 5x| < 1$ and write the solution in interval notation.

20. Solve $|3x - 2| \ge 6$ and write the solution in interval notation.

21. Solve $x^2 - x - 20 = 0$ by factoring.

22. Solve $6x^2 - 11x + 4 = 0$ by factoring.

23. Solve $x^2 - 2x + 6 = 0$ by completing the square.

24. Solve $2x^2 - 4x + 3 = 0$ by the quadratic formula.

25. Solve $3x^2 + 2x - 1 = 0$ by the quadratic formula.

In Exercises 26–28 solve for x.

26. $49x^2 - 9 = 0$

27. $kx^2 - 3\pi = 0$

28. $x^2 + x = 12$

In Exercises 29–31 determine the nature of the roots of the quadratic equation without solving.

29. $3r^2 = 2r + 5$ 30. $4x^2 + 20x + 25 = 0$ 31. $6y^2 - 2y = -7$

In Exercises 32–35 solve the given equation.

32. $\sqrt{x + 2} = x$

33. $\sqrt{x + 3} + \sqrt{2x - 3} = 6$

34. $3x^4 + 5x^2 - 5 = 0$

35. $\left(1 - \dfrac{2}{x}\right)^2 - 8\left(1 - \dfrac{2}{x}\right) + 15 = 0$

36.   A charitable organization rented an auditorium at a cost of $420 and split the cost among the attendees. If 10 additional persons had attended the meeting, the cost per person would have decreased by $1. How many attendees were there in the original group?

37.   Find the values of x for which $\sqrt{2x^2 - x - 6}$ has real values.

38.   Write the solution set of the inequality $x^2 + 4x - 5 \geq 0$ in interval notation.

39.   Write the solution set for $\dfrac{2x + 1}{x + 5} \geq 0$ in interval notation.

40.   Write the solution set for $(3 - x)(2x + 3)(x + 2) < 0$ in interval notation.

PROGRESS TEST 2A

In Problems 1–2 solve for y.

1. $5 - 4y = 2$

2. $\dfrac{2 + 5y}{3y - 1} = 6$

3.   One side of a triangle is 2 meters shorter than the base, and the other side is 3 meters longer than half the base. If the perimeter is 15 meters, find the length of each side.

4.   A trust fund invested a certain amount of money at 6.5% simple annual interest, a second amount $200 more than the first amount at 7.5%, and a third amount $300 more than twice the first amount at 9%. If the total annual income from these investments is $1962, how much was invested at each rate?

5.   Indicate whether the statement is true (T) or false (F): The equation $(2x - 1)^2 = 4x^2 - 4x + 1$ is an identity.

6.   Solve $-1 \leq 2x + 3 < 5$ and graph the solution set.

In Problems 7–8 solve and express the solution set in interval notation.

7. $3(2a - 1) - 4(a + 2) \leq 4$ 8. $-2 \leq 2 - x \leq 6$

9.   Solve $|4x - 1| = 9$.

10.   Solve $|2x - 1| \leq 5$ and graph the solution set.

11.   Solve $|1 - 3x| < 5$ and write the solution in interval notation.

12.   Solve $x^2 - 5x = 14$ by factoring.

13.   Solve $5x^2 - x + 4 = 0$ by completing the square.

14.   Solve $12x^2 + 5x - 3 = 0$ by the quadratic formula.

In Problems 15–16 solve for x.

15. $(2x - 5)^2 + 9 = 0$

16. $2 + \dfrac{1}{x} - \dfrac{3}{x^2} = 0$

In Problems 17–18 determine the nature of the roots of the quadratic equation without solving.

17. $6x^2 + x - 2 = 0$ 18. $3x^2 - 2x = -6$

In Problems 19–20 solve the given equation.

19. $x - \sqrt{4 - 3x} = -8$ 20. $3x^4 + 5x^2 - 2 = 0$

21.   The area of a rectangle is 96 square meters. If the length and the width are each increased by 2 meters, the area of the newly formed rectangle is 140 square meters. Find the dimensions of the original rectangle.

22.   Find the values of x for which $\sqrt{3x^2 - 4x + 1}$ has real values.

In Problems 23–24, write the solution set in interval notation.

23. $-2x^2 + 3x - 1 \le 0$ 24. $(x - 1)(2 - 3x)(x + 2) \le 0$

PROGRESS TEST 2B

In Problems 1–2 solve for x.

1. $3(2x + 5) = 5 - (3x - 1)$ 2. $3x - k^2 = -kx$

3.   A silver alloy which is 60% silver is to be combined with a silver alloy which is 80% silver to produce 120 ounces of an alloy which is 75% silver. How many ounces of each alloy must be used?

4.   Solve $\dfrac{ax + b}{cx + b} = d$ for x.

5.   Indicate whether the statement is true (T) or false (F): $x = -1$ is a solution of the equation $\dfrac{x - 1}{x + 1} = 0$.

6.   Solve $-9 \le 1 - 5x \le -4$ and graph the solution set.

In Problems 7–8 solve and express the solution set in interval notation.

7. $\dfrac{x}{4} - \dfrac{1}{2} \le \dfrac{1}{2} - x$ 8. $\dfrac{-2}{3 - x} \ge 0$

9.   Solve $|1 - 3x| = 7$.

10.   Solve $\dfrac{|x - 4|}{2} \ge 1$ and graph the solution set.

11.   Solve $|5x + 2| > 3$ and write the solution set in interval notation.
12.   Solve $6x^2 + 13x - 5 = 0$ by factoring.
13.   Solve $2x^2 - 5x + 2 = 0$ by completing the square.
14.   Solve $3x^2 - x = -7$ by the quadratic formula.

In Problems 15–16 solve for x.

15. $(x - 3)^2 + 2 = 0$ 16. $k_1 + \dfrac{k_2}{x^2} = k_3$

In Problems 17–18 determine the nature of the roots of the quadratic equation without solving.

17. $6z^2 - 4z = -2$ 18. $4y^2 - 20y + 25 = 0$

In Problems 19–20 solve the given equation.

19. $\sqrt{x - 1} - \sqrt{3x - 2} = -1$ 20. $\dfrac{8}{x^{4/3}} + \dfrac{9}{x^{2/3}} + 1 = 0$

21.   If the price of a large candy bar rose by 10 cents, a buyer would receive 2 fewer candy bars for $6.00 than she would at the lower price. What was the lower price?

22.   Find the values of x for which $x/\sqrt{2x - 1}$ has real values.

In Problems 23–24 write the solution set in interval notation.

23. $\dfrac{x^2}{x + 5} \le 0$ 24. $(3x - 2)(x + 4)(1 - x) > 0$

CHAPTER
THREE
FUNCTIONS

What is the effect of increased fertilization on the growth of an azalea? If the minimum wage is increased, what will be the impact upon the number of unemployed workers? When a submarine dives, can we calculate the water pressure against the hull at a given depth?

Each of the questions posed above seeks a relationship between phenomena. The search for relationships, or correspondence, is a central concept in our attempts to study the universe and is used in mathematics, engineering, the physical and biological sciences, the social sciences, and in business and economics.

The concept of a function has been developed as a means of organizing and assisting in the study of relationships. Since graphs are powerful means of exhibiting relationships, we begin with a study of the Cartesian or rectangular coordinate system. We will then formally define a function and will provide a number of ways of viewing the function concept. Function notation will be introduced to provide a convenient means of writing functions.

We will also explore some special types of functional relationships (increasing and decreasing functions) and will see that variation can be viewed as a functional relationship.

3.1
RECTANGULAR COORDINATE SYSTEMS

In Chapter One we associated the system of real numbers with points on the real number line. That is, we saw that there is a one-to-one correspondence between the system of real numbers and points on the real number line.

We will now develop an analogous way to handle points in a plane. We begin by drawing a pair of perpendicular lines intersecting at a point O called the **origin**. One of the lines, called the **x-axis**, is usually drawn in a horizontal position. The other line, called the **y-axis**, is usually drawn vertically.

If we think of the x-axis as a real number line, we may mark off some convenient unit of length, with positive numbers to the right of the origin and negative numbers to the left of the origin. Similarly, we may think of the y-axis as a real number line. Again, we may mark off a convenient unit of length (usually the same as the unit of length on the x-axis) with the upward direction representing positive numbers and the downward direction negative numbers. The x and y axes are called **coordinate axes**, and together they constitute a **rectangular coordinate system** or a **Cartesian coordinate system**. The coordinate axes divide the plane into four **quadrants** which we label I, II, III, and IV as in Figure 1.

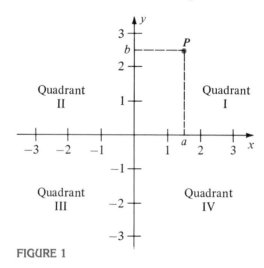

FIGURE 1

By using the coordinate axes, we can outline a procedure for labeling a point P in the plane. From P, draw a perpendicular to the x-axis and note that it meets the x-axis at $x = a$. Now draw a perpendicular from P to the y-axis and note that it meets the y-axis at $y = b$. We say that the **coordinates** of P are given by the **ordered pair** (a,b). The term "ordered pair" means that the order is significant, that is, the ordered pair (a,b) is different from the ordered pair (b,a).

The first number of the ordered pair (a,b) is sometimes called the **abscissa** or **x-coordinate** of P. The second number is called the **ordinate** or **y-coordinate** of P.

We have now developed a procedure for associating with each point P in the plane a unique ordered pair of real numbers (a,b). We usually write the point P as $P(a,b)$. Conversely, every ordered pair of real numbers (a,b) determines a unique point P in the plane. The point P is located at the intersection of the lines perpendicular to the x-axis and to the y-axis at the points on the axes having coordinates a and b, respectively. We have thus established a one-to-one correspondence between the set of all points in the plane and the set of all ordered pairs of real numbers.

We have indicated a number of points in Figure 2. Note that all points on

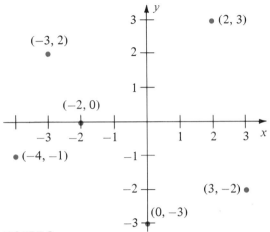

FIGURE 2

the x-axis have a y-coordinate of 0 and all points on the y-axis have an x-coordinate of 0. It is important to observe that the x-coordinate of a point P is the distance of P from the y-axis; the y-coordinate is its distance from the x-axis. The point $(2,3)$ in Figure 2 is 2 units from the y-axis and 3 units from the x-axis.

THE DISTANCE FORMULA

There is a very useful formula which gives the distance $\overline{PQ}$ between two points $P(x_1, y_1)$ and $Q(x_2, y_2)$. In Figure 3a we have shown the x-coordinate of a point as the distance of the point from the y-axis, and the y-coordinate as its distance from the x-axis. Thus we labeled the horizontal segments x_1 and x_2 and the vertical segments y_1 and y_2. In Figure 3b we use the lengths from Figure 3a to indicate that $\overline{PR} = x_2 - x_1$ and $\overline{QR} = y_2 - y_1$. Since triangle PRQ is a right triangle, we can apply the Pythagorean theorem.

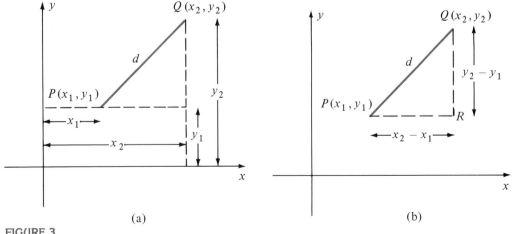

(a)

(b)

FIGURE 3

$$d^2 = (x_2 - x_1)^2 + (y_2 - y_1)^2$$

Although the points in Figure 3 are both in Quadrant I, the same result will be obtained for any two points. Since distance cannot be negative, we have

The Distance Formula

The distance $\overline{PQ}$ between the points $P(x_1,y_1)$ and $Q(x_2,y_2)$ in a plane is

$$\overline{PQ} = \sqrt{(x_2 - x_1)^2 + (y_2 - y_1)^2}$$

It is also clear from the distance formula that $\overline{PQ} = \overline{QP}$.

EXAMPLE 1
Find the distance between the points $P(-2,-3)$ and $Q(1,2)$.

Solution
Using the distance formula, we have

$$\overline{PQ} = \sqrt{[1 - (-2)]^2 + [2 - (-3)]^2} = \sqrt{3^2 + 5^2} = \sqrt{34}$$

PROGRESS CHECK
Find the distance between the points $P(-3,2)$ and $Q(4,-2)$.

Answer
$\sqrt{65}$

EXAMPLE 2
Show that the triangle with vertices $A(-2,3)$, $B(3,-2)$, and $C(6,1)$ is a right triangle.

Solution
It is a good idea to draw a diagram as in Figure 4. We compute the lengths of the three sides.

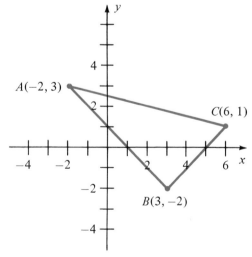

FIGURE 4

$$\overline{AB} = \sqrt{(3+2)^2 + (-2-3)^2} = \sqrt{50}$$
$$\overline{BC} = \sqrt{(6-3)^2 + (1+2)^2} = \sqrt{18}$$
$$\overline{AC} = \sqrt{(6+2)^2 + (1-3)^2} = \sqrt{68}$$

If the Pythagorean theorem holds, then triangle ABC is a right triangle. We see that

$$(\overline{AC})^2 = (\overline{AB})^2 + (\overline{BC})^2 \qquad \text{since } 68 = 50 + 18$$

and we conclude that triangle ABC is a right triangle whose hypotenuse is AC.

GRAPHS OF EQUATIONS

By the **graph of an equation in two variables** x and y we mean the set of all points $P(x,y)$ in the plane, whose coordinates (x,y) satisfy the given equation. We say that the ordered pair (a,b) is a **solution** of the equation if substitution of a for x and b for y yields a true statement.

To graph $y = x^2 - 4$, an equation in the variables x and y, we note that we can obtain solutions by assigning arbitrary values to x and computing corresponding values of y. Thus, if $x = 3$, then $y = 3^2 - 4 = 5$, and the ordered pair $(3,5)$ is a solution of the equation. Table 1 shows a number of solutions. We now plot the points corresponding to these ordered pairs. Since the equation has an infinite number of solutions, the plotted points represent only a portion of the graph. We assume that the curve behaves nicely between the plotted points and connect these points by a smooth curve (Figure 5). We must plot enough points to feel reasonably certain of the outline of the curve.

TABLE 1

x	-3	$-\dfrac{3}{2}$	-1	0	1	2	$\dfrac{5}{2}$
y	5	$-\dfrac{7}{4}$	-3	-4	-3	0	$\dfrac{9}{4}$

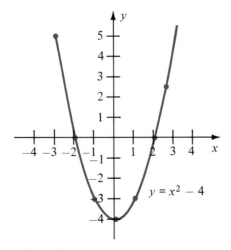

$$y = x^2 - 4$$

FIGURE 5

The abscissa of a point at which a graph meets the x-axis is called an **x-intercept.** Since the graph in Figure 5 meets the x-axis at the points $(2,0)$ and $(-2,0)$, we see that 2 and -2 are the x-intercepts. Similarly, we define the **y-intercept** as the ordinate of a point at which the graph meets the y-axis. In Figure 5 the y-intercept is -4. Intercepts are often easy to calculate and are useful in sketching a graph.

EXAMPLE 3
Graph the equation $y = |x + 1|$.

Solution
We apply the definition of absolute value to obtain

$$y = |x + 1| = \begin{cases} x + 1 & \text{if} \quad x + 1 \geq 0 \\ -(x + 1) & \text{if} \quad x + 1 < 0 \end{cases}$$

or

$$y = \begin{cases} x + 1 & \text{if} \quad x \geq -1 \\ -x - 1 & \text{if} \quad x < -1 \end{cases}$$

We then form a table by assigning values to x and calculating the corresponding values of y.

x	-3	-2	-1	0	1	2	3
y	-2	1	0	1	2	3	4

The points are joined by a smooth curve (Figure 6), which consists of two rays or half-lines intersecting at $(-1,0)$.

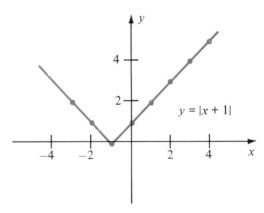

FIGURE 6

SYMMETRY

The mathematical definition of symmetry provides a precise means of testing for a property that you have always dealt with intuitively. In Figure 7a we have sketched a graph that appears to be symmetric about the x-axis; that is, each point above the x-axis appears to be reflected in a point below the x-axis. Thus, if (x_1, y_1) is a point on the upper portion of the curve, then $(x_1, -y_1)$ must lie on the lower portion of the curve. Similarly, using the graph in Figure 7b, we can argue that symmetry about the y-axis

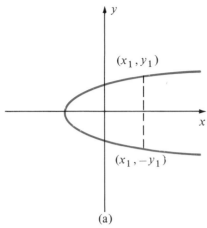

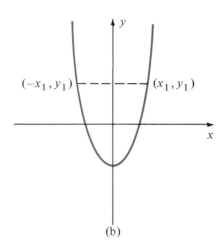

(a)   (b)

FIGURE 7

occurs if, for every point (x_1, y_1) on the curve, $(-x_1, y_1)$ also lies on the curve. In summary,

> If the equation of a curve is unchanged when x is replaced by $-x$, then the curve is **symmetric with respect to the y-axis.**
> If the equation of a curve is unchanged when y is replaced by $-y$, then the curve is **symmetric with respect to the x-axis.**

EXAMPLE 4
Use symmetry to assist in graphing the equations.

(a) $y = x^2 - 1$ (b) $x = y^2 + 1$

Solution
(a)   Replacing x by $-x$ in $y = x^2 - 1$, we obtain

$$y = (-x)^2 - 1 = x^2 - 1$$

Since the equation is unaltered, the curve is symmetric with respect to the y-axis. Now, replacing y by $-y$,

$$-y = x^2 - 1$$

is *not* the same equation. Thus, the curve is not symmetric with respect to the x-axis. We now form a table of values and use the symmetry to help sketch the graph of Figure 8a.
(b)   Replacing x by $-x$ in $x = y^2 + 1$, we obtain

$$-x = y^2 + 1$$

which is not the same equation. Replacing y by $-y$, we find that

$$x = (-y)^2 + 1 = y^2 + 1$$

which is the same equation. Thus, the curve is symmetric with respect to the x-axis. Note that in forming the table of values of $x = y^2 + 1$ shown in Figure 8b we assigned values to y and calculated the corresponding values of x from the given equation. Solving the equation for y yields $y = \pm \sqrt{x - 1}$, which

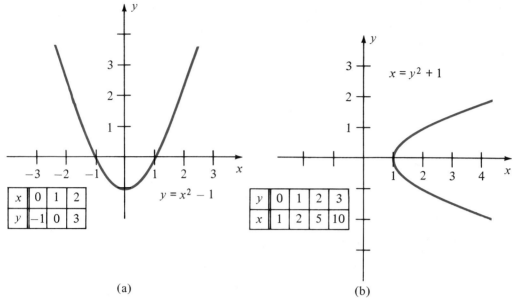

FIGURE 8

confirms the symmetry about the x-axis. We can think of the upper half of Figure 8b as the graph of $y = \sqrt{x - 1}$ and the lower half as the graph of $y = -\sqrt{x - 1}$.

EXAMPLE 5
Without sketching the graph, determine symmetry with respect to the x- and y-axes.

(a) $x^2 + 4y^2 - y = 1$ 　　　　　　　　　(b) $xy = 5$

Solution
(a) Replacing x by $-x$ in

$$x^2 + 4y^2 - y = 1$$
$$(-x)^2 + 4y^2 - y = 1$$
$$x^2 + 4y^2 - y = 1$$

Since the equation is unaltered, the curve is symmetric with respect to the y-axis. Now, replacing y by $-y$,

$$x^2 + 4(-y)^2 - (-y) = 1$$
$$x^2 + 4y^2 + y = 1$$

is *not* the same equation. Thus, the curve is not symmetric with respect to the x-axis.

(b) Replacing x by $-x$, we have $-xy = 5$, which is not the same equation. Replacing y by $-y$, we again have $-xy = 5$. Thus, the curve is not symmetric with respect to either axis.

PROGRESS CHECK
Without graphing, determine symmetry with respect to the coordinate axes.

(a) $x^2 - y^2 = 1$ 　　　　(b) $x + y = 10$ 　　　　(c) $y = \dfrac{1}{x^2 + 1}$

Answers
(a) Symmetric with respect to both x- and y-axes.
(b) Not symmetric with respect to either axis.
(c) Symmetric with respect to the y-axis.

There is another type of symmetry that is easy to verify—symmetry with respect to the origin. In Figure 9 we have sketched the graph of $y = x^3$,

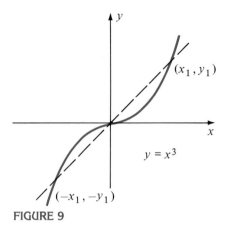

FIGURE 9

which appears to be symmetric with respect to the origin. We see from the graph that if (x_1, y_1) lies on the curve, so does $(-x_1, -y_1)$.

> If the equation of a curve is unchanged when x is replaced by $-x$ and y is replaced by $-y$, then the curve is **symmetric with respect to the origin.**

EXAMPLE 6
Determine symmetry with respect to the origin.

(a) $y = x^3 - 1$ (b) $y^2 = \dfrac{x^2 + 1}{x^2 - 1}$

Solution
(a) Replacing x by $-x$ and y by $-y$ in $y = x^3 - 1$

$$-y = (-x)^3 - 1$$
$$-y = -x^3 - 1$$
$$y = x^3 + 1$$

Since the equation is altered, it is not symmetric with respect to the origin.
(b) Replacing x by $-x$ and y by $-y$ in $y^2 = \dfrac{x^2 + 1}{x^2 - 1}$

$$(-y)^2 = \frac{(-x)^2 + 1}{(-x)^2 - 1}$$
$$y^2 = \frac{x^2 + 1}{x^2 - 1}$$

The equation is unchanged and we conclude that the curve is symmetric with respect to the origin.

PROGRESS CHECK
Determine symmetry with respect to the origin.

(a) $x^2 + y^2 = 1$ (b) $y^2 = x - 1$ (c) $y = x + \dfrac{1}{x}$

Answers
(a) Symmetric with respect to the origin.
(b) Not symmetric with respect to the origin.
(c) Symmetric with respect to the origin.

Note that in Example 6b and in (a) of the last Progress Check, the curves are symmetric with respect to both the x- and y-axes, as well as the origin. In fact, we have the following general rule.

A curve that is symmetric with respect to both coordinate axes is also symmetric with respect to the origin. However, a curve that is symmetric with respect to the origin need not be symmetric with respect to the coordinate axes.

The curve $y = x^3$ in Figure 9 is symmetric with respect to the origin, but not with respect to the coordinate axes.

EXERCISE SET 3.1
In each of Exercises 1–2 plot the given points on the same coordinate axes.

1. $(2,3), (-3,-2), \left(-\dfrac{1}{2},\dfrac{1}{2}\right), \left(0, \dfrac{1}{4}\right), \left(-\dfrac{1}{2}, 0\right), (3,-2)$

2. $(-3,4), (5,-2), (-1,-3), \left(-1, \dfrac{3}{2}\right), (0,1.5)$

In Exercises 3–8 find the distance between each pair of points.

3. $(5,4), (2,1)$ 4. $(-4,5), (-2,3)$
5. $(-1,-5), (-5, -1)$ 6. $(-3,0), (2, -4)$
7. $\left(\dfrac{2}{3},\dfrac{3}{2}\right), (-2,-4)$ 8. $\left(-\dfrac{1}{2},3\right), \left(-1, -\dfrac{3}{4}\right)$

In Exercises 9–12 find the length of the shortest side of the triangle determined by the three given points.

9. $A(6,2), B(-1,4), C(0,-2)$ 10. $P(2,-3), Q(4,4), R(-1,-1)$
11. $R\left(-1,\dfrac{1}{2}\right), S\left(-\dfrac{3}{2}, 1\right), T(2,-1)$ 12. $F(-5,-1), G(0,2), H(1,-2)$

In Exercises 13–16 determine if the given points form a right triangle. (*Hint:* A triangle is a right triangle if and only if the lengths of the sides satisfy the Pythagorean theorem.)

13. $(1,-2), (5,2), (2,1)$ 14. $(2,-3), (-1,-1), (3,4)$
15. $(-4,1), (1,4), (4,-1)$ 16. $(1,-1), (-6,1), (1,2)$

In Exercises 17–20 show that the points lie on the same line. (*Hint:* Three points are collinear if and only if the sum of the lengths of two sides equals the length of the third side.)

17. $(-1,2), (1,1), (5,-1)$ 18. $(-1,-4), (1,10), (0,3)$

19. $(-1,2), (1,5), \left(-2, \dfrac{1}{2}\right)$ 20. $(-1,-5), (1,1), (-2,-8)$

21. Find the perimeter of the quadrilateral whose vertices are $(-2,-1), (-4,5),$ $(3,5), (4,-2).$

22. Show that the points $(-2,-1), (2,2), (5,-2)$ are the vertices of an isosceles triangle.

23. Show that the points $(9,2), (11,6), (3,5),$ and $(1,1)$ are the vertices of a parallelogram.

24. Show that the point $(-1,1)$ is the midpoint of the line segment whose endpoints are $(-5,-1)$ and $(3,3).$

25. The points $A(2,7), B(4,3),$ and $C(x,y)$ determine a right triangle whose hypotenuse is AB. Find x and y. (*Hint:* There is more than one answer.)

26. The points $A(2,6), B(4,6), C(4,8),$ and $D(x,y)$ form a rectangle. Find x and y.

In Exercises 27–32 sketch the graph of the given equation.

27. $y = 2x + 4$ 28. $y = -3x + 5$ 29. $y = \sqrt{x}$

30. $y = \sqrt{x - 1}$ 31. $y = |x + 3|$ 32. $y = 2 - |x|$

In Exercises 33–38 use symmetry where possible to assist in sketching the graph of the given equation.

33. $y = 3 - x^2$ 34. $y = 3x - x^2$ 35. $y = x^3 + 1$

36. $x = y^3 - 1$ 37. $x = y^2 - 1$ 38. $y = 3x$

Without graphing, determine whether each curve is symmetric with respect to the x-axis, the y-axis, the origin, or none of these.

39. $3x + 2y = 5$ 40. $y = 4x^2$ 41. $y^2 = x - 4$

42. $x^2 - y = 2$ 43. $y^2 = 1 + x^3$ 44. $y = (x - 2)^2$

45. $y^2 = (x - 2)^2$ 46. $y^2 x + 2x = 4$ 47. $y^2 x + 2x^2 = 4x^2 y$

48. $y^3 = x^2 - 9$ 49. $y = \dfrac{x^2 + 4}{x^2 - 4}$ 50. $y = \dfrac{1}{x^2 + 1}$

51. $y^2 = \dfrac{x^2 + 1}{x^2 - 1}$ 52. $4x^2 + 9y^2 = 36$ 53. $xy = 4$

3.2
FUNCTIONS AND FUNCTION NOTATION

The equation

$$y = 2x + 3$$

assigns a value to y for every value of x. If we let X denote the set of values that we can assign to x, and let Y denote the set of values that the equation assigns to y, we can show the correspondence schematically as in Figure 10.

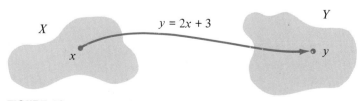

FIGURE 10

We are particularly interested in the situation when, for each element x in X, there corresponds one and only one element y in Y; that is, the rule assigns exactly one y for a given x. This type of correspondence plays a fundamental role in mathematics and is given a special name.

A **function** is a rule which, for each x in a set X, assigns exactly one y in a set Y. The element y is called the **image** of x. The set X is called the **domain** of the function and the set of all images is called the **range** of the function.

We can think of the rule defined by the equation $y = 2x + 3$ as a function machine (see Figure 11). Each time we drop a value of x from the domain

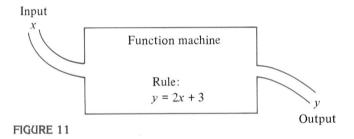

Input
x

Function machine

Rule:
$y = 2x + 3$

y
Output

FIGURE 11

into the input hopper, exactly one value of y falls out of the output hopper. If we drop in $x = 5$, the function machine follows the rule and produces $y = 13$. Since we are free to choose those values of x which we drop into the machine, we call x the **independent variable**; the value of y which drops out depends upon the choice of x so y is called the **dependent variable**. We say that the dependent variable is a function of the independent variable; that is, *the output is a function of the input*.

Let's look at a few schematic presentations. The correspondence in Figure 12a is a function; for each x in X there is exactly one corresponding

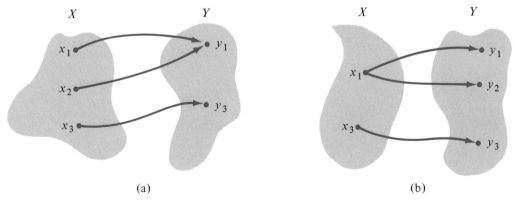

X Y X Y

x_1 y_1 x_1 y_1

x_2 y_2

y_3

x_3 x_3 y_3

(a)   (b)

FIGURE 12

value of y in Y. Although y_1 is the image of both x_1 and x_2, this does not violate the definition of a function. However, the correspondence in Figure 12b is not a function since x_1 has two images, y_1 and y_2, assigned to it, which violates the definition of a function.

VERTICAL LINE TEST

There is a simple graphic way to test if an equation determines a function. Let's graph the equations $y = x^2$ and $y^2 = x$. Now we draw vertical lines on both graphs in Figure 13. No vertical line intersects the graph of $y = x^2$ in more than one point. This means that the equation $y = x^2$ assigns exactly one y for each x and therefore determines y as a function of x. On the other hand, since some vertical lines intersect the graph of $y^2 = x$ in two points, the equation $y^2 = x$ assigns *two* values of y to some values of x and the resulting correspondence does not determine y as a function of x. Thus, *not every equation in the variables x and y determines y as a function of x.*

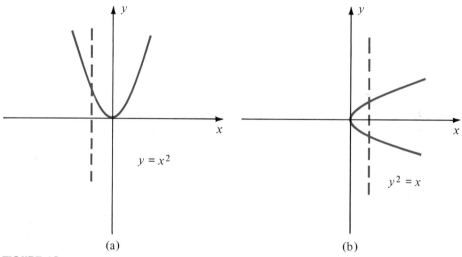

(a)   (b)

FIGURE 13

Vertical Line Test

An equation determines a function if and only if no vertical line meets the graph of the equation in more than one point.

EXAMPLE 1
Which of the following graphs of equations determine a function?

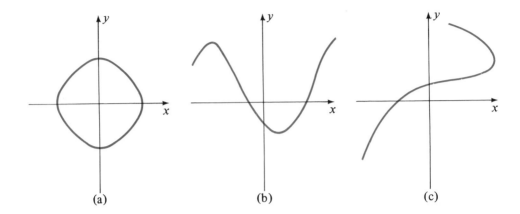

(a)   (b)   (c)

(a)   Not a function. Some vertical lines meet the graph in more than one point.
(b)   A function. Passes the vertical line test.
(c)   Not a function. Fails the vertical line test.

DOMAIN AND RANGE

We have defined the domain of a function to be the set of values assumed by the independent variable. In more advanced courses in mathematics, the domain may include complex numbers. For our work in this chapter, we will restrict the domain of a function to be those real numbers for which the function is defined. When a function is defined by an equation, we must always be alert to two potential problems.

(a)   *Division by zero.* For example, the domain of the function

$$y = \frac{2}{x-1}$$

is the set of all real numbers other than $x = 1$. When $x = 1$ the denominator is 0, and division by 0 is not defined.

(b)   *Even roots of negative numbers.* For example, the function

$$y = \sqrt{x-1}$$

is defined only for $x \geq 1$ since we exclude the square root of negative numbers. Hence the domain of the function consists of all real numbers $x \geq 1$.

The range of a function is, in general, not as easily determined as is the domain. The range is the set of all y-values which occur in the correspondence; that is, it is the set of all outputs of the function. For our purposes it will suffice to determine the range by examining the graph.

EXAMPLE 2
Graph the equation $y = \sqrt{x}$. If the correspondence determines a function, find the domain and range.

Solution

We obtain the graph of the equation by plotting points and connecting them to form a smooth curve. Applying the vertical line test to the graph as shown in Figure 14 we see that the equation determines a function. The domain of the function is the set $\{x | x \geq 0\}$ and the range is the set $\{y | y \geq 0\}$.

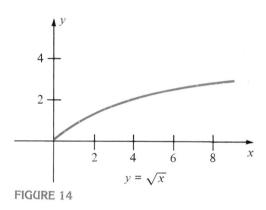

FIGURE 14

PROGRESS CHECK

Graph the equation $y = x^2 - 4$, $-3 \leq x \leq 3$. If the correspondence determines a function, find the domain and range.

Answer
The graph is that portion of the curve shown in Figure 5 (page 109) that lies between $x = -3$ and $x = 3$. The domain is $\{x | -3 \leq x \leq 3\}$; the range is $\{y | -4 \leq y \leq 5\}$.

FUNCTION NOTATION

It is customary to designate a function by a letter of the alphabet such as f, g, F, or C. We then denote the output corresponding to x by $f(x)$, which is read "f of x." Thus,

$$f(x) = 2x + 3$$

specifies a rule f for determining an output $f(x)$ for a given value of x. To find $f(x)$ when $x = 5$, we simply substitute 5 for x and obtain

$$f(5) = 2(5) + 3 = 13$$

The notation $f(5)$ is a convenient way of specifying "the value of the function f that corresponds to $x = 5$." The symbol f represents the function or rule; the notation $f(x)$ represents the output produced by the rule. For convenience, however, we will at times join in the common practice of designating the function f by $f(x)$.

EXAMPLE 3

(a) If $f(x) = 2x^2 - 2x + 1$, find $f(-1)$.
We substitute -1 for x.

$$f(-1) = 2(-1)^2 - 2(-1) + 1 = 5$$

(b) If $f(t) = 3t^2 - 1$, find $f(2a)$.
We substitute $2a$ for t.

$$f(2a) = 3(2a)^2 - 1 = 3(4a^2) - 1 = 12a^2 - 1$$

PROGRESS CHECK

(a) If $f(u) = u^3 + 3u - 4$, find $f(-2)$.
(b) If $f(t) = t^2 + 1$, find $f(t - 1)$.

Answers
(a) -18 (b) $t^2 - 2t + 2$

EXAMPLE 4

Let the function f be defined by $f(x) = x^2 - 1$. Find

(a) $f(-2)$ (b) $f(a)$ (c) $f(a + h)$ (d) $f(a + h) - f(a)$

Solution

(a) $f(-2) = (-2)^2 - 1 = 4 - 1 = 3$

(b) $f(a) = a^2 - 1$

(c) $f(a + h) = (a + h)^2 - 1 = a^2 + 2ah + h^2 - 1$

(d) $f(a + h) - f(a) = (a + h)^2 - 1 - (a^2 - 1)$
$$= a^2 + 2ah + h^2 - 1 - a^2 + 1$$
$$= 2ah + h^2$$

WARNING

(a) Note that $f(a + 3) \neq f(a) + f(3)$. Function notation is not to be confused with the distributive law.
(b) Note that $f(x^2) \neq f \cdot x^2$. The use of parentheses in function notation does not imply multiplication.
(c) Note that $f(x^2) \neq [f(x)]^2$. Squaring x is not the same as squaring $f(x)$.

EXAMPLE 5

The correspondence between Fahrenheit temperature F and Celsius temperature C is given by

$$F = \frac{9}{5}C + 32$$

(a) Find the Fahrenheit temperature corresponding to a Celsius reading of 37° (normal body temperature).
(b) Express C as a function of F.
(c) Find the Celsius temperature corresponding to a Fahrenheit reading of 212° (the boiling point of water).

Solution

(a) We substitute $C = 37$ to find $F(37)$.

$$F = \frac{9}{5}(37) + 32 = 98.6$$

Normal body temperature is 98.6° Fahrenheit.

(b) We solve the following equation for C:

$$F = \frac{9}{5}C + 32$$

$$F - 32 = \frac{9}{5}C$$

$$C = \frac{5}{9}(F - 32)$$

(c) We substitute $F = 212$ to find $C(212)$.

$$C = \frac{5}{9}(F - 32)$$

$$C = \frac{5}{9}(212 - 32) = \frac{5}{9}(180) = 100$$

Water boils at 100° Celsius.

EXERCISE SET 3.2

In Exercises 1–6 graph the equation. If the graph determines a function, find the domain and use the graph to determine the range of the function.

1. $y = 2x - 3$
2. $y = x^2 + x, -2 \le x \le 1$
3. $x = y + 1$
4. $x = y^2 - 1$
5. $y = \sqrt{x - 1}$
6. $y = |x|$

In Exercises 7–12 determine the domain of the function defined by the given rule.

7. $f(x) = \sqrt{2x - 3}$
8. $f(x) = \sqrt{5 - x}$
9. $f(x) = \dfrac{1}{\sqrt{x - 2}}$
10. $f(x) = \dfrac{-2}{x^2 + 2x - 3}$
11. $f(x) = \dfrac{\sqrt{x - 1}}{x - 2}$
12. $f(x) = \dfrac{x}{x^2 - 4}$

In Exercises 13–16 find the number (or numbers) whose image is 2.

13. $f(x) = 2x - 5$
14. $f(x) = x^2$
15. $f(x) = \dfrac{1}{x - 1}$
16. $f(x) = \sqrt{x - 1}$

Given the function f defined by $f(x) = 2x^2 + 5$, determine the following.

17. $f(0)$
18. $f(-2)$
19. $f(a)$
20. $f(3x)$
21. $3f(x)$
22. $-f(x)$

Given the function g defined by $g(x) = x^2 + 2x$, determine the following.

23. $g(-3)$
24. $g\left(\dfrac{1}{x}\right)$
25. $\dfrac{1}{g(x)}$
26. $g(-x)$
27. $g(a + h)$
28. $\dfrac{g(a + h) - g(a)}{h}$

Given the function F defined by $F(x) = \dfrac{x^2 + 1}{3x - 1}$, determine the following.

29. $F(-2.73)$ 30. $F(16.11)$ 31. $\dfrac{1}{F(x)}$

32. $F(-x)$ 33. $2F(2x)$ 34. $F(x^2)$

Given the function r defined by $r(t) = \dfrac{t - 2}{t^2 + 2t - 3}$, determine the following.

35. $r(-8.27)$ 36. $r(2.04)$ 37. $r(2a)$

38. $2r(a)$ 39. $r(a + 1)$ 40. $r(1 + h)$

41. If x dollars are borrowed at 7% simple annual interest, express the interest I at the end of 4 years as a function of x.
42. Express the area A of an equilateral triangle as a function of the length s of its side.
43. Express the diameter d of a circle as a function of its circumference C.
44. Express the perimeter P of a square as a function of its area A.

3.3
GRAPHS OF FUNCTIONS

We have used the graph of an equation to help us find out whether or not the equation determines a function. It is therefore natural that the **graph of a function** f is defined to be the graph of the equation $y = f(x)$. For example, the graph of the function f defined by the rule $f(x) = x^2 - 4$ is the graph of the equation $y = x^2 - 4$, which has been sketched in Figure 5 on page 109.

Equations are not the only way of defining a function. In many important applications, a function may be defined by a table or by several equations. We illustrate this idea by several examples.

EXAMPLE 1
The commission earned by a door-to-door cosmetics salesperson is determined as follows.

Weekly Sales	Commission
less than $300	20% of sales
$300 or more but less than $400	$60 + 35% of sales over $300
$400 or more	$95 + 60% of sales over $400

(a) Express the commission C as a function of sales s.
(b) Find the commission if the weekly sales are $425.
(c) Sketch the graph of the function.

Solution
(a) The function C can be described by three equations.

$$C(s) = \begin{cases} 0.20s & \text{if } s < 300 \\ 60 + 0.35\,(s - 300) & \text{if } 300 \le s < 400 \\ 95 + 0.60\,(s - 400) & \text{if } s \ge 400 \end{cases}$$

(b) When $s = 425$, we must use the third equation and substitute to determine $C(425)$.

$$C(425) = 95 + 0.60(425 - 400)$$
$$= 95 + 0.60(25)$$
$$= 110$$

The commission on sales of $425 is $110.

(c) The graph of the function C consists of three line segments (Figure 15).

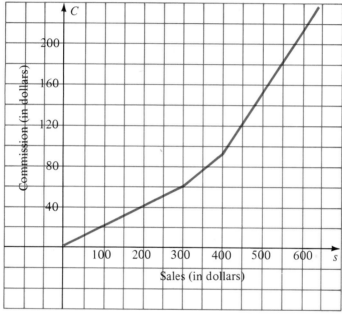

FIGURE 15

EXAMPLE 2
Sketch the graph of the function f defined by

$$f(x) = \begin{cases} x^2 & \text{if} & -2 \le x \le 2 \\ 2x + 1 & \text{if} & 2 < x \le 5 \end{cases}$$

Solution
We form a table of points to be plotted.

x	-2	-1	0	1	2	3	4	5
y	4	1	0	1	4	7	9	11

Note that the graph in Figure 16 has a gap. Also note that the point (2,5) has been marked with an open circle to indicate that it is not on the graph of the function.

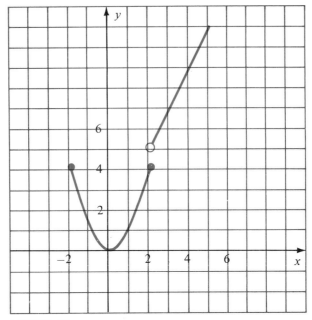

FIGURE 16

INCREASING AND DECREASING FUNCTIONS

We say that the straight line in Figure 17a is increasing or rising since the values of y increase as we move from left to right. Since the graph of a function f is obtained by sketching the equation $y = f(x)$, we can give a precise definition of increasing and decreasing functions.

> **Increasing and Decreasing Functions**
>
> A function f is **increasing** if $f(x_1) < f(x_2)$ whenever $x_1 < x_2$.
>
> A function f is **decreasing** if $f(x_1) > f(x_2)$ whenever $x_1 < x_2$.

In other words, if a function is increasing, the dependent variable y assumes larger values as we move from left to right (Figure 17a); for a decreasing function (Figure 17b), y takes on smaller values as we move from left to right. The function pictured in Figure 17c is *neither* increasing nor decreasing according to this definition. In fact, one portion of the graph is decreasing and another is increasing. We can modify our definition of increasing and decreasing functions to apply to *intervals* in the domain. A function may then be increasing in some intervals and decreasing in others.

Returning to Figure 17c, note that the function is decreasing when $x \leq -3$ and increasing when $x \geq -3$. This is the "usual" situation for a function; there are intervals in which the function is increasing and intervals in which it is decreasing. Of course, there is another possibility. The

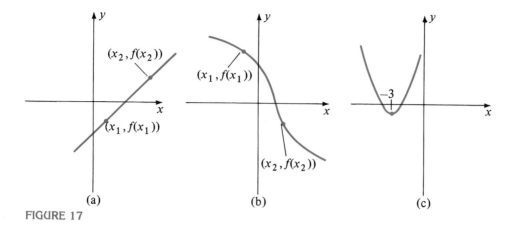

(a)   (b)   (c)

FIGURE 17

function may retain the same value on an interval, in which case we call it a **constant function** over that interval.

EXAMPLE 3
Given the function f defined by $f(x) = 1 - x^2$, use a graph to determine where the function is increasing and where it is decreasing.

Solution
We graph $y = 1 - x^2$ by plotting several points and noting that the graph is symmetric with respect to the y-axis. From the graph in Figure 18 we see that

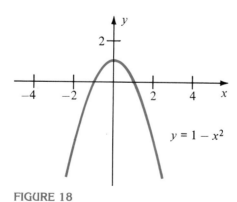

FIGURE 18

$$f \text{ is increasing when } x \leq 0$$
$$f \text{ is decreasing when } x \geq 0$$

EXAMPLE 4
The function f is defined by

$$f(x) = \begin{cases} |x| & \text{if } x \leq 2 \\ -3 & \text{if } x > 2 \end{cases}$$

Use a graph to find the values of x for which the function is increasing, decreasing, and constant.

Solution
We sketch the graph of f by plotting a number of points. From the graph in Figure 19 we determine that

$$f \text{ is increasing if } 0 \leq x \leq 2$$

$$f \text{ is decreasing if } x \leq 0$$

$$f \text{ is constant and has value } -3 \text{ if } x > 2.$$

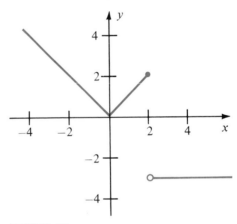

FIGURE 19

PROGRESS CHECK
The function f is defined by

$$f(x) = \begin{cases} 2x + 1 & \text{if } x < -1 \\ 0 & \text{if } -1 \leq x \leq 3 \\ -2x + 1 & \text{if } x > 3 \end{cases}$$

Use a graph to find the values of x for which the function is increasing, decreasing, and constant.

Answer
Increasing when $x < -1$.   Constant when $-1 \leq x \leq 3$ Decreasing when $x > 3$

POLYNOMIAL FUNCTIONS
The polynomial function of first-degree

$$f(x) = ax + b$$

is called a **linear function.** We will study this function in detail in a later section and will show that its graph is a straight line. For now, we sketch the graphs of a few linear functions to "convince" ourselves that they appear to be straight lines.

EXAMPLE 5

Sketch the graphs of $f(x) = 2x + 3$ and $g(x) = -2x$ on the same coordinate axes. Represent $f(5)$ by drawing the vertical line from the x-axis to the point $(5, f(5))$.

Solution

We form a table of values for $y = 2x + 3$ and $y = -2x$, plot the corresponding points, and connect these by "smooth curves" as in Figure 20.

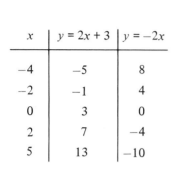

x	$y = 2x + 3$	$y = -2x$
-4	-5	8
-2	-1	4
0	3	0
2	7	-4
5	13	-10

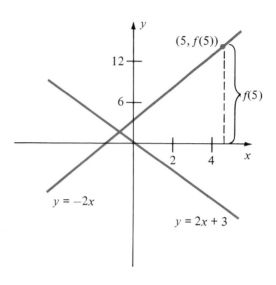

FIGURE 20

The polynomial function of second-degree

$$f(x) = ax^2 + bx + c, \ a \neq 0$$

is called a **quadratic function.** The graph of this function is called a **parabola** and will be studied in detail in a later chapter.

EXAMPLE 6

Sketch the graph of $f(x) = 2x^2 - 4x + 3$.

Solution

We need to graph $y = 2x^2 - 4x + 3$. We form a table of values, plot the corresponding points, and connect these by a smooth curve as in Figure 21.

An investigation of polynomials of any degree reveals that they are all functions. The graphs of higher degree polynomials are more complex than the straight line or parabola but are always "smooth curves." The exercises are intended to help you gain experience with the graphs of polynomial functions. The method of plotting points enables us to sketch a curve but doesn't reveal the turning points of the curve. More sophisticated techniques that aid in sketching curves by determining "maximum" and "minimum" points and other useful information are developed in calculus courses. We will return to the graphs of polynomial functions in a later chapter.

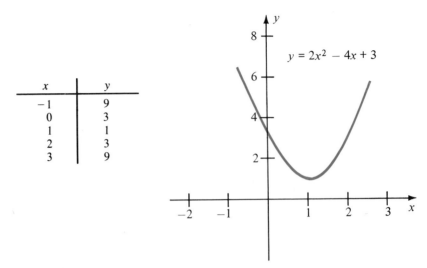

x	y
−1	9
0	3
1	1
2	3
3	9

FIGURE 21

EXERCISE SET 3.3

In Exercises 1–16 sketch the graph of the function and state where it is increasing, decreasing, and constant.

1. $f(x) = 3x + 1$

2. $f(x) = 3 - 2x$

3. $f(x) = x^2 + 1$

4. $f(x) = x^2 - 4$

5. $f(x) = 9 - x^2$

6. $f(x) = 4x - x^2$

7. $f(x) = |2x + 1|$

8. $f(x) = |1 - x|$

9. $f(x) = \begin{cases} 2x, & x > -1 \\ -x - 1, & x \le -1 \end{cases}$

10. $f(x) = \begin{cases} x + 1, & x > 2 \\ 1, & -1 \le x \le 2 \\ -x + 1, & x < -1 \end{cases}$

11. $f(x) = \begin{cases} x, & x < 2 \\ 2, & x \ge 2 \end{cases}$

12. $f(x) = \begin{cases} -x, & x \le -2 \\ x^2, & -2 < x \le 2 \\ -x, & 3 \le x \le 4 \end{cases}$

13. $f(x) = \begin{cases} -x^2, & -3 < x < 1 \\ 0, & 1 \le x \le 2 \\ -3x, & x > 2 \end{cases}$

14. $f(x) = \begin{cases} 2 & \text{if } x \text{ is an integer} \\ -1 & \text{if } x \text{ is not an integer} \end{cases}$

15. $f(x) = \begin{cases} -2, & x < -2 \\ -1, & -2 \le x \le -1 \\ 1, & x > -1 \end{cases}$

16. $f(x) = \begin{cases} \dfrac{x^2 - 1}{x - 1}, & x \ne 1 \\ 3, & x = 1 \end{cases}$

In Exercises 17–24 sketch the graphs of the given functions on the same coordinate axes.

17. $f(x) = x^2$, $g(x) = 2x^2$, $h(x) = \dfrac{1}{2}x^2$

18. $f(x) = \dfrac{1}{2}x^2$, $g(x) = \dfrac{1}{3}x^2$, $h(x) = \dfrac{1}{4}x^2$

19. $f(x) = 2x^2$, $g(x) = -2x^2$

20. $f(x) = x^2 - 2$, $g(x) = 2 - x^2$

21. $f(x) = x^3$, $g(x) = 2x^3$

22. $f(x) = \dfrac{1}{2}x^3$, $g(x) = \dfrac{1}{4}x^3$

23. $f(x) = x^3, g(x) = -x^3$ 24. $f(x) = -2x^3$, $g(x) = -4x^3$

25. The telephone company charges a fee of $6.50 per month for the first 100 message units and an additional fee of 6 cents for each of the next 100 message units. A reduced rate of 5 cents is charged for each message unit after the first 200 units. Express the monthly charge C as a function of the number of message units u.

26. The annual dues of a union are as shown in the table.

Employee's annual salary	Annual dues
less than $8000	$60
$8000 or more but less than $15,000	$60 + 1% of the salary in excess of $8000
$15,000 or more	$130 + 2% of the salary in excess of $15,000

Express the annual dues d as a function of the salary.

27. A tour operator who runs charter flights to Rome has established the following pricing schedule. For a group of no more than 100 people, the round trip fare per person is $300, with a minimum rental of $30,000 for the plane. For a group which has more than 100 but less than 150 people the fare for all passengers will be reduced by $1 for each passenger in excess of 100. Write the tour operator's total revenue R as a function of the number of people x in the group.

28. A firm packages and ships 1-pound jars of instant coffee. The cost C of shipping is 40 cents for the first pound and 25 cents for each additional pound.
     (a) Write C as a function of the weight w (in pounds) for $0 < w \le 30$.
     (b) What is the cost of shipping a package containing 24 jars of instant coffee?

29. The daily rates of a car rental firm are $14 plus 8 cents per mile.
     (a) Express the cost C of renting a car as a function of the number of miles m traveled.
     (b) What is the domain of the function?
     (c) How much would it cost to rent a car for a 100-mile trip?

30. In a wildlife preserve, the population P of eagles depends upon the population x of its basic food supply, rodents. Suppose that P is given by

$$P(x) = 0.002x + 0.004x^2$$

Find the eagle population when the rodent population is
(a) 500   (b) 2000

3.4
LINEAR FUNCTIONS

In the last section we said that a linear function is of the form $f(x) = ax + b$, and we observed that the graphs of such functions appear to be straight lines. In this section we show that this conjecture is true; that is, the graph of a linear function is always a straight line.

SLOPE OF THE STRAIGHT LINE

Consider a straight line L that is not parallel to the y-axis (see Figure 22) and let $P_1(x_1,y_1)$ and $P_2(x_2,y_2)$ by any two distinct points on L. We have indicated the increments or changes $x_2 - x_1$ and $y_2 - y_1$ in the x and y

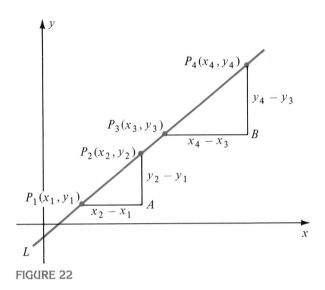

FIGURE 22

coordinates, respectively, from P_1 to P_2. Note that the increment $y_2 - y_1$ can be positive, negative, or zero, while the increment $x_2 - x_1$ can only be negative or positive. If we choose any other pair of points, say $P_3(x_3,y_3)$ and $P_4(x_4,y_4)$, we will, in general, obtain different increments in the x- and y-coordinates. However, since triangles P_1AP_2 and P_3BP_4 are similar, the ratio $\dfrac{y_2 - y_1}{x_2 - x_1}$ will be the same as the ratio $\dfrac{y_4 - y_3}{x_4 - x_3}$. We call this ratio the **slope of the line** L and denote it by m.

> **Slope of a Line**
>
> The slope of a line which is not vertical is given by
>
> $$m = \frac{y_2 - y_1}{x_2 - x_1}$$
>
> where $P_1(x_1, y_1)$ and $P_2(x_2, y_2)$ are any two points on the line.

For a *vertical* line, $x_1 = x_2$, so that $x_2 - x_1 = 0$. Since we cannot divide by 0, we say that a vertical line has no slope.

EXAMPLE 1
Find the slope of the line which passes through the points $(4,2)$, $(1,-2)$.

Solution

We may choose either point as (x_1, y_1) and the other as (x_2, y_2). Our choice is

$$(x_1, y_1) = (4, 2) \quad \text{and} \quad (x_2, y_2) = (1, -2)$$

Then

$$m = \frac{y_2 - y_1}{x_2 - x_1} = \frac{-2 - 2}{1 - 4} = \frac{-4}{-3} = \frac{4}{3}$$

The student should verify that reversing the choice of P_1 and P_2 produces the same result for the slope m. We may choose either point as P_1 and the other as P_2, but we must use this choice consistently once it has been made.

Slope is a means of measuring the steepness of a line. That is, slope specifies the number of units we must move up or down to reach the line after moving one unit to the left or right of the line. In Figure 23 we have indicated several lines with positive and negative slopes. We can summarize this way.

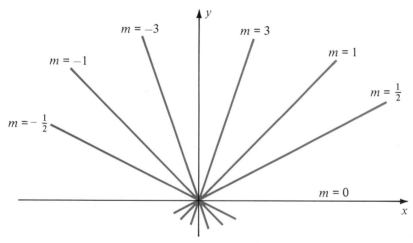

FIGURE 23

(a) When $m > 0$, the line is the graph of an increasing function.

(b) When $m < 0$, the line is the graph of a decreasing function.

(c) When $m = 0$, the line is the graph of a constant function.

(d) Slope does not exist for a vertical line, and a vertical line is not the graph of a function.

EQUATIONS OF THE STRAIGHT LINE

We can apply the concept of slope to develop two important forms of the equations of a straight line. In Figure 24 the point $P_1(x_1, y_1)$ lies on a line L

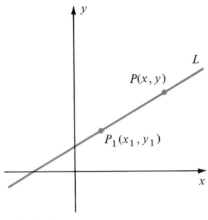

FIGURE 24

whose slope is m. If $P(x,y)$ is any other point on L, then we may use P and P_1 to compute m, that is,

$$m = \frac{y - y_1}{x - x_1}$$

This can be written in the form

$$y - y_1 = m(x - x_1)$$

Since (x_1, y_1) satisfies this equation, every point on L satisfies this equation. Conversely, any point satisfying this equation must lie on the line L since there is only one line through $P_1(x_1, y_1)$ with slope m. This equation is called the point-slope form of a line.

Point-Slope Form

$$y - y_1 = m(x - x_1)$$

is an equation of the line with slope m that passes through the point (x_1, y_1).

EXAMPLE 2
Find an equation of the line that passes through the points $(6, -2)$ and $(-4, 3)$.

Solution
We first find the slope. Letting $(x, y) = (6, -2)$ and $(x_2, y_2) = (-4, 3)$, then

$$m = \frac{y_2 - y_1}{x_2 - x_1} = \frac{3 - (-2)}{-4 - 6} = \frac{5}{-10} = -\frac{1}{2}$$

Next, the point-slope form is used with $m = -\frac{1}{2}$ and $(x_1, y_1) = (6, -2)$.

$$y - y_1 = m(x - x_1)$$

$$y - (-2) = -\frac{1}{2}(x - 6)$$

$$y = -\frac{1}{2}x + 1$$

The student should verify that using the point $(-4,3)$ and $m = -\frac{1}{2}$ in the point-slope form will yield the same equation.

PROGRESS CHECK
Find an equation of the line that passes through the points $(-5,0)$ and $(2,-5)$.

Answer
$7y = -5x - 25$

There is another form of the equation of the straight line that is very useful. In Figure 25 the line L meets the y-axis at the point $(0,b)$, and is assumed to have slope m. Then we can let $(x_1,y_1) = (0,b)$ and use the point-slope form.

$$y - y_1 = m(x - x_1)$$
$$y - b = m(x - 0)$$
$$y = mx + b$$

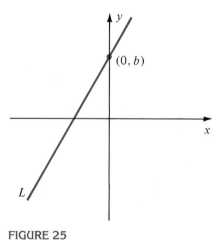

FIGURE 25

Recalling that b is the y-intercept, we call this equation the slope-intercept form of the line.

Slope-Intercept Form

The graph of the equation

$$y = mx + b$$

is a straight line with slope m and y-intercept b.

Since the graph of $y = mx + b$ is the graph of the function $f(x) = mx + b$, we arrive at the important conclusion that the graph of a linear function is a straight line.

EXAMPLE 3
Find the slope and y-intercept of the line $y - 3x + 1 = 0$.

Solution
The equation must be placed in the form $y = mx + b$. Thus, solving for y,

$$y = 3x - 1$$

and we find that $m = 3$ is the slope and $b = -1$ is the y-intercept.

PROGRESS CHECK
Find the slope and y-intercept of the line $2y + x - 3 = 0$.

Answer

$$slope = m = -\frac{1}{2}; \; y\text{-}intercept = b = \frac{3}{2}$$

The **general first-degree equation** in x and y can always be written in the form

$$Ax + By + C = 0$$

where A, B, and C are constants. We can rewrite this equation as

$$By = -Ax - C$$

If $B \neq 0$, the equation becomes

$$y = -\frac{A}{B}x - \frac{C}{B}$$

which we recognize as having a straight line graph with slope $-A/B$ and y-intercept $-C/B$. If $B = 0$, the original equation becomes $Ax + C = 0$ whose graph is a vertical line. We have thereby proved the following.

> The graph of the first-degree equation
>
> $$Ax + By + C = 0$$
>
> is a straight line.

HORIZONTAL AND VERTICAL LINES

In Figure 26a we have drawn a horizontal line through the point (a,b). Every point on this line has the form (x,b) since the y-coordinate remains constant. If $P(x_1,b)$ and $Q(x_2,b)$ are any two distinct points on the line, then the slope is

$$m = \frac{b - b}{x_2 - x_1} = 0$$

We have established the following.

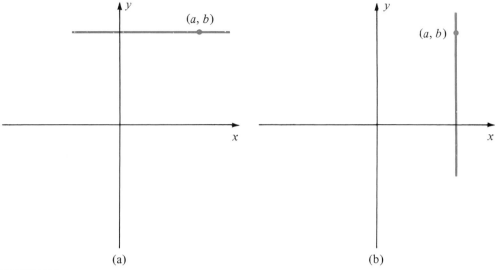

FIGURE 26

Horizontal Lines

The equation of the horizontal line through the point (a,b) is

$$y = b$$

The slope of a horizontal line is 0.

In Figure 26b, every point on the vertical line through the point (a,b) has the form (a,y) since the x-coordinate remains constant. The slope computation using any two points $P(a,y_1)$ and $Q(a,y_2)$ on the line produces

$$m = \frac{y_2 - y_1}{a - a} = \frac{y_2 - y_1}{0}$$

Since we cannot divide by 0, slope is not defined for a vertical line.

Vertical Lines

The equation of the vertical line through the point (a,b) is

$$x = a$$

The slope of a vertical line is undefined.

EXAMPLE 4

Find the equations of the horizontal and vertical lines through $(-4,7)$.

Solution

The horizontal line has the equation $y = 7$. The vertical line has the equation $x = -4$.

WARNING

Don't confuse "no slope" and "zero slope." A horizontal line has zero slope. A vertical line has no slope, by which we mean that the slope is undefined.

PARALLEL AND PERPENDICULAR LINES

The concept of slope of a line can be used to determine when two lines are parallel or perpendicular. Since parallel lines have the same "steepness," we intuitively recognize that they must have the same slope.

> Two lines with slopes m_1 and m_2 are parallel if and only if $m_1 = m_2$.

The criterion for perpendicular lines can be stated in this way.

> Two lines with slopes m_1 and m_2 are perpendicular if and only if
> $$m_2 = -\frac{1}{m_1}.$$

These two theorems do not apply to vertical lines since the slope of a vertical line is undefined. The proofs of these theorems are geometric in nature and are outlined in Exercises 54 and 56 of Exercise Set 3.4.

EXAMPLE 5
Given the line $y = 3x - 2$, find an equation of the line passing through the point $(-5,4)$ that is (a) parallel to the given line and (b) perpendicular to the given line.

Solution
We first note that the line $y = 3x - 2$ has slope $m_1 = 3$.

(a) Every line parallel to the line $y = 3x - 2$ must have slope $m_2 = m_1 = 3$. We therefore seek a line with slope 3 that passes through the point $(-5,4)$. Using the point-slope formula,

$$y - y_1 = m(x-x_1)$$
$$y - 4 = 3(x+5)$$
$$y = 3x + 19$$

(b) Every line perpendicular to the line $y = 3x - 2$ has slope $m_2 = -1/m_1 = -1/3$. The line we seek has slope $-1/3$ and passes through the point $(-5,4)$. We can again apply the point-slope formula to obtain

$$y - y_1 = m(x-x_1)$$
$$y - 4 = -\frac{1}{3}(x+5)$$
$$y = -\frac{1}{3}x + \frac{7}{3}$$

The three lines are shown in Figure 27.

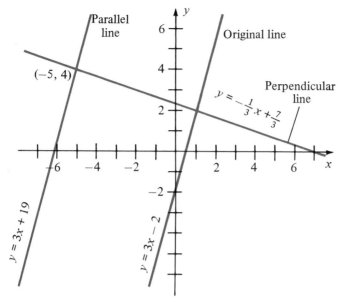

FIGURE 27

We have seen that slope tells us a great deal about the character of a line. The concept of slope can be extended to curves other than straight lines. In calculus we use this extended definition to show that the straight line is the only curve whose slope is constant. Thus, slope not only helps to characterize lines but also serves to distinguish straight lines from all other curves.

EXERCISE SET 3.4

In Exercises 1–6 determine the slope of the line through the given points. State whether the line is the graph of an increasing function, a decreasing function, or a constant function.

1. $(2, 3), (-1, -3)$ 2. $(1, 2), (-2, 5)$ 3. $(-2, 3), (0, 0)$

4. $(2, 4), (-3, 4)$ 5. $\left(\frac{1}{2}, 2\right), \left(\frac{3}{2}, 1\right)$ 6. $(-4, 1), (-1, -2)$

7. Use slopes to show that the points $A(-1, -5)$, $B(1, -1)$, and $C(3, 3)$ are collinear (lie on the same line).

8. Use slopes to show that the points $A(-3, 2)$, $B(3, 4)$, $C(5, -2)$ and $D(-1, -4)$, are the vertices of a parallelogram.

In Exercises 9–12 determine an equation of the line with the given slope that passes through the given point.

9. $m = 2, (-1, 3)$ 10. $m = -\frac{1}{2}, (1, -2)$

11. $m = 3, (0, 0)$ 12. $m = 0, (-1, 3)$

In Exercises 13–18 determine an equation of the line through the given points.

13. $(2, 4), (-3, -6)$ 14. $(-3, 5), (1, 7)$ 15. $(0, 0), (3, 2)$

16. $(-2, 4), (3, 4)$ 17. $\left(-\frac{1}{2}, -1\right), \left(\frac{1}{2}, 1\right)$ 18. $(-8, -4), (3, -1)$

In Exercises 19–24 determine an equation of the line with the given slope m and the given y-intercept b.

19. $m = 3, b = 2$ 20. $m = -3, b = -3$ 21. $m = 0, b = 2$

22. $m = -\frac{1}{2}, b = \frac{1}{2}$ 23. $m = \frac{1}{3}, b = -5$ 24. $m = -2, b = -\frac{1}{2}$

In Exercises 25–30 determine the slope m and y-intercept b of the given line.

25. $3x + 4y = 5$ 26. $2x - 5y + 3 = 0$

27. $y - 4 = 0$ 28. $x = -5$

29. $3x + 4y + 2 = 0$ 30. $x = -\frac{1}{2}y + 3$

In Exercises 31–36 write an equation of (a) the horizontal line passing through the given point and (b) the vertical line passing through the given point.

31. $(-6, 3)$ 32. $(-5, -2)$ 33. $(-7, 0)$

34. $(0, 5)$ 35. $(9, -9)$ 36. $\left(-\frac{3}{2}, 1\right)$

In Exercises 37–40 determine the slope of every line that is (a) parallel to the given line and (b) perpendicular to the given line.

37. $y = -3x + 2$ 38. $2y - 5x + 4 = 0$

39. $3y = 4x - 1$ 40. $5y + 4x = -1$

In Exercises 41–44 determine an equation of the line through the given point that (a) is parallel to the given line and (b) is perpendicular to the given line.

41. $(1,3); y = -3x + 2$ 42. $(-1,2); 3y + 2x = 6$

43. $(-3,2); 3x + 5y = 2$ 44. $(-1,-3); 3y + 4x - 5 = 0$

45. The Celsius (C) and Fahrenheit (F) temperature scales are related by a linear equation. Water boils at 212°F or 100°C, and freezes at 32°F or 0°C.
    (a) Write a linear equation expressing F in terms of C.
    (b) What is the Fahrenheit temperature when the Celsius temperature is 20°?

46. The college bookstore sells a textbook that costs $10 for $13.50, and a textbook that costs $12 for $15.90. If the markup policy of the bookstore is linear, write a linear function that relates sales price S and cost C. What is the cost of a book that sells for $22?

47. An appliance manufacturer finds that it had sales of $200,000 five years ago and sales of $600,000 this year. If the growth in sales is assumed to be linear, what will the sales amount be five years from now?

48. A product which cost $2.50 three years ago sells for $3 this year. If price increases are assumed to be linear, how much will the product cost six years from now?

49. Find a real number c such that $P(-2,2)$ is on the line $3x + cy = 4$.

50. Find a real number c such that the line $cx - 5y + 8 = 0$ has x-intercept 4.

51. If the points $(-2,-3)$ and $(-1,5)$ are on the graph of a linear function f, find $f(x)$.

52. If $f(1) = 4$ and $f(-1) = 3$ and the function f is linear, find $f(x)$.

53. Prove that the linear function $f(x) = ax + b$ is an increasing function if $a > 0$ and is a decreasing function if $a < 0$.

54. In the accompanying figure, lines L_1 and L_2 are parallel. Points A and D are selected on lines L_1 and L_2, respectively. Lines parallel to the x-axis are constructed through A and D which intersect the y-axis at points B and E. Supply a reason for each of the steps in the following proof.

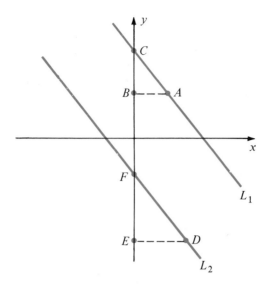

(a) Angles *ABC* and *DEF* are equal.

(b) Angles *ACB* and *DFE* are equal.

(c) Triangles *ABC* and *DEF* are similar.

(d) $\dfrac{\overline{CB}}{\overline{BA}} = \dfrac{\overline{FE}}{\overline{ED}}$

(e) $m_1 = \dfrac{\overline{CB}}{\overline{BA}}, \quad m_2 = \dfrac{\overline{FE}}{\overline{ED}}$

(f) $m_1 = m_2$

(g) Parallel lines have the same slope.

55. Prove that if two lines have the same slope, then they are parallel.

56. In the accompanying figure, lines perpendicular to each other, with slopes m_1 and m_2, intersect at a point Q. A perpendicular from Q to the x-axis intersects the x-axis at the point C. Supply a reason for each of the steps in the following proof.

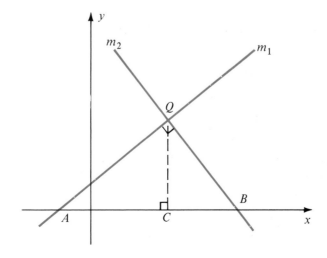

    (a) Angles CAQ and BQC are equal.

    (b) Triangles ACQ and BCQ are similar.

    (c) $\dfrac{\overline{CQ}}{\overline{AC}} = \dfrac{\overline{CB}}{\overline{CQ}}$

    (d) $m_1 = \dfrac{\overline{CQ}}{\overline{AC}}, \quad m_2 = -\dfrac{\overline{CQ}}{\overline{CB}}$

    (e) $m_2 = -\dfrac{1}{m_1}$

57.   Prove that if two lines have slopes m_1 and m_2 such that $m_2 = -\dfrac{1}{m_1}$, then the lines are perpendicular.

3.5
DIRECT AND INVERSE VARIATION

Two functional relationships occur so frequently that they are given distinct names. They are direct and inverse variation. We say that two quantities **vary directly** if an increase in one causes a proportional increase in the other. In the table

x	1	2	3	4
y	3	6	9	12

we see that an increase in x causes a proportional increase in y. If we look at the ratios y/x we have

$$\frac{y}{x} = \frac{3}{1} = \frac{6}{2} = \frac{9}{3} = \frac{12}{4} = 3$$

or $y = 3x$. The ratio y/x remains constant for all values of y and $x \neq 0$. This is an example of the

> ### Principle of Direct Variation
> y varies directly as x means $y = kx$ for some constant k.

As another example, y varies directly as the square of x means $y = kx^2$ for some constant k. The constant k is called the **constant of variation.**

EXAMPLE 1
Suppose that y varies directly as the cube of x, and $y = 24$ when $x = -2$. Write the appropriate equation, solve for the constant of variation k, and use this k to relate the variables.

Solution
Using the principle of direct variation, the functional relationship is

$$y = kx^3 \qquad \text{for some constant } k$$

Substituting the values $y = 24$ and $x = -2$, we have

$$24 = k \cdot (-2)^3 = -8k$$

$$k = -3$$

Thus,

$$y = -3x^3$$

PROGRESS CHECK

(a) If P varies directly as the square of V, and $P = 64$ when $V = 16$, find the constant of variation.

(b) The circumference C of a circle varies directly as the radius r. If $C = 25.13$ when $r = 4$, express C as a function of r, that is, use the constant of variation to relate the variables C and r.

Answers

(a) $\dfrac{1}{4}$ (b) $C = 6.2825r$

Two quantities are said to **vary inversely** if an increase in one causes a proportional decrease in the other. In the table

x	1	2	3	4
y	24	12	8	6

we see that an increase in x causes a proportional decrease in y. If we look at the product xy we have

$$xy = 1 \cdot 24 = 2 \cdot 12 = 3 \cdot 8 = 4 \cdot 6 = 24$$

or

$$y = \frac{24}{x}$$

In general, we have the

Principle of Inverse Variation

y varies inversely as x means $y = \dfrac{k}{x}$ for some constant k.

Once again, k is called the constant of variation.

EXAMPLE 2

Suppose that y varies inversely as x^2, and $y = 10$ when $x = 10$. Write the appropriate equation, solve for the constant of variation k, and use this k to relate the variables.

Solution

The functional relationship is

$$y = \frac{k}{x^2} \qquad \text{for some constant } k$$

Substituting $y = 10$ and $x = 10$,

$$10 = \frac{k}{(10)^2} = \frac{k}{100}$$

$$k = 1000$$

Thus,

$$y = \frac{1000}{x^2}$$

PROGRESS CHECK
If v varies inversely as the cube of w, and $v = 2$ when $w = -2$, find the constant of variation.

Answer
−16

An equation of variation can involve more than two variables. We say that a quantity **varies jointly** as two or more other quantities if it varies directly as their product.

EXAMPLE 3
Express as an equation: P varies jointly as R, S, and the square of T.

Solution
Since P must vary directly as $R \cdot S \cdot T^2$, we have $P = k \cdot R \cdot S \cdot T^2$ for some constant k.

EXAMPLE 4
A snow removal firm finds that the annual profit P varies jointly as the number of available plows p and the square of the total inches of snowfall s, and inversely as the price per gallon of gasoline g. If the profit is $15,000 when the snowfall is 6 inches, 5 plows are used, and the price of gasoline is $1.50 per gallon, express the profit P as a function of s, p, and g.

Solution
We are given that

$$P = k\frac{s^2 p}{g}$$

for some constant k. To determine k, we substitute $P = 15,000$, $p = 5$, $s = 6$, and $g = 1.5$. Thus,

$$15{,}000 = k\frac{(5)(6)^2}{1.5} = 120k$$

$$k = \frac{15{,}000}{120} = 125$$

Thus,

$$P = 125\frac{ps^2}{g}$$

EXERCISE SET 3.5
1. In the following table, y varies directly with x.

x	2	3	4	6	8	12		
y	8	12	16	24			80	120

(a) Find the constant of variation.
(b) Write an equation showing that y varies directly with x.
(c) Fill the blanks in the table.

2. In the accompanying table, y varies inversely with x.

x	1	2	3	6	9	12	15	18		
y	6	3	2	1	$\frac{2}{3}$				$\frac{1}{4}$	$\frac{1}{10}$

(a) Find the constant of variation,
(b) Write an equation showing that y varies inversely with x.
(c) Fill the blanks in the table.

3. If y varies directly as x, and $y = -\frac{1}{4}$ when $x = 8$,
(a) find the constant of variation,
(b) find y when $x = 12$.

4. If C varies directly as the square of s, and $C = 12$ when $s = 6$,
(a) find the constant of variation,
(b) find C when $s = 9$.

5. If s varies directly as the square of t, and $s = 10$ when $t = 10$,
(a) find the constant of variation,
(b) find s when $t = 5$.

6. If V varies as the cube of T, and $V = 16$ when $T = 4$,
(a) find the constant of variation,
(b) find V when $T = 6$.

7. If y varies inversely as x, and $y = -\frac{1}{2}$ when $x = 6$,
(a) find the constant of variation,
(b) find y when $x = 12$.

8. If V varies inversely as the square of p, and $V = \frac{2}{3}$ when $p = 6$,
(a) find the constant of variation,
(b) find V when $p = 8$.

9. If K varies inversely as the cube of r, and $K = 8$ when $r = 4$,
(a) find the constant of variation,
(b) find K when $r = 5$.

10. If T varies inversely as the cube of u, and $T = 2$ when $u = 2$,
(a) find the constant of variation,
(b) find T when $u = 5$.

11. If M varies directly as the square of r and inversely as the square of s, and $M = 4$ when $r = 4$ and $s = 2$,
(a) write the appropriate equation relating M, r, and s,
(b) find M when $r = 6$ and $s = 5$.

12. If f varies jointly as u and v, and $f = 36$ when $u = 3$ and $v = 4$,
(a) write the appropriate equation connecting f, u, and v,
(b) find f when $u = 5$ and $v = 2$.

13. If T varies jointly as p and the cube of v, and inversely as the square of u, and $T = 24$ when $p = 3$, $v = 2$, and $u = 4$,
(a) write the appropriate equation connecting T, p, v, and u,
(b) find T when $p = 2$, $v = 3$, and $u = 36$.

14. If A varies jointly as the square of b and the square of c, and inversely as the cube of d, and $A = 18$ when $b = 4$, $c = 3$, $d = 2$,
(a) write the appropriate equation relating A, b, c, and d,
(b) find A when $b = 9$, $c = 4$, and $d = 3$.

15. The distance s an object falls from rest in t seconds varies directly as the square of t. If an object falls 144 feet in 3 seconds,
(a) how far does it fall in 5 seconds?
(b) how long does it take to fall 400 feet?

16. In a certain state the income tax paid by a person varies directly as the income. If the tax is $20 per month when the monthly income is $1600, find the tax due when the monthly income is $900.

17. The resistance R of a conductor varies inversely as the area A of its cross section. If $R = 20$ ohms when $A = 8$ square centimeters, find R when $A = 12$ square centimeters.

18. The pressure P of a certain enclosed gas varies directly as the temperature T and inversely as the volume V. Suppose that 300 cubic feet of gas exert a pressure of 20 pounds per square foot when the temperature is 500°K (absolute temperature measured on the Kelvin scale). What is the pressure of this gas when the temperature is lowered to 400°K and the volume is increased to 500 cubic feet?

19. The intensity of illumination I from a source of light varies inversely as the square of the distance d from the source. If the intensity is 200 candlepower when the source is 4 feet away,
    (a) what is the intensity when the source is 6 feet away?
    (b) how close should the source be to provide an intensity of 50 candlepower?

20. The weight of a body in space varies inversely as the square of its distance from the center of the earth. If a body weighs 400 pounds on the surface of the earth, how much does it weigh 1000 miles from the surface of the earth? (Assume that the radius of the earth is 4000 miles.)

21. The equipment cost of a printing job varies jointly as the number of presses and the number of hours that the presses are run. When 4 presses are run for 6 hours, the equipment cost is $1200. If the equipment cost for 12 hours of running is $3600, how many presses are being used?

22. The current I in a wire varies directly as the electromotive force E and inversely as the resistance R. In a wire whose resistance is 10 ohms a current of 36 amperes is obtained when the electromotive force is 120 volts. Find the current produced when $E = 220$ volts and $R = 30$ ohms.

23. The illumination from a light source varies directly as the intensity of the source and inversely as the square of the distance from the source. If the illumination is 50 candlepower per square foot on a screen 2 feet away from a light source whose intensity is 400 candlepower, what is the illumination 4 feet away from a source whose intensity is 3840 candlepower?

24. If f varies directly as u and inversely as the square of v, what happens to f if both u and v are doubled?

3.6
COMBINING FUNCTIONS;
INVERSE FUNCTIONS

We can combine two functions such as

$$f(x) = x^2 \qquad g(x) = x - 1$$

by the usual operations of addition, subtraction, multiplication, and division. Using these functions f and g, we can form

$$(f + g)(x) = f(x) + g(x) = x^2 + x - 1$$
$$(f - g)(x) = f(x) - g(x) = x^2 - (x - 1) = x^2 - x + 1$$

$$(f \cdot g)(x) = f(x) \cdot g(x) = x^2 (x - 1) = x^3 - x^2$$

$$\frac{f}{g}(x) = \frac{f(x)}{g(x)} = \frac{x^2}{x - 1}$$

In each case we have combined two functions f and g to form a new function. Note, however, that the domain of the new functions need not be the same as the domain of either of the original functions. The function formed by division in the above example has as its domain the set of all real numbers x except $x = 1$, since we cannot divide by 0. On the other hand, the original functions $f(x) = x^2$ and $g(x) = x - 1$ are both defined at $x = 1$.

EXAMPLE 1

Given $f(x) = x - 4$, $g(x) = x^2 - 4$, find the following.

(a) $(f + g)(x)$ (b) $(f - g)(x)$ (c) $(f \cdot g)(x)$

(d) $\left(\dfrac{f}{g}\right)(x)$ (e)   the domain of $\left(\dfrac{f}{g}\right)(x)$

Solution

(a) $(f + g)(x) = f(x) + g(x) = x - 4 + x^2 - 4 = x^2 + x - 8$

(b) $(f - g)(x) = f(x) - g(x) = x - 4 - (x^2 - 4) = -x^2 + x$

(c) $(f \cdot g)(x) = f(x) \cdot g(x) = (x - 4)(x^2 - 4) = x^3 - 4x^2 - 4x + 16$

(d) $\left(\dfrac{f}{g}\right)(x) = \dfrac{f(x)}{g(x)} = \dfrac{x - 4}{x^2 - 4}$

(e)   The domain of $\left(\dfrac{f}{g}\right)(x)$ must exclude values of x for which $x^2 - 4 = 0$. Thus, the domain of $\left(\dfrac{f}{g}\right)(x)$ consists of the set of all real numbers except 2 and -2.

PROGRESS CHECK

Given $f(x) = 2x^2$, $g(x) = x^2 - 5x + 6$, find the following.

(a) $(f + g)(x)$ (b) $(f - g)(x)$ (c) $(f \cdot g)(x)$

(d) $\left(\dfrac{f}{g}\right)(x)$ (e)   the domain of $\left(\dfrac{f}{g}\right)(x)$

Answers

(a) $3x^2 - 5x + 6$ (b) $x^2 + 5x - 6$

(c) $2x^4 - 10x^3 + 12x^2$ (d) $\dfrac{2x^2}{x^2 - 5x + 6}$

(e)   *The set of all real numbers except 2 and 3.*

There is another, important way in which two functions f and g can be combined to form a new function. In Figure 28, the function f assigns the value y in set Y to x in set X; then, function g assigns the value z in set Z to y in Y. The net effect of this combination of f and g is a new function h, called the **composite function of g and f, $g \circ f$**, which assigns z in Z to x in X.

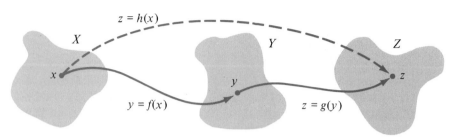

FIGURE 28

We write this new function as

$$h(x) = (g \circ f)(x) = g[f(x)]$$

which is read "g of f of x."

EXAMPLE 2
Given $f(x) = x^2$, $g(x) = x - 1$, find the following.

(a) $f[g(3)]$ (b) $g[f(3)]$ (c) $f[g(x)]$ (d) $g[f(x)]$

Solution

(a) We begin by evaluating $g(3)$.
$$g(x) = x - 1$$
$$g(3) = 3 - 1 = 2$$

Therefore, $f[g(3)] = f(2)$

Since $\quad\quad f(x) = x^2$

then $\quad\quad\quad f(2) = 2^2 = 4$

Thus, $\quad\quad\ f[g(3)] = 4$

(b) Beginning with $f(3)$, we have

$f(3) = 3^2 = 9$

Then we find by substituting $f(3) = 9$ that

$g[f(3)] = g(9) = 9 - 1 = 8$

(c) Since $g(x) = x - 1$, we make the substitution

$f[g(x)] = f(x - 1) = (x - 1)^2 = x^2 - 2x + 1$

(d) Since $f(x) = x^2$, we make the substitution

$g[f(x)] = g(x^2) = x^2 - 1$

Note that $f[g(x)] \neq g[f(x)]$.

PROGRESS CHECK
Given $f(x) = x^2 - 2x$, $g(x) = 3x$, find the following.

(a) $f[g(-1)]$ (b) $g[f(-1)]$ (c) $f[g(x)]$

(d) $g[f(x)]$ (e) $(f \circ g)(2)$ (f) $(g \circ f)(2)$

Answers
(a) 15   (b) 9   (c) $9x^2 - 6x$
(d) $3x^2 - 6x$ (e) 24   (f) 0

An element in the range of a function may correspond to more than one element in the domain of the function. In Figure 29 we see that y in Y corresponds to both x_1 and x_2 in X. If we demand that every element in the domain be assigned to a *different* element of the range, then the function is called **one-to-one.** More formally,

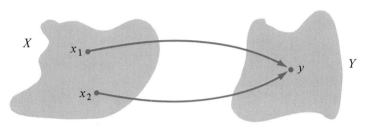

FIGURE 29

A function f is one-to-one if $f(a) = f(b)$ only when $a = b$.

There is a simple means of determining if a function f is one-to-one by examining the graph of the function. In Figure 30a we see that a horizontal line meets the graph in more than one point. Thus, $f(a) = f(b)$ although $a \neq b$; hence the function is not one-to-one. On the other hand, no horizontal line meets the graph in Figure 30b in more than one point; the graph thus determines a one-to-one function. In summary, we have the following test.

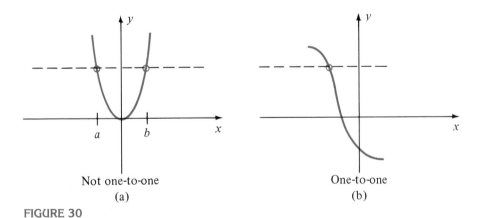

Not one-to-one   One-to-one
(a)   (b)

FIGURE 30

Horizontal Line Test

If no horizontal line meets the graph of a function in more than one point, then the function is one-to-one.

EXAMPLE 4
Which of the graphs in Figure 31 are graphs of one-to-one functions?

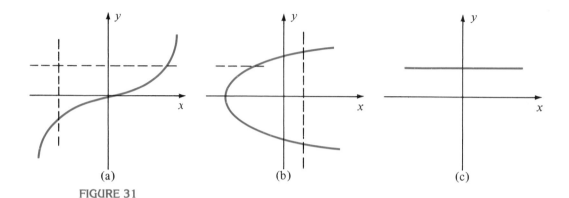

(a)   (b)   (c)

FIGURE 31

Solution
(a) No *vertical* line meets the graph in more than one point; hence, it is the graph
     of a function. No *horizontal* line meets the graph in more than one point;
     hence, it is the graph of a one-to-one function.
(b) No *horizontal* line meets the graph in more than one point. But *vertical* lines
     do meet the graph in more than one point. It is therefore not the graph of a
     function and consequently cannot be the graph of a one-to-one function.
(c) No *vertical* line meets the graph in more than one point; hence, it is the graph
     of a function. But a *horizontal* line does meet the graph in more than one
     point. This is the graph of a function but not of a one-to-one function.

PROGRESS CHECK
Which of the graphs in Figure 32 are graphs of one-to-one functions?

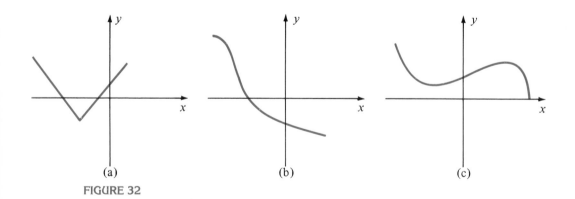

(a)   (b)   (c)

FIGURE 32

Answer
(b)

Suppose the function f in Figure 33a is a one-to-one function and that $y = f(x)$. Since f is one-to-one, we know that the correspondence is unique, that is, x in X is the *only* element of the domain for which $y = f(x)$. It is

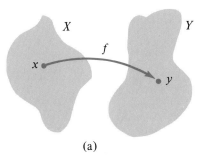

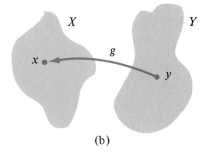

(a)   (b)

FIGURE 33

then possible to define a function g (Figure 33b) with domain Y and range X which reverses the correspondence, that is,

$$g(y) = x \quad \text{for every } x \text{ in } X$$

If we substitute $y = f(x)$, we have

$$g[f(x)] = x \quad \text{for every } x \text{ in } X \tag{1}$$

Substituting $g(y) = x$ in the equation $f(x) = y$ yields

$$f[g(y)] = y \quad \text{for every } y \text{ in } Y \tag{2}$$

The functions f and g of Figure 33 are therefore seen to satisfy the properties of Equations (1) and (2). Such a pair of functions are called inverse functions.

> **Inverse Functions**
>
> If f is a one-to-one function with domain X and range Y then the function g with domain Y and range X satisfying
>
> $$g[f(x)] = x \quad \text{for every } x \text{ in } X$$
> $$f[g(y)] = y \quad \text{for every } y \text{ in } Y$$
>
> is called the **inverse function** of f.

It is not difficult to show that the inverse of a one-to-one function is unique (see Exercise 51).

Since the inverse (reciprocal) of a real number x ($\neq 0$) can be written as x^{-1}, it is natural to write the inverse of a function f as f^{-1}. Thus we have

$$f^{-1}[f(x)] = x \quad \text{for every } x \text{ in } X$$
$$f[f^{-1}(y)] = y \quad \text{for every } y \text{ in } Y$$

See Figure 34 for a graphical representation.

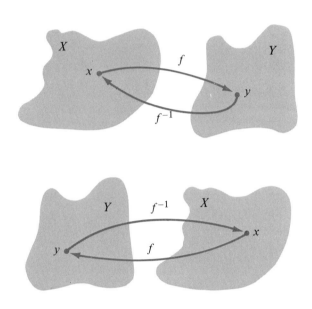

FIGURE 34

In the following chapter we will study a very important class of inverse functions, the exponential and logarithmic functions. Always remember that *we can define the inverse function of f only if f is one-to-one*.

EXAMPLE 5
Let f be the function defined by

$$f(x) = x^2 - 4, \quad x \geq 0$$

Verify that the inverse of f is given by

$$f^{-1}(x) = \sqrt{x + 4}$$

Solution
We must verify that $f[f^{-1}(x)] = x$ and $f^{-1}[f(x)] = x$. Thus,

$$f[f^{-1}(x)] = f(\sqrt{x + 4})$$
$$= (\sqrt{x + 4})^2 - 4$$
$$= x + 4 - 4 = x$$

and

$$f^{-1}[f(x)] = f^{-1}(x^2 - 4)$$
$$= \sqrt{(x^2 - 4) + 4}$$
$$= \sqrt{x^2} = |x|$$

Since $x \geq 0$,

$$f^{-1}[f(x)] = |x| = x$$

We have verified that the equations defining inverse functions hold, and conclude that the inverse of f is as given.

We may also think of the function f defined by $y = f(x)$ as the set of all ordered pairs $(x, f(x))$ where x assumes all values in the domain of f. Since the inverse function reverses the correspondence, the function f^{-1} is the set of all ordered pairs $(f(x), x)$ where $f(x)$ assumes all values in the range of f. With this approach, we see that the graphs of inverse functions are related in a distinct manner. First, note that the points (a,b) and (b,a) in Figure 35a are located symmetrically with respect to the graph of the line $y = x$. That is, if we fold the paper along the line $y = x$, the two points will

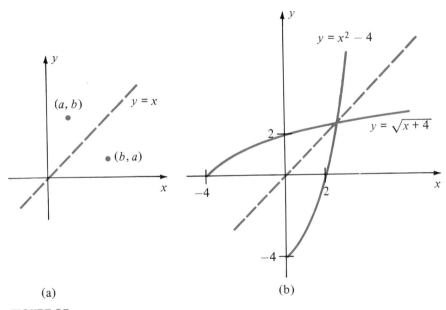

(a)   (b)

FIGURE 35

coincide. But if (a,b) lies on the graph of a function f, then (b,a) must lie on the graph of f^{-1}. Thus, the graphs of a pair of inverse functions are reflections of each other about the line $y = x$. In Figure 35b we have sketched the graphs of the functions from Example 5 on the same coordinate axes to demonstrate this interesting relationship.

It is sometimes possible to find an inverse by algebraic methods, as is shown by the following example.

EXAMPLE 6
Find the inverse function of $f(x) = 2x - 3$.

Inverse of f	
Step 1. Write $y = f(x)$.	*Step 1.* $y = 2x - 3$
Step 2. Solve for x.	*Step 2.* $y + 3 = 2x$
	$x = \dfrac{y + 3}{2}$
Step 3. $f^{-1}(y)$ is the expression for x found in Step 2.	*Step 3.*
	$f^{-1}(y) = \dfrac{y + 3}{2}$
Step 4. To obtain $f^{-1}(x)$, replace y by x. (Note that y and x are simply symbols for a variable.)	*Step 4.*
	$f^{-1}(x) = \dfrac{x + 3}{2}$

PROGRESS CHECK
Given $f(x) = 3x + 5$, find f^{-1}.

Answer

$f^{-1}(x) = \dfrac{x - 5}{3}$

WARNING

(a) In general, $f^{-1}(x) \neq \dfrac{1}{f(x)}$.

If $g(x) = x - 1$, then

$$g^{-1}(x) \neq \dfrac{1}{x - 1}$$

Use the methods of this section to show that

$$g^{-1}(x) = x + 1$$

(b) The inverse function notation is *not* to be thought of as a power.

EXERCISE SET 3.6
In Exercises 1–10 $f(x) = x^2 + 1$ and $g(x) = x - 2$. Determine the following.

1. $(f + g)(x)$ 2. $(f + g)(2)$ 3. $(f - g)(x)$
4. $(f - g)(3)$ 5. $(f \cdot g)(x)$ 6. $(f \cdot g)(-1)$
7. $\left(\dfrac{f}{g}\right)(x)$ 8. $\left(\dfrac{f}{g}\right)(-2)$
9. the domains of f and of g
10. the domains of $\dfrac{f}{g}$ and of $\dfrac{g}{f}$

In Exercises 11–18 $f(x) = 2x + 1$ and $g(x) = 2x^2 + x$. Determine the following.
11. $(f \circ g)(x)$ 12. $(g \circ f)(x)$ 13. $(f \circ g)(2)$

14. $(g \circ f)(3)$ 15. $(f \circ g)(x + 1)$ 16. $(f \circ f)(-2)$
17. $(g \circ f)(x - 1)$ 18. $(g \circ g)(x)$

In Exercises 19–24 $f(x) = x^2 + 4$ and $g(x) = \sqrt{x + 2}$. Determine the following.

19. $(f \circ g)(x)$ 20. $(g \circ f)(x)$ 21. $(f \circ f)(-1)$
22. the domain of $(f \circ g)(x)$
23. the domain of $(g \circ f)(x)$
24. the domain of $(g \circ g)(x)$

In Exercises 25–28 determine $(f \circ g)(x)$ and $(g \circ f)(x)$.

25. $f(x) = x - 1$, $g(x) = x + 2$ 26. $f(x) = \sqrt{x + 1}$, $g(x) = x + 2$

27. $f(x) = \dfrac{1}{x + 1}$, $g(x) = \dfrac{1}{x - 1}$ 28. $f(x) = \dfrac{x + 1}{x - 1}$, $g(x) = x$

In Exercises 29–34 verify that $g = f^{-1}$ for the given functions f and g by showing that $f[g(x)] = x$ and $g[f(x)] = x$.

29. $f(x) = 2x + 4$ $g(x) = \dfrac{1}{2}x - 2$

30. $f(x) = 3x - 2$ $g(x) = \dfrac{1}{3}x + \dfrac{2}{3}$

31. $f(x) = 2 - 3x$ $g(x) = -\dfrac{1}{3}x + \dfrac{2}{3}$

32. $f(x) = x^3$ $g(x) = \sqrt[3]{x}$

33. $f(x) = \dfrac{1}{x}$ $g(x) = \dfrac{1}{x}$

34. $f(x) = \dfrac{1}{x - 2}$ $g(x) = \dfrac{1}{x} - 2$

In Exercises 35–42 find $f^{-1}(x)$. Sketch the graphs of $y = f(x)$ and $y = f^{-1}(x)$ on the same coordinate axes.

35. $f(x) = 2x + 3$ 36. $f(x) = 3x - 4$ 37. $f(x) = 3 - 2x$

38. $f(x) = \dfrac{1}{2}x + 1$ 39. $f(x) = \dfrac{1}{3}x - 5$ 40. $f(x) = 2 - \dfrac{1}{5}x$

41. $f(x) = x^3 + 1$ 42. $f(x) = \dfrac{1}{x + 1}$

In Exercises 43–50 use the horizontal line test to determine whether the given function is a one-to-one function.

43. $f(x) = 2x - 1$ 44. $f(x) = 3 - 5x$
45. $f(x) = x^2 - 2x + 1$ 46. $f(x) = x^2 + 4x + 4$
47. $f(x) = -x^3 + 1$ 48. $f(x) = x^3 - 2$

49. $f(x) = \begin{cases} 2x, & x \le -1 \\ x^2, & -1 < x \le 0 \\ 3x - 1, & x > 0 \end{cases}$ 50. $f(x) = \begin{cases} x^2 - 4x + 4, & x \le 2 \\ x, & x > 2 \end{cases}$

51. Prove that a one-to-one function can have at most one inverse function. (*Hint:* Assume that the functions g and h are both inverses of the function f. Show that $g(x) = h(x)$ for all real values x in the range of f.)

52. Prove that the linear function $f(x) = ax + b$ is a one-to-one function if $a \ne 0$, and is not a one-to-one function if $a = 0$.

53. Find the inverse of the linear function $f(x) = ax + b$, $a \ne 0$.

TERMS AND SYMBOLS

origin (p. 106)	symmetry (p. 110)	quadratic function (p. 127)
x-axis (p. 106)	symmetric with respect to the x-axis (p. 111)	parabola (p. 127)
y-axis (p. 106)		polynomial function (p. 127)
coordinate axes (p. 106)	symmetric with respect to the y-axis (p. 111)	slope (p. 130)
rectangular coordinate system (p. 106)		point-slope form (p. 132)
	symmetric with respect to the origin (p. 113)	
Cartesian coordinate system (p. 106)		slope-intercept form (p. 133)
quadrant (p. 106)	function (p. 116)	general first-degree equation (p. 134)
coordinates of a point (p. 106)	image (p. 116)	
	domain (p. 116)	direct variation (p. 140)
ordered pair (p. 106)	range (p. 116)	
abscissa (p. 106)	independent variable (p. 116)	constant of variation (p. 140)
x-coordinate (p. 106)	dependent variable (p. 116)	inverse variation (p. 141)
ordinate (p. 106)	vertical line test (p. 117)	joint variation (p. 142)
y-coordinate (p. 106)	$f(x)$ (p. 119)	composite function (p. 145)
distance formula (p. 108)	graph of a function (p. 122)	$f[g(x)]$ (p. 146)
graph of an equation in two variables (p. 109)	increasing function (p. 124)	$f \circ g$ (p. 146)
	decreasing function (p. 124)	one-to-one function (p. 147)
solution of an equation in two variables (p. 109)	constant function (p. 125)	horizontal line test (p. 147)
x-intercept (p. 110)	linear function (p. 126)	inverse function (p. 149)
y-intercept (p. 110)		f^{-1} (p. 149)

KEY IDEAS FOR REVIEW

◼ In a rectangular coordinate system, every ordered pair of real numbers (a,b) corresponds to a point in the plane, and every point in the plane corresponds to an ordered pair of real numbers.

◼ The distance $\overline{PQ}$ between points $P(x_1, y_1)$ and $Q(x_2, y_2)$ is given by the distance formula

$$\overline{PQ} = \sqrt{(x_2 - x_1)^2 + (y_2 - y_1)^2}$$

◼ An equation in two variables can be graphed by plotting points which satisfy the equation and joining the points to form a smooth curve.

◼ A function is a rule which assigns exactly one element y of a set Y to each element x of a set X. The domain is the set of inputs and the range is the set of outputs.

◼ A graph represents a function if no vertical line meets the graph in more than one point.

- The domain of a function is the set of all real numbers for which the function is defined. Beware of division by zero and square roots of negative numbers.

- Function notation gives the definition of the function and also the value or expression at which to evaluate the function. Thus, if the function f is defined by $f(x) = x^2 + 2x$, then the notation $f(3)$ denotes the result of replacing the independent variable x by 3 wherever it appears.

$$f(x) = x^2 + 2x$$

$$f(3) = 3^2 + 2(3) = 15$$

- To graph $f(x)$, simply graph the equation $y = f(x)$.

- An equation is not the only way to define a function. Sometimes a function is defined by a table or chart, or by several equations. Moreover, not every equation determines a function.

- As we move from left to right, the graph of an increasing function rises and the graph of a decreasing function falls. The graph of a constant function neither rises nor falls; it is horizontal.

- The graph of a function can have holes or gaps, and can be defined in "pieces."

- Polynomials in one variable are all functions and have "smooth" curves as their graphs.

- The graph of the linear function $f(x) = ax + b$ is a straight line.

- Any two points on a line can be used to find its slope m.

$$m = \frac{y_2 - y_1}{x_2 - x_1}$$

- Positive slope indicates that a line is rising; negative slope indicates that a line is falling.

- The slope of a horizontal line is 0; the slope of a vertical line is undefined.

- The point slope form of a line is $y - y_1 = m(x - x_1)$.

- The slope-intercept form of a line is $y = mx + b$.

- The graphs of the linear function $f(x) = ax + b$ and of the general first-degree equation $Ax + By = C$ are always straight lines.

- The equation of the horizontal line through the point (a,b) is $y = b$; the equation of the vertical line through the point (a,b) is $x = a$.

- Parallel lines have the same slope.

- The slopes of perpendicular lines are negative reciprocals of each other.

- Direct and inverse variation are functional relationships.

- We say that y varies directly as x if $y = kx$ for some constant k; we say that y varies inversely as x if $y = k/x$ for some constant k.

- Joint variation is a term for direct variation involving more than two quantities.

- Functions can be combined by the usual operations of addition, subtraction, multiplication and division. However, the domain of the resulting function need not correspond with the domain of either of the original functions.

- A composite function is a function of a function.

- We say a function is one-to-one if every element of the range corresponds to precisely one element of the domain.

■ No horizontal line meets the graph of a one-to-one function in more than one point.

■ The inverse of a function, f^{-1}, reverses the correspondence defined by the function f. The domain of f becomes the range of f^{-1}, and the range of f becomes the domain of f^{-1}.

■ The inverse of a function f is defined only if f is a one-to-one function.

REVIEW EXERCISES

1. Find the distance between the points $(-4,-6)$ and $(2,-1)$.
2. Find the length of the longest side of the triangle whose vertices are $(3,-4)$, $(-2,-6)$ and $(-1,2)$.

In Exercises 3–4 sketch the graph of the given equation.

3. $y = 1 - |x|$　　　　　　　　　　4. $y = \sqrt{x-2}$

In Exercises 5–6 analyze the given equation for symmetry with respect to the x-axis, y-axis, and origin.

5. $y^2 = 1 - x^3$　　　　　　　　6. $y^2 = \dfrac{x^2}{x^2 - 5}$

In Exercises 7–8 determine if the graph is that of a function.

7.

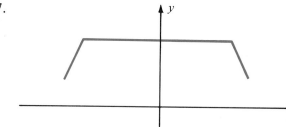

8.

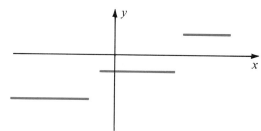

In Exercises 9–10 determine the domain of the given function.

9. $f(x) = \sqrt{3x-5}$　　　　　　10. $f(x) = \dfrac{x}{x^2 + 2x + 1}$

11. If $f(x) = \sqrt{x-1}$ find a real number whose image is 15.
12. If $f(t) = t^2 + 1$ find a real number whose image is 10.

In Exercises 13–15 $f(x) = x^2 - x$. Evaluate the following.

13. $f(-3)$　　　　　　14. $f(y-1)$　　　　　　15. $\dfrac{f(2+h) - f(2)}{h}$

Exercises 16–19 refer to the function f defined by

$$f(x) = \begin{cases} x - 1, & x \leq -1 \\ x^2, & -1 < x \leq 2 \\ -2, & x > 2 \end{cases}$$

16. Sketch the graph of the function f.
17. Determine where the function f is increasing, decreasing, and constant.
18. Evaluate $f(-4)$.
19. Evaluate $f(4)$.

In Exercises 20–25 the points A and B have coordinates $(-4,-6)$ and $(-1,3)$, respectively.

20. Find the slope of the line through A and B.
21. Find an equation of the line through the points A and B.
22. Find an equation of the line through A that is parallel to the y-axis.
23. Find an equation of the horizontal line through B.
24. Find an equation of the line through A that is parallel to the line $2x - y - 3 = 0$.
25. Find an equation of the line through B that is perpendicular to the line $3y + x - 5 = 0$.

26. If R varies directly as q, and $R = 20$ when $q = 5$, find R when $q = 40$.

27. If S varies inversely as the cube of t, and $S = 8$ when $t = -1$, find S when $t = -2$.

28. If P varies jointly as q and r and inversely as the square of t, and $P = -3$ when $q = 2$, $r = -3$, and $t = 4$, find P when $q = -1$, $r = \frac{1}{2}$, and $t = 4$.

In Exercises 29–34 $f(x) = x + 1$ and $g(x) = x^2 - 1$. Determine the following.

29. $(f + g)(x)$ 30. $(f \cdot g)(-1)$ 31. $\left(\dfrac{f}{g}\right)(x)$

32. the domain of $\left(\dfrac{f}{g}\right)(x)$ 33. $(g \circ f)(x)$ 34. $(f \circ g)(2)$

In Exercises 35–38 $f(x) = \sqrt{x} - 2$ and $g(x) = x^2$. Determine the following.

35. $(f \circ g)(x)$ 36. $(g \circ f)(x)$
37. $(f \circ g)(-2)$ 38. $(g \circ f)(-2)$

In Exercises 39–40 $f(x) = 2x + 4$ and $g(x) = \dfrac{x}{2} - 2$.

39. Prove that f and g are inverse functions of each other.
40. Sketch the graphs of $y = f(x)$ and $y = g(x)$ on the same coordinate axes.

PROGRESS TEST 3A

1. Find the perimeter of the triangle whose vertices are $(2,5)$, $(-3,1)$, and $(-3,4)$.
2. Use symmetry to assist in sketching the graph of the equation $y = 2x^2 - 1$.
3. Analyze the equation $y = 1/x^3$ for symmetry with respect to the axes and origin.
4. Determine the domain of the function $\dfrac{1}{\sqrt{x} - 1}$.
5. If $f(x) = \sqrt{x} - 1$, find a real number whose image is 4.
6. If $f(x) = 2x^2 + 3$, find $f(2t)$.

Problems 7–9 refer to the function f defined by

$$f(x) = \begin{cases} 0, & x < -2 \\ |x|, & -2 \le x \le 3 \\ x^2 - x, & x > 3 \end{cases}$$

7. Determine where the function f is increasing, decreasing, and constant.
8. Evaluate $f(-5)$ 9.   Evaluate $f(-2)$
10. Find an equation of the line through the points $(-3,5)$ and $(-5,2)$.
11. Find an equation of the vertical line through the point $(-3,4)$.
12. Find the slope m and y-intercept b of the line whose equation is $2y - x = 4$.
13. Find an equation of the line through the point $(4,-1)$ that is parallel to the x-axis.
14. Find an equation of the line through the point $(-2,3)$ that is perpendicular to the line $y - 3x - 2 = 0$.
15. If h varies directly as the cube of r, and $h = 2$ when $r = -\frac{1}{2}$, find h when $r = 4$.
16. If T varies jointly as a and the square of b, and inversely as the cube of c, and $T = 64$ when $a = -1$, $b = \frac{1}{2}$, and $c = 2$, find T when $a = 2$, $b = 4$, and $c = -1$.

In Problems 17–19 $f(x) = \dfrac{1}{x - 1}$ and $g(x) = x^2$. Find the following.

17. $(f - g)(2)$ 18. $\left(\dfrac{g}{f}\right)(x)$ 19. $(g \circ f)(3)$

20. Prove that $f(x) = -3x + 1$ and $g(x) = -\frac{1}{3}(x - 1)$ are inverse functions of each other.

PROGRESS TEST 3B

1. Find the length of the shorter diagonal of the parallelogram whose vertices are $(-3,2)$, $(-5,-4)$, $(3,-4)$, and $(5,2)$.
2. Use symmetry to assist in sketching the graph of the equation $y^2 = -3x + 4$.
3. Analyze the equation $x^2 - xy + 2 = 0$ for symmetry with respect to the axes and the origin.
4. Determine the domain of the function

$$f(x) = \frac{x^2}{16 - x^2}$$

5. If $f(x) = x^2 - 2x$, find a real number whose image is -1.
6. If $f(x) = \sqrt{x} - 1$, find $f(4)$.

Problems 7–9 refer to the function f defined by

$$f(x) = \begin{cases} x^2 - 1, & x \le -3 \\ 10, & -3 < x \le 3 \\ \sqrt{x}, & x > 3 \end{cases}$$

7. Determine where the function f is increasing, decreasing, and constant.
8. Evaluate $f(2)$.   9.   Evaluate $f(-5)$.
10. Find the slope of the line through the points $(-2,-3)$ and $(-4,6)$.
11. Find an equation of the horizontal line through the point $(-6,-5)$.
12. Find the y-intercept of the line through the points $(4,-3)$ and $(-1,2)$.
13. Determine the slope of every line that is perpendicular to the line $6y - 2x = 5$.
14. Determine an equation of the line through the point $(3,-2)$ that is parallel to the line $3y + x - 4 = 0$.

15. If A varies inversely as the square of b, and $A = -2$ when $b = 4$, find A when $b = 3$.

16. If R varies jointly as x and the square root of y, and inversely as the square of z, and $R = 3/8$ when $x = 2$, $y = 9$, and $z = 4$, find the constant of variation.

In Problems 17–19 $f(x) = \dfrac{1}{\sqrt{x+1}}$ and $g(x) = x - 1$. Find the following.

17. $(f \cdot g)(3)$ 18. $(f + g)(1)$ 19. $(f \circ g)(x)$

20. Prove that $f(x) = \dfrac{1}{x}$ and $g(x) = \dfrac{1}{x}$ are inverse functions.

CHAPTER FOUR
EXPONENTIAL AND LOGARITHMIC FUNCTIONS

Thus far in our study of algebra we have dealt primarily with functions that are polynomials, or sums, differences, products, quotients, and powers of polynomials. In this chapter, we introduce a new type of function, the exponential function, and its inverse, the logarithmic function.

Exponential functions apply in nature and are useful in chemistry, biology, and economics, as well as in mathematics and engineering. We will study applications of exponential functions in calculating such quantities as compound interest and the growth rate of bacteria in a culture medium.

Logarithms can be viewed as another way of writing exponents. Historically, logarithms have been used to simplify calculations; in fact, the slide rule, a device long used by engineers, is based on logarithmic scales. In today's world of inexpensive hand calculators, the need for manipulating logarithms is reduced. The section on computing with logarithms will provide enough background to allow you to use this powerful tool but omits some of the detail found in older textbooks.

The function $f(x) = 2^x$ is very different from any of the functions we have worked with thus far. Previously, we defined functions by using the basic algebraic operations (addition, subtraction, multiplication, division, powers, and roots). However, $f(x) = 2^x$ has a variable in the exponent and doesn't fall into the class of algebraic functions. Rather, it is our first example of an exponential function.

> An **exponential function** has the form
>
> $$f(x) = a^x$$
>
> where $a > 0$, $a \neq 1$. The constant a is called the **base**, and the independent variable x may assume any real value.

The simplest way to become familiar with exponential functions is to sketch their graphs.

EXAMPLE 1
Sketch the graph of $f(x) = 2^x$.

Solution
We let $y = 2^x$ and we form a table of values of x and y. Then we plot these points and sketch the smooth curve as in Figure 1.

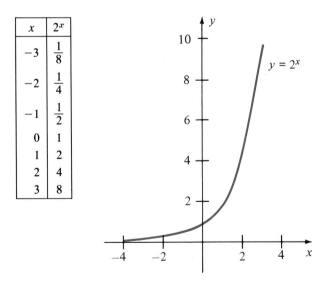

x	2^x
-3	$\frac{1}{8}$
-2	$\frac{1}{4}$
-1	$\frac{1}{2}$
0	1
1	2
2	4
3	8

FIGURE 1

In a sense, we have cheated in our definition of $f(x) = 2^x$ and in sketching the graph in Figure 1. Since we have not explained the meaning of 2^x when x is irrational, we have no right to plot values such as $2^{\sqrt{2}}$. For our purposes, however, it will be adequate to think of $2^{\sqrt{2}}$ as the value we approach by taking successively closer approximations to $\sqrt{2}$ such as $2^{1.4}$, $2^{1.41}$, $2^{1.414}$,…. A precise definition is given in more advanced mathematics courses where it is also shown that the laws of exponents hold for irrational exponents.

We now look at $f(x) = a^x$ when $0 < a < 1$.

EXAMPLE 2
Sketch the graph of $f(x) = \left(\dfrac{1}{2}\right)^x = 2^{-x}$.

Solution
We form a table, plot points, and sketch the graph. See Figure 2.

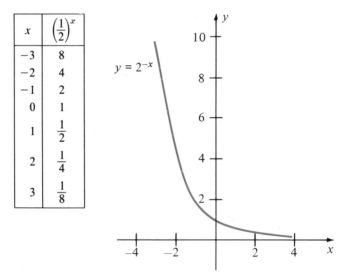

x	$\left(\dfrac{1}{2}\right)^x$
-3	8
-2	4
-1	2
0	1
1	$\dfrac{1}{2}$
2	$\dfrac{1}{4}$
3	$\dfrac{1}{8}$

FIGURE 2

In Figure 3, we have sketched the graphs of $f(x) = 2^x$, $g(x) = 3^x$, $h(x) = \left(\dfrac{1}{2}\right)^x$, and $k(x) = \left(\dfrac{1}{3}\right)^x$ on the same coordinate axes.

The graphs in Figure 3 illustrate the following. Recall that the definition of an exponential function requires $a > 0$ and $a \neq 1$.

- Since $a^0 = 1$, the graph of $f(x) = a^x$ passes through the point $(0,1)$ for any value of a.
- The domain of $f(x) = a^x$ consists of the set of all real numbers; the range is the set of all positive real numbers.
- If $a > 1$, a^x is an increasing function; if $a < 1$, a^x is a decreasing function.
- If $a < b$, then $a^x < b^x$ for all $x > 0$, and $a^x > b^x$ for all $x < 0$. Note in Figure 3 that $y = 3^x$ lies above $y = 2^x$ when $x > 0$ and below when $x < 0$.

Since a^x is either increasing or decreasing, it never assumes the same value twice. (Recall that $a \neq 1$.) This leads to a useful conclusion.

If $a^x = a^y$, then $x = y$.

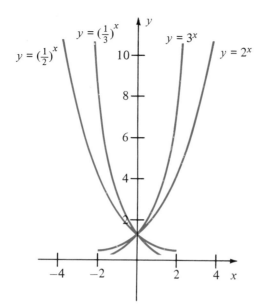

FIGURE 3

The graphs of a^x and b^x intersect only at $x = 0$. This observation provides us with the following result.

$$\text{If } a^x = b^x \text{ for all } x \neq 0, \text{ then } a = b.$$

EXAMPLE 3
Solve for x.

(a) $3^{10} = 3^{5x}$. Since $a^x = a^y$ implies $x = y$, we have

$$10 = 5x$$
$$2 = x$$

(b) $2^7 = (x - 1)^7$. Since $a^x = b^x$ implies $a = b$, we have

$$2 = x - 1$$
$$3 = x$$

PROGRESS CHECK
Solve for x.

(a) $2^8 = 2^{x + 1}$ (b) $4^{2x + 1} = 4^{11}$

Answer
(a) 7   *(b) 5*

There is an irrational number, denoted by the letter e, which plays an important role in mathematics. In calculus we show that the expression

$$\left(1 + \frac{1}{m}\right)^m \tag{1}$$

gets closer and closer to the number e as m gets larger and larger. We can evaluate this expression for different values of m as shown in Table 1.

TABLE 1

m	1	2	10	100	1000	10,000	100,000	1,000,000
$\left(1 + \dfrac{1}{m}\right)^m$	2.0	2.25	2.5937	2.7048	2.7169	2.7181	2.7182	2.71828

From this table we see that as m gets larger and larger the expression $\left(1 + \dfrac{1}{m}\right)^m$ gets closer and closer to the number 2.71828, which is an approximation to e. The graphs of e^x and e^{-x} are shown in Figure 4.

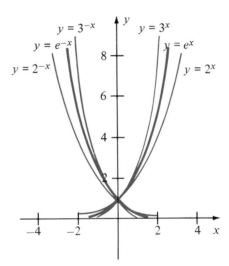

FIGURE 4

APPLICATIONS

Exponential functions occur in a wide variety of applied problems. We will look at problems dealing with population growth, such as the growth of bacteria in a culture medium; radioactive decay, such as determining the half-life of strontium 90; and interest earned when the interest rate is compounded.

The function Q defined by

$$Q(t) = q_0 e^{kt} \qquad (k > 0)$$

in which the variable t represents time, is called an **exponential growth model**; k is a constant and t is the independent variable. We may think of Q as the quantity of a substance available at any given time t. Note when $t = 0$ we have

$$Q(0) = q_0 e^0 = q_0$$

which says that q_0 is the initial quantity. (It is customary to use the subscript 0 to denote an initial value.) The constant k is called the **growth constant.**

EXAMPLE 4

The number of bacteria in a culture after t hours is described by the exponential growth model

$$Q(t) = 50e^{0.7t}$$

(a)   Find the initial number of bacteria, q_0, in the culture.
(b)   How many bacteria are in the culture after 10 hours?

Solution

(a)   To find q_0 we need to evaluate $Q(t)$ at $t = 0$.

$$Q(0) = 50e^{0.7(0)} = 50e^0 = 50 = q_0$$

Thus, initially there are 50 bacteria in the culture.

(b)   The number of bacteria in the culture after 10 hours is given by $Q(10)$.

$$Q(10) = 50e^{0.7(10)} = 50e^7 = 50(1096.6) = 54,830$$

Thus, there are 54,830 bacteria after 10 hours. (The value $e^7 = 1096.6$ can be found by using Table I in the Appendix; it can also be found by using a calculator with a "y^x" key with $y = e = 2.71828$ and $x = 7$.)

PROGRESS CHECK

The number of bacteria in a culture after t minutes is described by the exponential growth model $Q(t) = q_0 e^{0.005t}$. If there were 100 bacteria present initially, how many bacteria will be present after one hour has elapsed?

Answer
135

The model defined by the function

$$Q(t) = q_0 e^{-kt} \quad (k > 0)$$

is called an **exponential decay model;** k is a constant, called the **decay constant,** and t is the independent variable denoting time. Here is an application of this model.

EXAMPLE 5

A substance has a decay rate of 5% per hour. If 500 grams are present initially, how much of the substance remains after 4 hours?

Solution

The general equation of an exponential decay model is

$$Q(t) = q_0 e^{-kt}$$

In our model, $q_0 = 500$ grams (since the quantity available initially is 500 grams) and $k = 0.05$ (since the decay rate is 5% per hour). After 4 hours,

$$Q(4) = 500e^{-0.05(4)} = 500e^{-0.2} = 500(0.8187) = 409.4$$

($e^{-0.2} = 0.8187$ is obtained from Table I in the Appendix). Thus, there remain 409.4 grams of the substance.

PROGRESS CHECK

The number of grams Q of a certain radioactive substance present after t seconds is given by the exponential decay model $Q(t) = q_0 e^{-0.4t}$. If 200 grams of the substance are present initially, find how much remains after 6 seconds.

Answer
18.1 grams

In Section 2.2, we studied simple interest as an application of linear equations. Recall that if the principal P is invested at a simple annual interest rate r then the amount or sum S that we have on hand after t years is given by

$$S = P + Prt$$

or

$$S = P(1 + rt)$$

In many business transactions the interest that is added to the principal at regular time intervals also earns interest. This is called the **compound interest** process.

The time period between successive additions of interest is known as the **conversion period**. Thus, if interest is compounded quarterly, the conversion period is three months; if interest is compounded semiannually, the conversion period is six months.

Suppose now that a principal P is invested at an annual interest rate r, compounded k times a year. Then each conversion period lasts $1/k$ years. Thus, the amount S_1 at the end of the first conversion period is

$$S_1 = P + P \cdot r \cdot \frac{1}{k} = P\left(1 + \frac{r}{k}\right)$$

The amount S_2 at the end of the second conversion period is

$$S_2 = P\left(1 + \frac{r}{k}\right) + P\left(1 + \frac{r}{k}\right) \cdot r \cdot \frac{1}{k}$$

$$= \left[P\left(1 + \frac{r}{k}\right)\right]\left(1 + \frac{r}{k}\right)$$

or

$$S_2 = P\left(1 + \frac{r}{k}\right)^2$$

In this way, we see that the amount S after n conversion periods is given by

$$S = P\left(1 + \frac{r}{k}\right)^n \tag{2}$$

which is usually written

$$S = P(1 + i)^n$$

where $i = r/k$. Table IV in the Appendix gives values of $(1 + i)^n$ for a number of values of i and n.

EXAMPLE 6

Suppose that $6000 is invested at an annual interest rate of 8%, compounded quarterly. What is the value of the investment after 3 years?

Solution

We are given $P = 6000$, $r = 0.08$, $k = 4$, and $n = 12$ (since there are four conversion periods per year for three years). Thus,

$$i = \frac{r}{k} = \frac{0.08}{4} = 0.02$$

and

$$S = P(1 + i)^n = 6000(1 + 0.02)^{12}$$

Table IV in the Appendix with $i = 0.02$ and $n = 12$ yields

$$S = 6000(1.26824179) = 7609.45$$

Thus, the sum at the end of the three year period is $7609.45.

PROGRESS CHECK

Suppose that $5000 is invested at an annual interest rate of 6% compounded semiannually. What is the value of the investment after 12 years?

Answer
$10,163.97

CONTINUOUS COMPOUNDING

When P, r, and t are held fixed and the frequency of compounding is increased, the return on the investment is increased. We wish to determine the effect of making the number of conversions per year larger and larger.

Suppose a principal P is invested at an annual rate r, compounded k times per year. After t years, the number of conversions is $n = tk$. Then, the value of the investments after t years is given by Equation (3) as

$$S = P\left(1 + \frac{r}{k}\right)^{tk} \tag{3}$$

Letting $m = \dfrac{k}{r}$, we can rewrite Equation (3) as

$$S = P\left(1 + \frac{1}{m}\right)^{tmr}$$

or

$$S = P\left[\left(1 + \frac{1}{m}\right)^{m}\right]^{rt}$$

If the number of conversions per year k gets larger and larger, then m gets larger and larger. Since the expression

$$\left(1 + \frac{1}{m}\right)^{m}$$

was seen in Table 1 to get closer and closer to e as m gets larger and larger, we conclude that

$$S = Pe^{rt} \tag{4}$$

As the number of conversions increases, so does the value of the investment. But there is a limit or bound to this value and it is given by Equation (4). We say that Equation (4) represents the result of **continuous compounding.**

EXAMPLE 7
Suppose that $20,000 is invested at an annual interest rate of 7% compounded continuously. What is the value of the investment after 4 years?

Solution
We have $P = 20,000$, $r = 0.07$, and $t = 4$, and we substitute in Equation (4).

$$S = Pe^{rt}$$
$$= 20,000e^{0.07(4)} = 20,000e^{0.28}$$
$$= 20,000(1.3231) \qquad \text{from Table I, Appendix}$$
$$= 26,462$$

The sum available after 4 years is $26,462.

PROGRESS CHECK
Suppose that $10,000 is invested at an annual interest rate of 10% compounded continuously. What is the value of the investment after 6 years?

Answer
$18,221

By solving Equation (4) for P, we can determine the principal P that must be invested at continuous compounding to have a certain amount S at some future time. The values of e^{-x} from Table I in the Appendix will be used in this connection.

EXAMPLE 8
Suppose that a principal P is to be invested at continuous compound interest of 8% per year to yield $10,000 in five years. Approximately how much should be invested?

Solution
Using Equation (4) with $S = 10,000$, $r = 0.08$, and $t = 5$, we have

$$10,000 = Pe^{0.08(5)} = Pe^{0.40}$$
$$P = \frac{10,000}{e^{0.40}}$$
$$= 10,000e^{-0.40}$$
$$= 10,000(0.6703) \qquad \text{from Table I, Appendix}$$
$$= 6703$$

Thus, approximately $6703 should be invested initially.

PROGRESS CHECK

Approximately how much money should a 35-year-old woman invest now at continuous compound interest of 10% per year to obtain the sum of $20,000 upon her retirement at age 65?

Answer
$996

EXERCISE SET 4.1

In Exercises 1–12 sketch the graph of the given function f.

1. $f(x) = 4^x$
2. $f(x) = 4^{-x}$
3. $f(x) = 10^x$
4. $f(x) = 10^{-x}$
5. $f(x) = 2^{x+1}$
6. $f(x) = 2^{x-1}$
7. $f(x) = 2^{|x|}$
8. $f(x) = 2^{-|x|}$
9. $f(x) = 2^{2x}$
10. $f(x) = 3^{-2x}$
11. $f(x) = e^{x+1}$
12. $f(x) = e^{-2x}$

In Exercises 13–20 solve for x.

13. $2^x = 2^3$
14. $2^{x-1} = 2^4$
15. $3^x = 9^{x-2}$
16. $2^x = 8^{x+2}$
17. $2^{3x} = 4^{x+1}$
18. $3^{4x} = 9^{x-1}$
19. $e^{x-1} = e^3$
20. $e^{x-1} = 1$

In Exercises 21–24 solve for a.

21. $(a + 1)^x = (2a - 1)^x$
22. $(2a + 1)^x = (a + 4)^x$
23. $(a + 1)^x = (2a)^x$
24. $(2a + 3)^x = (3a + 1)^x$

In Exercises 25–29 use Table I in the Appendix to evaluate e^x and e^{-x}.

25. The number of bacteria in a culture after t hours is described by the exponential growth model $Q(t) = 200e^{0.25t}$.
    (a) What is the initial number of bacteria in the culture?
    (b) Find the number of bacteria in the culture after 20 hours.
    (c) Use Table I in the Appendix to complete the following table.

t	1	4	8	10
Q				

26. The number of bacteria in a culture after t hours is described by the exponential growth model $Q(t) = q_0e^{0.01t}$. If there were 400 bacteria present initially, how many bacteria will be present after two *days*?

27. At the beginning of 1975 the world population was approximately 4 billion. Suppose that the population is described by an exponential growth model, and that the rate of growth is 2% per year. Give the approximate world population in the year 2000.

28. The number of grams of potassium 42 present after t hours is given by the exponential decay model $Q(t) = q_0e^{-0.055t}$. If 400 grams of the substance were present initially, how much remains after 10 hours?

29. A radioactive substance has a decay rate of 4% per hour. If 1000 grams are present initially, how much of the substance remains after 10 hours?

In Exercises 30–33 use Table IV in the Appendix to assist in the computation.

30. An investor purchases a $12,000 savings certificate paying 10% annual interest compounded semiannually. Find the amount received when the savings certificate is redeemed at the end of 8 years.

31. The parents of a newborn infant place $10,000 in an investment which pays 8% annual interest compounded quarterly. What sum is available at the end of 18 years to finance the child's college education?

32. A widow is offered a choice of two investments. Investment A pays 8% annual interest compounded quarterly, and investment B pays 9% compounded annually. Which investment will yield a greater return?

33. A firm intends to replace its present computer in five years. The treasurer suggests that $25,000 be set aside in an investment paying 12% compounded monthly. What sum will be available for the purchase of the new computer?

In Exercises 34–38 use Tables I and IV in the Appendix to assist in the computations.

34. If $5000 is invested at an annual interest rate of 9% compounded continuously, how much is available after 5 years?

35. If $100 is invested at an annual interest rate of 5.5% compounded continuously, how much is available after 10 years?

36. A principal P is to be invested at continuous compound interest of 9% to yield $50,000 in 20 years. What is the approximate value of P to be invested?

37. A 40-year-old executive plans to retire at age 65. How much should be invested at 12% annual interest compounded continuously to provide the sum of $50,000 upon retirement?

38. Investment A offers 8% annual interest compounded semiannually, and investment B offers 8% annual interest compounded continuously. If $1000 were invested in each, what would be the approximate difference in value after 10 years?

4.2
LOGARITHMIC FUNCTIONS

We have previously noted that $f(x) = a^x$ is an increasing function if $a > 1$ and is a decreasing function if $0 < a < 1$. It is then clear that no horizontal line can meet the graph of $f(x) = a^x$ in more than one point. We conclude that the exponential function is a one-to-one function.

In Figure 5a, we see the function $f(x) = 2^x$ assigning values in the set Y for various values of x in the domain X. Since $f(x) = 2^x$ is a one-to-one

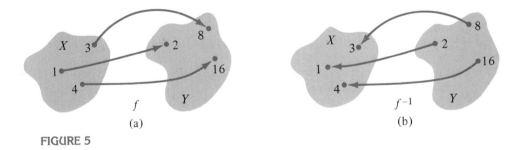

FIGURE 5

function, it makes sense to seek a function f^{-1} which will return the values of the range of f back to their origin as in Figure 5b. That is,

f maps 3 into 8, f^{-1} maps 8 into 3

f maps 4 into 16, f^{-1} maps 16 into 4

and so on. Since 2^x is always positive, we see that the domain of f^{-1} is the set of all positive real numbers. Its range is the set of all real numbers.

The function f^{-1} of Figure 5b has a special name, the **logarithmic function base 2,** which we write as $\log_2$. It is also possible to generalize and define the logarithmic function as the inverse of the exponential function with any base $a > 0$ and $a \neq 1$.

Logarithmic Function Base a

$y = \log_a x$ if and only if $x = a^y$.

The notation $\ln x$ is used to indicate logarithms to the base e. We call $\ln x$ the **natural logarithm** of x. Thus,

---

$$\ln x = \log_e x$$

The exponential form $x = a^y$ and the logarithmic form $y = \log_a x$ are two ways of expressing the same relationship between x, y, and a. Further, it is always possible to convert from one form to the other. The natural question, then, is why bother to create a logarithmic form when we already have an equivalent exponential form? We will show later that the logarithmic function has a number of very useful properties that make its study well worthwhile.

EXAMPLE 1

Write in exponential form.

(a) $\log_3 9 = 2$ The exponential form is $3^2 = 9$.

(b) $\log_2 \frac{1}{8} = -3$ The exponential form is $2^{-3} = \frac{1}{8}$.

(c) $\log_{16} 4 = \frac{1}{2}$ The exponential form is $16^{1/2} = 4$.

(d) $\ln 7.39 = 2$ The exponential form is $e^2 = 7.39$.

PROGRESS CHECK

Write in exponential form.

(a) $\log_4 64 = 3$

(b) $\log_{10} \left(\frac{1}{10,000} \right) = -4$

(c) $\log_{25} 5 = \frac{1}{2}$

(d) $\ln 0.3679 = -1$

Answers

(a) $4^3 = 64$

(b) $10^{-4} = \frac{1}{10,000}$

(c) $25^{1/2} = 5$

(d) $e^{-1} = 0.3679$

EXAMPLE 2
Write in logarithmic form.

(a) $36 = 6^2$ The logarithmic form is $\log_6 36 = 2$.

(b) $7 = \sqrt{49}$ The logarithmic form is $\log_{49} 7 = \dfrac{1}{2}$.

(c) $\dfrac{1}{16} = 4^{-2}$ The logarithmic form is $\log_4 \dfrac{1}{16} = -2$.

(d) $0.1353 = e^{-2}$ The logarithmic form is $\ln 0.1353 = -2$.

PROGRESS CHECK
Write in logarithmic form.

(a) $64 = 8^2$ (b) $6 = 36^{1/2}$

(c) $\dfrac{1}{7} = 7^{-1}$ (d) $20.09 = e^3$

Answers

(a) $\log_8 64 = 2$ (b) $\log_{36} 6 = \dfrac{1}{2}$

(c) $\log_7 \dfrac{1}{7} = -1$ (d) $\ln 20.09 = 3$

Logarithmic equations can often be solved by changing them to equivalent exponential forms.

EXAMPLE 3
Solve for x.

(a) $\log_3 x = -2$. The equivalent exponential form is
$$x = 3^{-2}$$
Thus,
$$x = \dfrac{1}{9}$$

(b) $\log_5 125 = x$. In exponential form we have
$$5^x = 125$$
Writing 125 to the base 5,
$$5^x = 5^3$$
and since $a^x = a^y$ implies $x = y$, we conclude that
$$x = 3$$

(c) $\log_x 81 = 4$. The equivalent exponential form is
$$x^4 = 81 = 3^4$$
and thus
$$x = 3$$

(d) $\ln x = \dfrac{1}{2}$. The equivalent exponential form is

$$x = e^{1/2}$$

or

$$x = 1.65$$

which we obtain from Table I in the Appendix, or by using a calculator with a "y^x" key.

PROGRESS CHECK
Solve for x.

(a) $\log_x 1000 = 3$ (b) $\log_2 x = 5$ (c) $x = \log_7 \dfrac{1}{49}$

Answers
(a) 10   *(b)* 32   *(c)* -2

If $f(x) = a^x$, then $f^{-1}(x) = \log_a x$. Recall that inverse functions have the property that

$$f[f^{-1}(x)] = x \quad \text{and} \quad f^{-1}[f(x)] = x$$

Substituting $f(x) = a^x$ and $f^{-1}(x) = \log_a x$, we have

$$f[f^{-1}(x)] = x \qquad\qquad f^{-1}[f(x)] = x$$
$$f(\log_a x) = x \qquad\qquad f^{-1}(a^x) = x$$
$$a^{\log_a x} = x \qquad\qquad \log_a a^x = x$$

The following two identities are useful in simplifying expressions and should be remembered.

$$a^{\log_a x} = x$$
$$\log_a a^x = x$$

The following pair of identities can be established by converting to the equivalent exponential form.

$$\log_a a = 1$$
$$\log_a 1 = 0$$

EXAMPLE 4
Evaluate.

(a) $8^{\log_8 5}$ (b) $\log_{10} 10^{-3}$

(c) $\log_7 7$ (d) $\log_4 1$

Solution
(a) 5   (b) -3 (c) 1   (d) 0

PROGRESS CHECK

Evaluate.

(a) $\log_3 3^4$ (b) $6^{\log_6 9}$ (c) $\log_5 1$ (d) $\log_8 8$

Answers

(*a*) *4*   (*b*) *9*   (*c*) *0*   (*d*) *1*

To sketch the graph of a logarithmic function, we convert to the equivalent exponential form. For example, to sketch the graph of $\log_2 x$ we form a table of values for the equivalent exponential equation $x = 2^y$.

y	-3	-2	-1	0	1	2	3
$x = 2^y$	$\dfrac{1}{8}$	$\dfrac{1}{4}$	$\dfrac{1}{2}$	1	2	4	8

We can now plot these points and sketch a smooth curve, as in Figure 6.

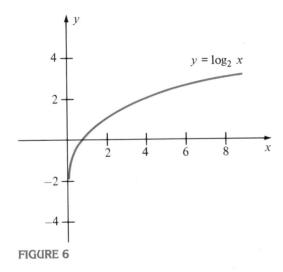

FIGURE 6

EXAMPLE 5

Sketch the graphs of $y = \log_3 x$ and $y = \log_{1/3} x$ on the same coordinate axes.

Solution

The graphs are shown in Figure 7.

The graphs in Figures 6 and 7 illustrate the following conclusions, which can be established in general.

- The point $(1,0)$ lies on the curve $y = \log_a x$ for any positive real number a. This is another way of saying $\log_a 1 = 0$.
- The domain of $f(x) = \log_a x$ is the set of all positive real numbers; the range is the set of all real numbers.
- When $a > 1$, $f(x) = \log_a x$ is an increasing function; when $0 < a < 1$, $f(x) = \log_a x$ is a decreasing function.

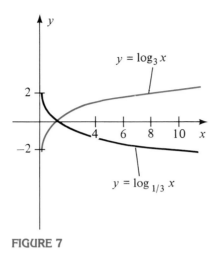

FIGURE 7

Since $\log_a x$ is either increasing or decreasing, the same value cannot be assumed more than once. Thus,

If $\log_a x = \log_a y$,   then $x = y$.

Since the graphs of $\log_a x$ and $\log_b x$ intersect only at $x = 1$, we have

If $\log_a x = \log_b x$ and $x \neq 1$,   then $a = b$.

EXAMPLE 6
Solve for x.

(a) $\log_5(x + 1) = \log_5 25$

$$x + 1 = 25$$
$$x = 24$$

(b) $\log_{x-1} 31 = \log_5 31$

$$x - 1 = 5$$
$$x = 6$$

PROGRESS CHECK
Solve for x.

(a) $\log_2 x^2 = \log_2 9$ (b) $\log_7 14 = \log_{2x} 14$

Answers

(a) *3, −3*   (b) $\dfrac{7}{2}$

The base 10 is used so frequently in work with logarithms that the notation $\log x$, with no base specified, is interpreted to mean $\log_{10} x$.

EXERCISE SET 4.2

Write each of the following in exponential form.

1. $\log_2 4 = 2$

2. $\log_5 125 = 3$

3. $\log_9 \dfrac{1}{81} = -2$

4. $\log_{64} 4 = \dfrac{1}{3}$

5. $\ln 20.09 = 3$

6. $\ln \dfrac{1}{7.39} = -2$

7. $\log_{10} 1000 = 3$

8. $\log_{10} \dfrac{1}{1000} = -3$

9. $\ln 1 = 0$

10. $\log_{10} 0.01 = -2$

11. $\log_3 \dfrac{1}{27} = -3$

12. $\log_{125} \dfrac{1}{5} = -\dfrac{1}{3}$

Write each of the following in logarithmic form.

13. $25 = 5^2$

14. $27 = 3^3$

15. $10,000 = 10^4$

16. $\dfrac{1}{100} = 10^{-2}$

17. $\dfrac{1}{8} = 2^{-3}$

18. $\dfrac{1}{27} = 3^{-3}$

19. $1 = 2^0$

20. $1 = e^0$

21. $6 = \sqrt{36}$

22. $2 = \sqrt[3]{8}$

23. $64 = 16^{3/2}$

24. $81 = 27^{4/3}$

25. $\dfrac{1}{3} = 27^{-1/3}$

26. $\dfrac{1}{2} = 16^{-1/4}$

Solve for x.

27. $\log_5 x = 2$

28. $\log_{16} x = \dfrac{1}{2}$

29. $\log_{25} x = -\dfrac{1}{2}$

30. $\log_{\frac{1}{2}} x = 3$

31. $\ln x = 2$

32. $\ln x = -3$

33. $\ln x = -\dfrac{1}{2}$

34. $\log_4 64 = x$

35. $\log_5 \dfrac{1}{25} = x$

36. $\log_x 4 = \dfrac{1}{2}$

37. $\log_x \dfrac{1}{8} = -\dfrac{1}{3}$

38. $\log_3(x - 1) = 2$

39. $\log_5(x + 1) = 3$

40. $\log_2(x - 1) = \log_2 10$

41. $\log_{x + 1} 24 = \log_3 24$

42. $\log_3 x^3 = \log_3 64$

43. $\log_{x + 1} 17 = \log_4 17$

44. $\log_{3x} 18 = \log_4 18$

Evaluate.

45. $3^{\log_3 6}$

46. $2^{\log_2(2/3)}$

47. $e^{\ln 2}$

48. $e^{\ln 1/2}$

49. $\log_5 5^3$

50. $\log_4 4^{-2}$

51. $\log_8 8^{1/2}$

52. $\log_{64} 64^{-1/3}$

53. $\log_7 49$

54. $\log_7 \sqrt{7}$

55. $\log_5 5$

56. $\ln e$

57. $\ln 1$

58. $\log_4 1$

59. $\log_2 \dfrac{1}{4}$

60. $\log_{16} 4$

61. $\log 10,000$

62. $\log 0.001$

63. $\ln e^2$

64. $\ln e^{-2/3}$

Sketch the graph of each given function.

65. $f(x) = \log_4 x$

66. $f(x) = \log_{\frac{1}{2}} x$

67. $f(x) = \log 2x$

68. $f(x) = \dfrac{1}{2} \log x$

69. $f(x) = \ln \dfrac{x}{2}$

70. $f(x) = \ln 3x$

71. $f(x) = \log_3 (x - 1)$

72. $f(x) = \log_3 (x + 1)$

PROPERTIES OF LOGARITHMS

There are three fundamental properties of logarithms that have made them a powerful computational aid.

Property 1. $\quad \log_a(x \cdot y) = \log_a x + \log_a y$

Property 2. $\quad \log_a\left(\dfrac{x}{y}\right) = \log_a x - \log_a y$

Property 3. $\quad \log_a x^n = n \log_a x$

These properties can be proved by using equivalent exponential forms. To prove the first property, $\log_a(x \cdot y) = \log_a x + \log_a y$, we let

$$\log_a x = u \quad \text{and} \quad \log_a y = v$$

Then the equivalent exponential forms are

$$a^u = x \quad \text{and} \quad a^v = y$$

Multiplying the left-hand and right-hand sides of these equations, we have

$$a^u \cdot a^v = x \cdot y$$

or

$$a^{u+v} = x \cdot y$$

Substituting a^{u+v} for $x \cdot y$ in $\log_a(x \cdot y)$ we have

$$\log_a(x \cdot y) = \log_a(a^{u+v})$$
$$= u + v \qquad \text{Since } \log_a a^x = x.$$

Substituting for u and v,

$$\log_a(x \cdot y) = \log_a x + \log_a y$$

Properties 2 and 3 can be established in much the same way.

It is these properties of logarithms that originally made their study worthwhile. Note the more complex operations of multiplication and division are converted to addition and subtraction, and exponentiation is converted to multiplication. We will first demonstrate these properties, and in the next section will apply them to realistic computational problems.

EXAMPLE 1

(a) $\log_{10}(225 \times 478) = \log_{10} 225 + \log_{10} 478$

(b) $\log_8 \left(\dfrac{422}{735}\right) = \log_8 422 - \log_8 735$

(c) $\log_2(2)^5 = 5 \log_2 2 = 5 \cdot 1 = 5$

(d) $\log_a \left(\dfrac{x \cdot y}{z}\right) = \log_a x + \log_a y - \log_a z$

PROGRESS CHECK

Write in terms of simpler logarithmic forms.

(a) $\log_4(1.47 \times 22.3)$

(b) $\log_a \dfrac{x-1}{\sqrt{x}}$

Answers

(a) $log_4\,1.47 + log_4\,22.3$

(b) $log_a\,(x-1) - \dfrac{1}{2}\,log_a\,x$

The following rules will assist in manipulating logarithmic forms more rapidly.

Rule	Example
1. Rewrite the expression so that each factor has a positive exponent.	$\log_a \dfrac{(x-1)^{-2}(y+2)^3}{\sqrt{x}}$ $= \log_a \dfrac{(y+2)^3}{(x-1)^2\sqrt{x}}$
2. Apply Property 1 and Property 2 for multiplication and division of logarithms. Each factor in the numerator will yield a term with a plus sign. Each factor in the denominator will yield a term with a minus sign.	$= \log_a (y+2)^3 - \log_a (x-1)^2 - \log_a \sqrt{x}$ $= 3\log_a (y+2) - 2\log_a (x-1) - \dfrac{1}{2}\log_a x$
3. Apply Property 3 to simplify.	

EXAMPLE 2

$$\log_a \frac{x^{-1/2}y^{3/2}}{(z+1)^5} = \log_a \frac{y^{3/2}}{x^{1/2}(z+1)^5}$$

$$= \log_a y^{3/2} - \log_a x^{1/2}(z+1)^5 \qquad \text{Property 2}$$

$$= \log_a y^{3/2} - \log_a x^{1/2} - \log_a (z+1)^5 \qquad \text{Property 1}$$

$$= \frac{3}{2}\log_a y - \frac{1}{2}\log_a x - 5\log_a (z+1) \qquad \text{Property 3}$$

PROGRESS CHECK

Simplify $\log_a \dfrac{(2x-3)^{1/2}(y+2)^{2/3}}{z^4}$

Answer

$\dfrac{1}{2}\,log_a\,(2x-3) + \dfrac{2}{3}\,log_a\,(y+2) - 4\,log_a\,z$

EXAMPLE 3

If $\log_a 1.5 = 0.37$, $\log_a 2 = 0.63$, and $\log_a 5 = 1.46$, find the following.

(a) $\log_a 7.5$

Since $\qquad\qquad\qquad\qquad 7.5 = 1.5 \times 5$

$$\log_a 7.5 = \log_a (1.5 \times 5)$$

$$= \log_a 1.5 + \log_a 5 \quad \text{Property 1}$$
$$= 0.37 + 1.46 \quad \text{Substitution}$$
$$= 1.83$$

(b) $\log_a \left[(1.5)^3 \cdot \sqrt[5]{\dfrac{2}{5}} \right]$

Write this as $\qquad \log_a (1.5)^3 + \log_a \left(\dfrac{2}{5} \right)^{1/5} \qquad$ Property 1

$$= 3 \log_a 1.5 + \frac{1}{5} \log_a \left(\frac{2}{5} \right) \qquad \text{Property 3}$$

$$= 3 \log_a 1.5 + \frac{1}{5} [\log_a 2 - \log_a 5] \quad \text{Property 2}$$

$$= 3(0.37) + \frac{1}{5} (0.63 - 1.46) \qquad \text{Substitution}$$

$$= 0.944$$

PROGRESS CHECK

If $\log_a 2 = 0.43$ and $\log_a 3 = 0.68$, find the following.

(a) $\log_a 18$

(b) $\log_a \sqrt[3]{\dfrac{9}{2}}$

Answers
(a) *1.79*

(b) *0.31*

 WARNING

(a)   Note that

$$\log_a (x + y) \neq \log_a x + \log_a y$$

Property 1 tells us that

$$\log_a (x \cdot y) = log_a x + \log_a y$$

Don't try to apply this property to $\log_a (x + y)$, which cannot be simplified.

(b)   Note that

$$\log_a x^n \neq (\log_a x)^n$$

By Property 3,

$$\log_a x^n = n \log_a x$$

We can also apply the properties of logarithms to combine terms involving logarithms.

EXAMPLE 4
Write as a single logarithm.

$$2 \log_a x - 3 \log_a (x + 1) + \log_a \sqrt{x - 1}$$
$$= \log_a x^2 - \log_a (x + 1)^3 + \log_a \sqrt{x - 1} \qquad \text{Property 3}$$

$$= \log_a x^2 \sqrt{x - 1} - \log_a (x + 1)^3 \qquad \text{Property 1}$$

$$= \log_a \frac{x^2 \sqrt{x - 1}}{(x + 1)^3} \qquad \text{Property 2}$$

PROGRESS CHECK

Write as a single logarithm.

$$\frac{1}{3} [\log_a (2x - 1) - \log_a (2x - 5)] + 4 \log_a x$$

Answer

$$\log_a x^4 \sqrt[3]{\frac{2x - 1}{2x - 5}}$$

 WARNING ———————————————————————————

(a)   Note that

$$\frac{\log_a x}{\log_a y} \neq \log_a (x - y)$$

Property 2 tells us that

$$\log_a\!\left(\frac{x}{y}\right) = \log_a x - \log_a y$$

Don't try to apply this property

to $\dfrac{\log_a x}{\log_a y}$, which cannot be simplified.

(b)   The expressions

$$\log_a x + \log_b x$$

and

$$\log_a x - \log_b x$$

cannot be simplified. Logarithms with different bases do not readily combine except in special cases.

EXERCISE SET 4.3

Express each of the following in terms of simpler logarithmic forms.

1. $\log_{10}(120 \times 36)$
2. $\log_6\left(\dfrac{187}{39}\right)$
3. $\log_3(3^4)$
4. $\log_3(4^3)$
5. $\log_a(2xy)$
6. $\ln (4xyz)$
7. $\log_a\left(\dfrac{x}{yz}\right)$
8. $\ln\left(\dfrac{2x}{y}\right)$
9. $\ln x^5$
10. $\log_3 y^{2/3}$
11. $\log_a(x^2y^3)$
12. $\log_a(xy)^3$
13. $\log_a \sqrt{xy}$
14. $\log_a \sqrt[3]{xy^4}$
15. $\ln(x^2y^3z^4)$

16. $\log_a(xy^3z^2)$ 17. $\ln(\sqrt{x}\,\sqrt[3]{y})$ 18. $\ln\sqrt[3]{xy^2}\,\sqrt[4]{z}$

19. $\log_a\left(\dfrac{x^2y^3}{z^4}\right)$ 20. $\ln\dfrac{x^4y^2}{z^{1/2}}$

If $\log 2 = 0.30$, $\log 3 = 0.47$, and $\log 5 = 0.70$, find the following.

21. $\log 6$ 22. $\log\dfrac{2}{3}$ 23. $\log 9$

24. $\log\sqrt{5}$ 25. $\log 12$ 26. $\log\dfrac{6}{5}$

27. $\log\dfrac{15}{2}$ 28. $\log 0.3$ 29. $\log\sqrt{7.5}$

30. $\log\sqrt[4]{30}$

Write each of the following as a simple logarithm.

31. $2\log x + \dfrac{1}{2}\log y$ 32. $3\log_a x - 2\log_a z$

33. $\dfrac{1}{3}\ln x + \dfrac{1}{3}\ln y$ 34. $\dfrac{1}{3}\ln x - \dfrac{2}{3}\ln y$

35. $\dfrac{1}{3}\log_a x + 2\log_a y - \dfrac{3}{2}\log_a z$ 36. $\dfrac{2}{3}\log_a x + \log_a y - 2\log_a z$

37. $\dfrac{1}{2}(\log_a x + \log_a y)$ 38. $\dfrac{2}{3}(4\ln x - 5\ln y)$

39. $\dfrac{1}{3}(2\ln x + 4\ln y) - 3\ln z$ 40. $\ln x - \dfrac{1}{2}(3\ln x + 5\ln y)$

41. $\dfrac{1}{2}\log_a(x-1) - 2\log_a(x+1)$

42. $2\log_a(x+2) - \dfrac{1}{2}(\log_a y + \log_a z)$

43. $3\log_a x - 2\log_a(x-1) + \dfrac{1}{2}\log_a\sqrt[3]{x+1}$

44. $4\ln(x-1) + \dfrac{1}{2}\ln(x+1) - 3\ln y$

4.4
COMPUTING WITH LOGARITHMS
(Optional)

We have earlier indicated that logarithms can be used to simplify complex calculations. In this section, we will demonstrate the power of logarithms in computational work.

We will use 10 as the base for our computations with logarithms since 10 is the base of our number system. We call logarithms to the base 10 **common logarithms** and it is customary to write $\log x$ in place of $\log_{10} x$.

We begin with the observation that any positive real number can be written as a product of a number c, $1 \le c < 10$, and a power of 10, say 10^k. This format is often referred to as **scientific notation**. Here are some examples.

$$643 = 6.43 \times 10^2 \qquad 4629 = 4.629 \times 10^3$$

$$754000 = 7.54 \times 10^5 \qquad\qquad 1.76 = 1.76 \times 10^0$$

$$0.0423 = 4.23 \times 10^{-2} \qquad 0.0000926 = 9.26 \times 10^{-5}$$

Let's apply properties of logarithms to the number 643 expressed in scientific notation. Thus,

$$643 = 6.43 \times 10^2$$

$$\begin{aligned}
\log 643 &= \log(6.43 \times 10^2) \\
&= \log 6.43 + \log 10^2 \qquad \text{Property 1} \\
&= \log 6.43 + 2\log 10 \qquad \text{Property 3} \\
&= \log 6.43 + 2 \qquad\qquad \log 10 = 1
\end{aligned}$$

In general,

> If x is a positive real number and $x = c \cdot 10^k$, then
>
> $$\log x = \log c + k$$

When $x = c \cdot 10^k$, with $1 \le c < 10$, the number $\log c$ is called the **mantissa** and the number k is called the **characteristic** of the number x. Since

$$x = c \cdot 10^k \quad \text{where } 1 \le c < 10$$

and since the function $y = \log x$ is an increasing function, we see that

$$\log 1 \le \log c < \log 10$$

or

$$0 \le \log c < 1$$

We may conclude that the mantissa is always a number between 0 and 1.

Table II in the Appendix can be used to approximate the logarithm of any three-digit number between 1.00 and 9.99 at intervals of 0.01. The following example shows how to proceed.

EXAMPLE 1

Find (a) log 73.5　　　(b) log 0.00451.

Solution

(a)　Since $73.5 = 7.35 \times 10^1$, the characteristic of 7.35 is 1. Using Table II in the Appendix, we find that log 7.35 is (approximately) 0.8663. Then

$$\log 73.5 = \log(7.35 \times 10^1) = \log 7.35 + \log 10^1$$

$$= 0.8663 + 1 = 1.8663$$

(b)　Since $0.00451 = 4.51 \times 10^{-3}$, the characteristic of 0.00451 is -3. From Table II we have

$$\log 0.00451 = 0.6542 - 3$$

Here we have a positive mantissa and a negative characteristic. For reasons that will be clear later, we always leave the answer in this form.

PROGRESS CHECK

Find the following.

(a) log 69,700   (b) log 0.000697   (c) log 0.697

Answers
(*a*) *4.8432*   (*b*) *0.8432 − 4*   (*c*) *0.8432 − 1*

WARNING

Note that
$$\log 0.00547 = 0.7380 - 3$$

but

$$\log 0.00547 \neq -3.7380$$

since this is algebraically incorrect. In fact, $0.7380 - 3 = -2.2620$

Table II in the Appendix can also be used in the reverse process, that is, to find x if $\log x$ is known. Since the entries in the body of Table II are numbers between 0 and 1, we must first write the number $\log x$ in the form

$$\log x = \log c + k$$

where $\log c$ is the mantissa, $0 \leq \log c < 1$, and k, the characteristic, is an integer. (This is why we insisted in Example 1b that the number be left in the form $0.6542 - 3$.)

EXAMPLE 2

(a) Find x if $\log x = 2.8351$.
     We write

$$\log x = 2.8351$$
$$= 0.8351 + 2$$
$$= \log c + k$$

We seek the mantissa 0.8351 in the body of Table II in the Appendix and find that it corresponds to log 6.84. Since the characteristic $k = 2$, we have

$$x = 6.84 \times 10^2 = 684$$

(b) Find x if $\log x = -6.6478$.
     We have to proceed carefully to insure that the mantissa is between 0 and 1. If we add and subtract 7, we have

$$\log x = (7 - 6.6478) - 7$$
$$= 0.3522 - 7$$
$$= \log c + k$$

We seek the mantissa 0.3522 in the body of Table II in the Appendix and find

$$0.3522 = \log 2.25$$

Since the characteristic $k = -7$, we have

$$x = 2.25 \times 10^{-7} = 0.000000225$$

PROGRESS CHECK
Find x in the following.

(a) $\log x = 3.8457$ (b) $\log x = 0.6201 - 2$ (c) $\log x = -2.0487$

Answers
(a)   7010   (b)   0.0417   (c)   0.00894

The following example shows how to use logarithms to simplify computations.

EXAMPLE 3
Approximate 478×0.0345 by using logarithms.

Solution

If $\hspace{6em} N = 478 \times 0.0345$

then $\hspace{5em} \log N = \log(478 \times 0.0345)$

$\hspace{9em} = \log 478 + \log 0.0345 \hspace{2em}$ Property 1

Using Table II in the Appendix,

$$\log 478 = 2.6794$$

$$\underline{\log 0.0345 = 0.5378 - 2}$$

$$\log N = 3.2172 - 2 \hspace{2em} \text{Adding the logarithms}$$

$$= 1.2172$$

Looking in the body of Table II, we find that the mantissa 0.2172 does not appear. However, 0.2175 does appear and corresponds to log 1.65. Thus,

$$N \approx 1.65 \times 10^1 = 16.5$$

Inexpensive calculators have reduced the importance of logarithms as a computational device. Still, many calculators cannot, for example, handle $\sqrt[5]{14.2}$ directly. If you know how to compute with logarithms and combine this knowledge with a calculator that can handle logarithms, you will have enhanced the power of your calculator. Many additional applications of logarithms occur in more advanced mathematics, especially in calculus.

EXAMPLE 4
Approximate $\dfrac{\sqrt{47.4}}{(2.3)^3}$ by using logarithms.

Solution

If $\hspace{9em} N = \dfrac{\sqrt{47.4}}{(2.3)^3}$

then $\hspace{5em} \log N = \dfrac{1}{2} \log 47.4 - 3 \log 2.3$

$$\dfrac{1}{2} \log 47.4 = \dfrac{1}{2}(1.6758) = 0.8379$$

$$\underline{3 \log 2.3 \hspace{1em} = 3(0.3617) = 1.0851}$$

$$\log N = 0.8379 - 1.0851 = -0.2472 \hspace{2em} \text{Subtracting the logarithms}$$

or

$$\log N = 0.7528 - 1 \qquad \text{Adding and subtracting 1}$$

From the body of Table II in the Appendix,

$$N \approx 5.66 \times 10^{-1} = 0.566$$

PROGRESS CHECK

Approximate $\dfrac{(4.64)^{3/2}}{\sqrt{7.42 \times 165}}$ by logarithms.

Answer
0.286

Problems in compound interest provide us with an opportunity to demonstrate the power of logarithms in computational work. In Section 4.1 we showed that the amount S available when a principal P is invested at an annual interest rate r compounded k times a year is given by

$$S = P(1 + i)^n$$

where $i = r/k$, and n is the number of conversion periods.

EXAMPLE 5

If $1000 is left on deposit at an interest rate of 8% per year compounded quarterly, how much money is in the account at the end of 6 years?

Solution

We have $P = 1000$, $r = 0.08$, $k = 4$, and $n = 24$ (since there are 24 quarters in 6 years). Thus,

$$S = P(1 + i)^n = 1000 \left(1 + \frac{0.08}{4}\right)^{24}$$

$$= 1000(1 + 0.02)^{24} = 1000(1.02)^{24}$$

Then

$$\log S = \log 1000 + 24 \log 1.02$$

$$= 3 + 24(0.0086) = 3.2064$$

From the body of Table II in the Appendix,

$$S \approx 1.61 \times 10^3 = 1610$$

The account contains (approximately) $1610 at the end of 6 years.

PROGRESS CHECK

If $1000 is left on deposit at an interest rate of 6% per year compounded semiannually, approximately how much is in the account at the end of four years?

Answer
$1267

EXERCISE SET 4.4

Write each of the following in scientific notation.

1. 2725
2. 493
3. 0.0084
4. 0.000914
5. 716,000
6. 527,600,000
7. 296.2
8. 32.767

Compute the following logarithms using Tables II and III in the Appendix.

9. log 3.56
10. ln 3.2
11. log 37.5
12. log 85.3
13. ln 4.7
14. ln 60
15. log 74
16. log 4230
17. log 48,200
18. log 7,890,000
19. log 0.342
20. log 0.00532

Using Tables II and III in the Appendix, find x.

21. $\log x = 0.4014$
22. $\ln x = -0.5108$
23. $\ln x = 1.0647$
24. $\log x = 2.7332$
25. $\ln x = 2.0669$
26. $\log x = 0.1903 - 2$
27. $\log x = 0.4099 - 1$
28. $\log x = 0.7024 - 2$
29. $\log x = 0.7832 - 4$
30. $\log x = 0.9320 - 2$
31. $\log x = -1.6599$
32. $\log x = -3.9004$

Find an approximate answer using logarithms.

33. $(320)(0.00321)$
34. $(8780)(2.13)$
35. $\dfrac{679}{321}$
36. $\dfrac{88.3}{97.2}$
37. $(3.19)^4$
38. $(42.3)^3(71.2)^2$
39. $\dfrac{(87.3)^2(0.125)^3}{(17.3)^3}$
40. $\sqrt[3]{(66.9)^4(0.781)^2}$
41. $\dfrac{\sqrt{7870}}{(46.3)^4}$
42. $\dfrac{(7.28)^{2/3}}{\sqrt[3]{(87.3)(16.2)^4}}$
43. $\dfrac{(32.870)(0.00125)}{(12.8)(124,000)}$

44. The period T (in seconds) of a simple pendulum of length L (in feet) is given by the formula

$$T = 2\pi \sqrt{\dfrac{L}{g}}$$

Using common logarithms, find the approximate value of T if $L = 4.72$ feet, $g = 32.2$, and $\pi = 3.14$.

45. Use logarithms to find the approximate amount that accumulates if $6000 is invested for 8 years in a bank paying 7% interest per year compounded quarterly.

46. Use logarithms to find the approximate sum if $8000 is invested for 6 years in a bank paying 8% interest per year compounded monthly.

47. If $10,000 is invested at 7.8% interest per year compounded semiannually, what sum is available after 5 years?

48. Which of the following offers will yield a greater return: 8% annual interest compounded annually, or 7.75% annual interest compounded quarterly?

49. Which of the following offers will yield a greater return: 9% annual interest compounded annually or 8.75% annual interest compounded quarterly?

50. The area of a triangle whose sides are a, b, and c in length is given by the

formula

$$A = \sqrt{s(s - a)(s - b)(s - c)}$$

where $s = \frac{1}{2}(a + b + c)$. Use logarithms to find the approximate area of a triangle whose sides are 12.86 feet, 13.72 feet, and 20.3 feet.

4.5
EXPONENTIAL AND LOGARITHMIC EQUATIONS

The following approach will often help in solving exponential and logarithmic equations.

> To solve an exponential equation, take logarithms of both sides of the equation.
>
> To solve a logarithmic equation, convert to the equivalent exponential form.

At times it may be necessary to simplify before converting to an equivalent form. Here are some examples.

EXAMPLE 1
Solve $3^{2x - 1} = 17$.

Solution
Taking logarithms to the base 10 of both sides of the equation,

$$\log 3^{2x - 1} = \log 17$$

$$(2x - 1)\log 3 = \log 17 \qquad \text{Property 3}$$

$$2x - 1 = \frac{\log 17}{\log 3}$$

$$2x = 1 + \frac{\log 17}{\log 3}$$

$$x = \frac{1}{2} + \frac{\log 17}{2 \log 3}$$

If a numerical value is required, Table II in the Appendix, or a calculator, can be used to approximate log 17 and log 3.

PROGRESS CHECK
Solve $2^{x + 1} = 3^{2x - 3}$.

Answer

$$\frac{\log 2 + 3 \log 3}{2 \log 3 - \log 2}$$

EXAMPLE 2
Solve $\log(2x + 8) = 1 + \log(x - 4)$.

Solution
If we rewrite the equation in the form

$$\log(2x + 8) - \log(x - 4) = 1$$

then we can apply Property 2 to form a single logarithm.

$$\log \frac{2x + 8}{x - 4} = 1$$

Now we convert to the equivalent exponential form.

$$\frac{2x + 8}{x - 4} = 10^1 = 10$$

$$2x + 8 = 10x - 40$$

$$x = 6$$

PROGRESS CHECK
Solve $\log(x + 1) = 2 + \log(3x - 1)$.

Answer
$\dfrac{101}{299}$

EXAMPLE 3
Solve $\log_2 x = 3 - \log_2(x + 2)$

Solution
Rewriting the equation with a single logarithm, we have

$$\log_2 x + \log_2(x + 2) = 3$$

$$\log_2[x(x + 2)] = 3 \qquad \text{Why?}$$

$$x(x + 2) = 2^3 = 8 \qquad \text{Equivalent exponential form}$$

$$x^2 + 2x - 8 = 0$$

$$(x - 2)(x + 4) = 0 \qquad \text{Factor}$$

$$x = 2 \quad \text{or} \quad x = -4$$

The "solution" $x = -4$ must be rejected since the original equation contains $\log_2 x$, which requires that x be positive.

PROGRESS CHECK
Solve $\log_3(x - 8) = 2 - \log_3 x$

Answer
$x = 9$

EXAMPLE 4
World population is increasing at an annual rate of 2.5%. If we assume an exponential growth model, in how many years will the population double?

Solution
The exponential growth model

$$Q(t) = q_0 e^{0.025t}$$

describes the population Q as a function of time t. Since the initial population is $Q(0) = q_0$, we seek the time t required for the population to double or become $2q_0$. We wish to solve the equation

$$Q(t) = 2q_0 = q_0 e^{0.025t}$$

for t. We then have

$$2q_0 = q_0 e^{0.025t}$$

$$2 = e^{0.025t} \qquad \text{Divide by } q_0$$

$$\ln 2 = \ln e^{0.025t} \qquad \text{Take natural logs of both sides}$$

$$= 0.025t \qquad \text{Since } \ln e^x = x$$

$$t = \frac{\ln 2}{0.025} = \frac{0.6931}{0.025} = 27.7$$

or approximately 28 years.

EXAMPLE 5
A trust fund invests $8000 at an annual interest rate of 8% compounded continuously. How long does it take for the initial investment to grow to $12,000?

Solution
Using Equation (4) of Section 4.1,

$$S = Pe^{rt}$$

we have $S = 12,000$, $P = 8000$, $r = 0.08$, and we must solve for t. Thus,

$$12,000 = 8000 e^{0.08t}$$

$$\frac{12,000}{8000} = e^{0.08t}$$

$$e^{0.08t} = 1.5$$

Taking natural logarithms of both sides we have

$$0.08t = \ln 1.5$$

$$t = \frac{\ln 1.5}{0.08} \approx \frac{0.4055}{0.08} \qquad \text{from Table III}$$

$$\approx 5.07$$

It takes approximately 5.07 years for the initial $8000 to grow to $12,000.

EXERCISE SET 4.5
In Exercises 1–29 solve for x.

1. $5^x = 18$
2. $2^x = 24$
3. $2^{x-1} = 7$
4. $3^{x-1} = 12$
5. $3^{2x} = 46$
6. $2^{2x-1} = 56$
7. $5^{2x-5} = 564$
8. $3^{3x-2} = 23.1$
9. $3^{x-1} = 2^{2x+1}$
10. $4^{2x-1} = 3^{2x+3}$
11. $2^{-x} = 15$
12. $3^{-x+2} = 103$
13. $4^{-2x+1} = 12$
14. $3^{-3x+2} = 2^{-x}$
15. $e^x = 18$
16. $e^{x-1} = 2.3$
17. $e^{2x+3} = 20$
18. $e^{-3x+2} = 40$
19. $\log x + \log 2 = 3$
20. $\log x - \log 3 = 2$
21. $\log_x(3 - 5x) = 1$
22. $\log_x(8 - 2x) = 2$

23. $\log x + \log(x - 3) = 1$

24. $\log x + \log(x + 21) = 2$

25. $\log(3x + 1) - \log(x - 2) = 1$

26. $\log(7x - 2) - \log(x - 2) = 1$

27. $\log_2 x = 4 - \log_2(x - 6)$

28. $\log_2(x - 4) = 2 - \log_2 x$

29. $\log_2(x + 4) = 3 - \log_2(x - 2)$

30. Suppose that world population is increasing at an annual rate of 2%. If we assume an exponential growth model, in how many years will the population double?

31. Suppose that the population of a certain city is increasing at an annual rate of 3%. If we assume an exponential growth model, in how many years will the population triple?

32. The population P of a certain city t years from now is given by

$$P = 20{,}000e^{0.05t}$$

How many years from now will the population be 50,000?

33. Potassium 42 has a decay rate of approximately 5.5% per hour. Assuming an exponential decay model, in how many hours will the original quantity of potassium 42 have been halved?

34. Consider an exponential decay model given by

$$Q = q_0 e^{-0.4t}$$

where t is in weeks. How many weeks does it take for Q to decay to $\frac{1}{4}$ of its original amount?

35. How long does it take an amount of money to double if it is invested at a rate of 8% per year compounded semiannually?

36. At what rate of annual interest, compounded semiannually, should a certain amount of money be invested so it will double in 8 years?

37. The number N of radios that an assembly line worker can assemble daily after t days of training is given by

$$N = 60 - 60e^{-0.04t}$$

After how many days of training does the worker assemble 40 radios daily?

38. The quantity Q (in grams) of a radioactive substance which is present after t days of decay is given by

$$Q = 400e^{-kt}$$

If $Q = 300$ when $t = 3$, find k, the decay rate.

39. A person on an assembly line produces P items per day after t days of training where

$$P = 400(1 - e^{-t})$$

How many days of training will it take this person to be able to produce 300 items per day?

40. Suppose that the number N of mopeds sold when x thousands of dollars are spent on advertising is given by

$$N = 4000 + 1000 \ln(x + 2)$$

How much advertising money must be spent to sell 6000 mopeds?

TERMS AND SYMBOLS

exponential function (p. 161)	**decay constant** (p. 165)	$\log_a x$ (p. 171)
base (p. 161)	**compound interest** (p. 166)	**ln** x (p. 171)
ax (p. 161)	**conversion period** (p. 166)	**natural logarithm** (p. 171)
e (p. 163)	**continuous compounding** (p. 168)	**common logarithm** (p. 181)
exponential growth model (p. 164)	**logarithmic function** (p. 171)	**scientific notation** (p. 181)
growth constant (p. 165)		**mantissa** (p. 182)
exponential decay model (p. 165)		**characteristic** (p. 182)

KEY IDEAS FOR REVIEW

■ An exponential function has a variable in the exponent and a base that is a positive constant.

■ The graph of the exponential function $f(x) = a^x$, $a > 0$, $a \neq 1$
- passes through the points $(0,1)$ and $(1, a)$ for any value of x.
- is increasing if $a > 1$ and decreasing if $0 < a < 1$.

■ The domain of the exponential function is the set of all real numbers; the range is the set of all positive numbers.

■ If $a^x = a^y$, then $x = y$ (assuming $a > 0$).

■ If $a^x = b^x$ for all $x \neq 0$, then $a = b$ (assuming $a, b > 0$)

■ Exponential functions play a key role in the following important applications.
- Exponential growth model: $Q(t) = q_0 e^{kt}$, $\quad k > 0$
- Exponential decay model: $\quad Q(t) = q_0 e^{-kt}$, $\quad k > 0$
- Compound interest: $\quad S = P(1 + i)^n$
- Continuous compounding: $\quad S = Pe^{rt}$

■ The logarithmic function $\log_a x$ is the inverse of the function a^x.

■ The logarithmic form $y = \log_a x$ and the exponential form $x = a^y$ are two ways of expressing the same relationship. In short, logarithms are exponents. Consequently, it is always possible to convert from one form to the other.

■ The following identities are useful in simplifying expressions and in solving equations.

$$a^{\log_a x} = x \qquad \log_a a = 1$$
$$\log_a a^x = x \qquad \log_a 1 = 0$$

■ The graph of the logarithmic function $f(x) = \log_a x$, $x > 0$
- passes through the points $(1,0)$ and $(a,1)$ for any $a > 0$.
- is increasing if $a > 1$ and decreasing if $0 < a < 1$.

■ The domain of the logarithmic function is the set of all positive real numbers; the range is the set of all real numbers.

■ If $\log_a x = \log_a y$, then $x = y$.

■ If $\log_a x = \log_b x$ and $x \neq 1$, then $a = b$.

■ The fundamental properties of logarithms are as follows.

Property 1. $\log_a(xy) = \log_a x + \log_a y$

Property 2. $\log_a \left(\dfrac{x}{y}\right) = \log_a x - \log_a y$

Property 3. $\log_a x^n = n \log_a x$

■ The fundamental properties of logarithms in conjunction with tables of logarithms are a powerful tool in performing calculations. It is these properties which make the study of logarithms worthwhile.

REVIEW EXERCISES

1. Sketch the graph of $f(x) = \left(\frac{1}{3}\right)^x$. Label the point $(-1, f(-1))$.
2. Solve $2^{2x} = 8^{x-1}$ for x.
3. Solve $(2a + 1)^x = (3a - 1)^x$ for a.
4. The sum of $8000 is invested in a certificate paying 12% annual interest compounded semiannually. What sum is available at the end of 4 years?

In Exercises 5–8 write each logarithmic form in exponential form and vice versa.

5. $27 = 9^{3/2}$

6. $\log_{64} 8 = \frac{1}{2}$

7. $\log_2 \frac{1}{8} = -3$

8. $6^0 = 1$

In Exercises 9–12 solve for x.

9. $\log_x 16 = 4$

10. $\log_5 \frac{1}{125} = x - 1$

11. $\ln x = -4$

12. $\log_3 (x + 1) = \log_3 27$

In Exercises 13–16 evaluate the given expression.

13. $\log_3 3^5$

14. $\ln e^{-1/3}$

15. $\log_3 \left(\frac{1}{3}\right)$

16. $e^{\ln 3}$

17. Sketch the graph of $f(x) = \log_3 x + 1$.

In Exercises 18–21 write the given expression in terms of simpler logarithmic forms.

18. $\log_a \frac{\sqrt{x-1}}{2x}$

19. $\log_a \frac{x(2-x)^2}{(y+1)^{1/2}}$

20. $\ln (x + 1)^4(y - 1)^2$

21. $\log \sqrt[5]{\frac{y^2 z}{z + 3}}$

In Exercises 22–25 use the values $\log 2 = 0.30$, $\log 3 = 0.50$, and $\log 7 = 0.85$ to evaluate the given expression.

22. $\log 14$

23. $\log 3.5$

24. $\log \sqrt{6}$

25. $\log 0.7$

In Exercises 26–29 write the given expression as a single logarithm.

26. $\frac{1}{3} \log_a x - \frac{1}{2} \log_a y$

27. $\frac{4}{3} [\log x + \log(x - 1)]$

28. $\ln 3x + 2(\ln y - \frac{1}{2} \ln z)$

29. $2 \log_a(x + 2) - \frac{3}{2} \log_a(x + 1)$

In Exercises 30–33 write the given number in scientific notation.

30. 476.5

31. 0.098

32. 26,475

33. 77.67

In Exercises 34–36 use logarithms to calculate the value of the given expression.

34. $(0.765)(32.4)^2$

35. $\sqrt{62.3}$

36. $\frac{2.1}{(32.5)^{5/2}}$

37. A substance is known to have a decay rate of 6% per hour. Approximately how many hours are required for the remaining quantity to be half of the original quantity?

In Exercises 38–40 solve for x.

38. $2^{3x-1} = 14$ 39. $2 \log x - \log 5 = 3$

40. $\log(2x - 1) = 2 + \log(x - 2)$

PROGRESS TEST 4A

1. Sketch the graph of $f(x) = 2^{x+1}$. Label the point $(1, f(1))$.

2. Solve $\left(\dfrac{1}{2}\right)^x = \left(\dfrac{1}{4}\right)^{2x+1}$

In Problems 3–4 convert from logarithmic form to exponential form and vice versa.

3. $\log_3 \dfrac{1}{9} = -2$ 4. $64 = 16^{3/2}$

In Problems 5–6 solve for x.

5. $\log_x 27 = 3$ 6. $\log_6 \left(\dfrac{1}{36}\right) = 3x + 1$

In Problems 7–8 evaluate the given expression.

7. $\ln e^{5/2}$ 8. $\log_5 \sqrt{5}$

In Problems 9–10 write the given expression in terms of simpler logarithmic forms.

9. $\log_a \dfrac{x^3}{y^2 z}$ 10. $\log \dfrac{x^2 \sqrt{2y-1}}{y^3}$

In Problems 11–12 use the values $\log 2.5 = 0.4$ and $\log 2 = 0.3$ to evaluate the given expression.

11. $\log 5$ 12. $\log 2\sqrt{2}$

In Problems 13–14 write the given expression as a single logarithm.

13. $2 \log x - 3 \log(y + 1)$ 14. $\dfrac{2}{3} [\log_a(x + 3) - \log_a(x - 3)]$

In Problems 15–16 write the given number in scientific notation.

15. 0.000273 16. 5.972

In Problems 17–18 use logarithms to evaluate the given expression.

17. $\dfrac{72.9}{(39.4)^2}$ 18. $\sqrt[3]{0.0176}$

19. The number of bacteria in a culture is described by the exponential growth model

$$Q(t) = q_0 e^{0.02t}$$

Approximately how many hours are required for the number of bacteria to double?

In Problems 20–21 solve for x.

20. $\log x - \log 2 = 2$ 21. $\log_4(x - 3) = 1 - \log_4 x$

PROGRESS TEST 4B

1. Sketch the graph of $f(x) = \left(\dfrac{1}{2}\right)^{x-1}$. Label the point $(2, f(2))$.

2. Solve $(a + 3)^x = (2a - 5)^x$ for a.

In Problems 3–4 convert from logarithmic form to exponential form and vice versa.

3. $\dfrac{1}{1000} = 10^{-3}$ 4. $\log_3 1 = 0$

In Problems 5–6 solve for x.

5. $\log_2(x - 1) = -1$ 6. $\log_{2x} 27 = \log_3 27$

In Problems 7–8 evaluate the given expression.

7. $\log_a 3^{10}$ 8. $e^{\ln 4}$

In Problems 9–10 write the given expression in terms of simpler logarithmic forms.

9. $\log_a(x - 1)(y + 3)^{5/4}$ 10. $\ln \sqrt{xy} \sqrt[4]{2z}$

In Problems 11–12 use the values $\log 2.5 = 0.4$, $\log 2 = 0.3$, and $\log 6 = 0.75$ to evaluate the given expression.

11. $\log 7.5$ 12. $\log 36$

In Problems 13–14 write the given expression as a single logarithm.

13. $\frac{3}{5} \ln(x - 1) + \frac{2}{5} \ln y - \frac{1}{5} \ln z$ 14. $\log \frac{x}{y} - \log \frac{y}{x}$

In Problems 15–16 write the given number in scientific notation.

15. 22,684,321   16. 0.297

In Problems 17–18 use logarithms to evaluate the given expression.

17. $(0.295)(31.7)^3$ 18. $\dfrac{\sqrt{42.9}}{(3.75)^2(747)}$

19. Suppose that $500 is invested in a certificate at an annual interest rate of 12% compounded monthly. What is the value of the investment after 6 months?

In Problems 20–21 solve for x.

20. $\log_x(x + 6) = 2$ 21. $\log (x - 9) = 1 - \log x$

CHAPTER FIVE
ANALYTIC GEOMETRY: THE CONIC SECTIONS

In 1637 the great French philosopher and scientist René Descartes developed an idea that the nineteenth century British philosopher John Stuart Mills described as "the greatest single step ever made in the progress of the exact sciences." Descartes combined the techniques of algebra with those of geometry and created a new field of study called **analytic geometry**. Analytic geometry enables us to apply algebraic methods and equations to the solution of problems in geometry and, conversely, to obtain geometric representations of algebraic equations.

We will first develop a formula for the coordinates of the midpoint of a line segment. We will then use the distance and midpoint formulas as tools to demonstrate the usefulness of analytic geometry by proving a number of general theorems from plane geometry.

The power of the methods of analytic geometry is also very well demonstrated, as we shall see in this chapter, in a study of the conic sections. We will find in the course of that study that (a) a geometric definition can be converted into an algebraic equation and (b) an algebraic equation can be classified by the type of graph it represents.

5.1
ANALYTIC GEOMETRY

We have previously seen that the length d of the line segment joining points $P_1(x_1, y_1)$ and $P_2(x_2, y_2)$ (see Figure 1) is given by

$$d = \sqrt{(x_2 - x_1)^2 + (y_2 - y_1)^2}$$

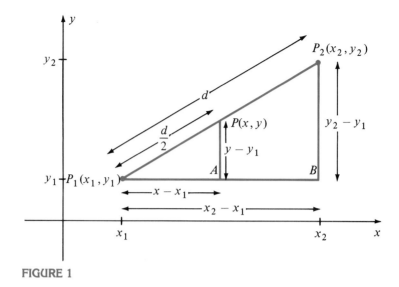

FIGURE 1

It is also possible to obtain a formula for the coordinates (x,y) of the midpoint P of the line segment whose endpoints are P_1 and P_2. Since P is the midpoint of P_1P_2, the length of P_1P is $d/2$. The lines PA and P_2B are parallel, so triangles P_1AP and P_1BP_2 are similar. Since corresponding sides of similar triangles are in proportion, we can write

$$\frac{\overline{P_1P_2}}{\overline{P_2B}} = \frac{\overline{P_1P}}{\overline{PA}}$$

or

$$\frac{d}{y_2 - y_1} = \frac{\dfrac{d}{2}}{y - y_1}$$

Solving for y we have

$$y = \frac{y_1 + y_2}{2}$$

Similarly,

$$\frac{\overline{P_1P_2}}{\overline{P_1B}} = \frac{\overline{P_1P}}{\overline{P_1A}} \quad \text{or} \quad \frac{d}{x_2 - x_1} = \frac{\dfrac{d}{2}}{x - x_1}$$

We solve for x to obtain

$$x = \frac{x_1 + x_2}{2}$$

We have established the following formula.

The Midpoint Formula

If $P(x,y)$ is the midpoint of the line segment whose
endpoints are $P_1(x_1,y_1)$ and $P_2(x_2,y_2)$, then

$$x = \frac{x_1 + x_2}{2} \qquad y = \frac{y_1 + y_2}{2}$$

EXAMPLE 1
Find the midpoint of the line segment whose endpoints are $P_1(3,4)$ and $P_2(-2, -6)$.

Solution
If $P(x,y)$ is the midpoint, then

$$x = \frac{x_1 + x_2}{2} = \frac{3 + (-2)}{2} = \frac{1}{2}$$

$$y = \frac{y_1 + y_2}{2} = \frac{4 + (-6)}{2} = -1$$

Thus, the midpoint is $(1/2, -1)$.

PROGRESS CHECK
Find the midpoint of the line segment whose endpoints are the following.

(a) $(0, -4), (-2, -2)$ (b) $(-10, 4), (7, -5)$

Answers
(a) *(-1, -3)*   *(b)* *(-3/2, -1/2)*

The formulas for distance, midpoint of a line segment, and slope of a line
are sufficient to allow us to demonstrate the beauty and power of analytic
geometry. With these tools, we can prove theorems from plane geometry
by placing the figures on a rectangular coordinate system.

EXAMPLE 2
Prove that the line joining the midpoints of two sides of a triangle is parallel to the
third side and equal to one half the third side.

Solution
We place the triangle OAB in a convenient location, namely, with one vertex at
the origin and one side on the positive x-axis (Figure 2). If Q and R are the
midpoints of OB and AB, then, by the midpoint formula, the coordinates of Q and
R are

$$Q\left(\frac{b}{2}, \frac{c}{2}\right) \qquad R\left(\frac{a + b}{2}, \frac{c}{2}\right)$$

We see that the line joining Q and R has slope 0 since the difference of the
y-coordinates is

$$\frac{c}{2} - \frac{c}{2} = 0$$

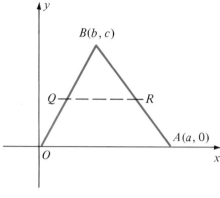

FIGURE 2

But side OA also has slope 0, which proves that QR is parallel to OA. Applying the distance formula to $\overline{QR}$ we have

$$\overline{QR} = \sqrt{(x_2 - x_1)^2 + (y_2 - y_1)^2}$$

$$= \sqrt{\left(\frac{a+b}{2} - \frac{b}{2}\right)^2 + \left(\frac{c}{2} - \frac{c}{2}\right)^2}$$

$$= \sqrt{\left(\frac{a}{2}\right)^2} = \frac{a}{2}$$

Since $\overline{OA}$ has length a, we have shown that $\overline{QR}$ is one half of $\overline{OA}$.

PROGRESS CHECK
Prove that the midpoint of the hypotenuse of a right triangle is equidistant from all three vertices.

Answer
(Hint: Place the triangle so that two legs coincide with the positive x- and y-axes. Find the coordinates of the midpoint of the hypotenuse by the midpoint formula. Finally, compute the distance from the midpoint to each vertex by the distance formula.)

EXERCISE SET 5.1
In Exercises 1–12, find the midpoint of the line segment whose endpoints are given.

1. (2,6), (3,4)   2. (1,1), (−2,5)   3. (2,0), (0,5)
4. (−3,0), (−5,2)   5. (−2,1), (−5,−3)   6. (2,3), (−1,3)
7. (0,−4), (0,3)   8. (1,−3), (3,2)   9. (−1,3), (−1,6)
10. (3,2), (0,0)   11. (1,−1), (−1,1)   12. (2,4), (2,−4)

13. Prove that the medians from the equal angles of an isosceles triangle are of equal length. (*Hint:* Place the triangle so that its vertices are at the points $A(-a,0)$, $B(a,0)$, and $C(0,b)$.)
14. Show that the midpoints of the sides of a rectangle are the vertices of a rhombus (a quadrilateral with four equal sides). (*Hint:* Place the rectangle so that its vertices are at the points (0,0), (a,0), (0,b), and (a,b).)
15. Prove that a triangle with two equal medians is isosceles.
16. Show that the sum of the squares of the lengths of the medians of a triangle equals three fourths the sum of the squares of the lengths of the sides. (*Hint:*

Place the triangle so that its vertices are at the points $(-a,0)$, $(b,0)$, and $(0,c)$.)

17. Prove that the diagonals of a rectangle are equal in length. (*Hint:* Place the rectangle so that its vertices are at the points $(0,0)$, $(a,0)$, $(0,b)$, and (a,b).)

5.2
THE CIRCLE

The conic sections provide us with an outstanding opportunity to demonstrate the double-edged power of analytic geometry. We will see that a geometric figure defined as a set of points can often be described analytically by an algebraic equation; conversely, we can start with an algebraic equation and use graphing procedures to study the properties of the curve.

First, let's see how the term conic section originates. If we pass a plane through a cone at various angles, the intersections shown in Figure 3 are called **conic sections**. In exceptional cases, the intersections include a point, a line, and a pair of lines.

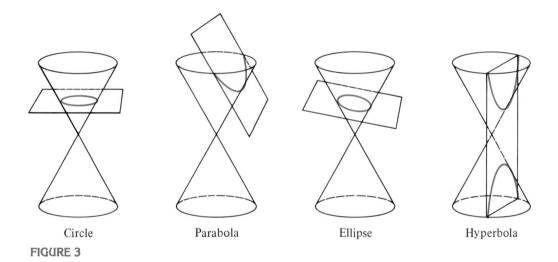

| Circle | Parabola | Ellipse | Hyperbola |

FIGURE 3

Let's begin with the geometric definition of a circle.

A **circle** is the set of all points in a plane that are at a given distance from a fixed point. The fixed point is called the **center** of the circle and the given distance is called the **radius**.

Using the methods of analytic geometry, we place the center at a point (h, k) as in Figure 4. If $P(x, y)$ is a point on the circle, then the distance from P to the center (h, k) must be equal to the radius r. By the distance formula

$$\sqrt{(x - h)^2 + (y - k)^2} = r$$

or

$$(x - h)^2 + (y - k)^2 = r^2$$

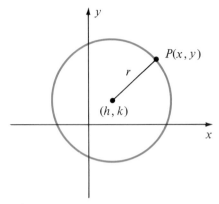

FIGURE 4

Since $P(x, y)$ is any point on the circle, we say that

$$(x - h)^2 + (y - k)^2 = r^2$$

is the **standard form of the equation of the circle** with center (h, k) and radius r.

EXAMPLE 1
Write the equation of the circle with center at $(2, -5)$ and radius 3.

Solution
Substituting $h = 2$, $k = -5$, and $r = 3$ in the equation

$$(x - h)^2 + (y - k)^2 = r^2$$

yields

$$(x - 2)^2 + (y + 5)^2 = 9$$

EXAMPLE 2
Find the center and radius of the circle whose equation is $(x + 1)^2 + (y - 3)^2 = 4$.

Solution
Since the standard form is

$$(x - h)^2 + (y - k)^2 = r^2$$

we must have

$$x - h = x + 1 \qquad y - k = y - 3 \qquad r^2 = 4$$

Solving, we find that

$$h = -1, \qquad k = 3, \qquad r = 2$$

Thus, the center is at $(-1, 3)$ and the radius is 2.

PROGRESS CHECK
Find the center and radius of the circle whose equation is $\left(x - \frac{1}{2}\right)^2 + (y + 5)^2 = 15$.

Answer

center $\left(\frac{1}{2}, -5\right)$, *radius* $\sqrt{15}$

When we are given the equation of a circle in the **general form**

$$Ax^2 + Ay^2 + Dx + Ey + F = 0, \qquad A \neq 0$$

in which the coefficients of x^2 and y^2 are the same, we may rewrite the equation in standard form. The process involves completing the square in each variable.

EXAMPLE 3

Write the equation of the circle $2x^2 + 2y^2 - 12x + 16y - 22 = 0$ in standard form.

Solution

Grouping the terms in x and y and factoring produces

$$2(x^2 - 6x) + 2(y^2 + 8y) = 22$$

Completing the square in both x and y, we have

$$2(x^2 - 6x + 9) + 2(y^2 + 8y + 16) = 22 + 18 + 32$$
$$2(x - 3)^2 + 2(y + 4)^2 = 72$$

Note that the quantities 18 and 32 were added to the right-hand side since each factor is mulitplied by 2. The last equation can be written as

$$(x - 3)^2 + (y + 4)^2 = 36$$

This is the standard form of the equation of the circle with center at $(3, -4)$ and radius of 6.

PROGRESS CHECK

Write the equation of the circle $4x^2 + 4y^2 - 8x + 4y = 103$ in standard form and determine the center and radius.

Answer

$$(x - 1)^2 + \left(y + \frac{1}{2}\right)^2 = 27, \; center \left(1, \; -\frac{1}{2}\right), \; radius \; \sqrt{27}$$

EXAMPLE 4

Write the equation $x^2 + y^2 - 6x + 13 = 0$ in standard form.

Solution

Regrouping, we have

$$(x^2 - 6x) + y^2 = -13$$

We now complete the square in x and y

$$(x^2 - 6x + 9) + y^2 = -13 + 9$$
$$(x - 3)^2 + y^2 = -4$$

Since $r^2 = -4$ is an impossible situation, the graph of the equation is not a circle. Note that the left-hand side of the equation in standard form is a sum of squares and is therefore nonnegative, while the right-hand side is negative. Thus, there are no real values of x and y that satisfy the equation. This is an example of an equation that does not have a graph!

PROGRESS CHECK

Write the equation $x^2 + y^2 - 12y + 36 = 0$ in standard form and analyze its graph.

Answer
The standard form is $x^2 + (y - 6)^2 = 0$. The equation is that of a "circle" with center at $(0, 6)$ and radius of 0. The "circle" is actually the point $(0, 6)$.

EXERCISE SET 5.2

In each of the following write the equation of the circle with center at (h,k) and radius r.

1. $(h,k) = (2,3)$, $r = 2$ 2. $(h,k) = (-3,0)$, $r = 3$

3. $(h,k) = (-2, -3)$, $r = \sqrt{5}$ 4. $(h,k) = (2, -4)$, $r = 4$

5. $(h,k) = (0,0)$, $r = 3$ 6. $(h,k) = (0, -3)$, $r = 2$

7. $(h,k) = (-1,4)$, $r = 2\sqrt{2}$ 8. $(h,k) = (2,2)$, $r = 2$

In each of the following find the center and radius of the circle with the given equation.

9. $(x - 2)^2 + (y - 3)^2 = 16$ 10. $(x + 2)^2 + y^2 = 9$

11. $(x - 2)^2 + (y + 2)^2 = 4$ 12. $\left(x + \dfrac{1}{2}\right)^2 + (y - 2)^2 = 8$

13. $(x + 4)^2 + \left(y + \dfrac{3}{2}\right)^2 = 18$ 14. $x^2 + (y - 2)^2 = 4$

15. $\left(x - \dfrac{1}{3}\right)^2 + y^2 = -\dfrac{1}{9}$ 16. $(x - 1)^2 + \left(y - \dfrac{1}{2}\right)^2 = 3$

Write the equation of each given circle in standard form and determine the center and radius if they exist.

17. $x^2 + y^2 + 4x - 8y + 4 = 0$ 18. $x^2 + y^2 - 2x + 6y - 15 = 0$

19. $2x^2 + 2y^2 - 6x - 10y + 6 = 0$ 20. $2x^2 + 2y^2 + 8x - 12y - 8 = 0$

21. $2x^2 + 2y^2 - 4x - 5 = 0$ 22. $4x^2 + 4y^2 - 2y + 7 = 0$

23. $3x^2 + 3y^2 - 12x + 18y + 15 = 0$ 24. $4x^2 + 4y^2 + 4x + 4y - 4 = 0$

Write the given equation in standard form and determine if the graph of the equation is a circle, a point, or neither.

25. $x^2 + y^2 - 6x + 8y + 7 = 0$ 26. $x^2 + y^2 + 4x + 6y + 5 = 0$

27. $x^2 + y^2 + 3x - 5x + 7 = 0$ 28. $x^2 + y^2 - 4x - 6y - 13 = 0$

29. $2x^2 + 2y^2 - 12x - 4 = 0$ 30. $2x^2 + 2y^2 + 4x - 4y + 25 = 0$

31. $2x^2 + 2y^2 - 6x - 4y - 2 = 0$ 32. $2x^2 + 2y^2 - 10y + 6 = 0$

33. $3x^2 + 3y^2 + 12x - 4y - 20 = 0$ 34. $x^2 + y^2 + x + y = 0$

35. $4x^2 + 4y^2 + 12x - 20y + 38 = 0$ 36. $4x^2 + 4y^2 - 12x - 36 = 0$

5.3
THE PARABOLA

We begin our study of the parabola with the geometric definition.

> A **parabola** is the set of all points that are
> equidistant from a given point and a given line.

The given point is called the **focus** and the given line is called the **directrix** of the parabola. In Figure 5, all points P on the parabola are equidistant from the focus F and the directrix L, that is, $\overline{PF} = \overline{PQ}$. The line through the focus that is perpendicular to the directrix is called the **axis of the parabola** (or simply the **axis**) and the parabola is seen to be symmetric with respect to the axis. The point V (Figure 5) where the parabola intersects its axis is called the **vertex** of the parabola. The vertex, then, is the point from which the parabola opens.

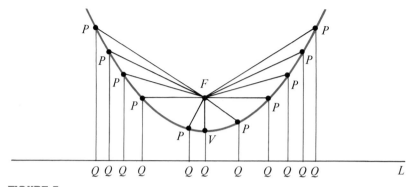

FIGURE 5

We can apply the methods of analytic geometry to find an equation of the parabola. We choose the y-axis as the axis of the parabola and the origin as the vertex (Figure 6). Since the vertex is on the parabola, it is

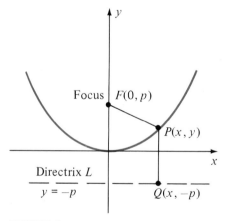

FIGURE 6

equidistant from the focus and the directrix. Thus, if the coordinates of the focus F are $(0,p)$, then the equation of the directrix is $y = -p$. We then let $P(x,y)$ be any point on the parabola and equate the distance from P to the focus F and the distance from P to the directrix L. Using the distance formula,

$$\overline{PF} = \overline{PQ}$$

$$\sqrt{(x - 0)^2 + (y - p)^2} = \sqrt{(x - x)^2 + (y + p)^2}$$

Squaring both sides,

$$x^2 + y^2 - 2py + p^2 = y^2 + 2py + p^2$$

$$x^2 = 4py$$

If we treat $4p$ as a constant and substitute $a = 4p$, we have the general equation of a parabola.

$$x^2 = ay$$

Note that substituting $-x$ for x leaves the equation unchanged, verifying symmetry with respect to the y-axis.

When the parabola has the x-axis as its axis and the vertex is at the origin, the general equation of the parabola is

$$y^2 = ax$$

Note that substituting $-y$ for y leaves this equation unchanged, verifying symmetry with respect to the x-axis. Graphs of parabolas illustrating both types of symmetry are shown in Figure 7.

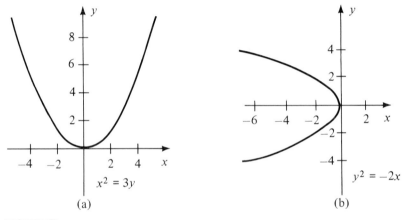

(a)   (b)

FIGURE 7

It is also possible to determine an equation of the parabola when the vertex is at some arbitrary point (h,k). The form of the equation depends upon whether the axis of the parabola is parallel to the x-axis or to the y-axis and is summarized in Table 1. Note that if the point (h,k) is the

TABLE 1

Standard Forms of the Equations of the Parabola			
Equation	Vertex	Axis	Direction
$(x - h)^2 = a(y - k)$	(h,k)	$x = h$	Up if $a > 0$ Down if $a < 0$
$(y - k)^2 = a(x - h)$	(h,k)	$y = k$	Right if $a > 0$ Left if $a < 0$

origin, then $h = k = 0$ and we arrive at the equations we derived previously, $x^2 = ay$ and $y^2 = ax$. Thus, in all cases, the sign of the constant a determines the direction in which the parabola opens. An equation of a parabola can always be written in the standard forms shown in Table 1 by completing the square, as in Example 2.

EXAMPLE 1

Determine the vertex, axis, and direction of the graph of the parabola

$$\left(x - \frac{1}{2}\right)^2 = -3(y + 4)$$

Solution

Comparison of the equation with the standard form

$$(x - h)^2 = a(y - k)$$

yields $h = \frac{1}{2}$, $k = -4$, $a = -3$. The axis of the parabola is always found by setting the square term equal to 0.

$$\left(x - \frac{1}{2}\right)^2 = 0$$

$$x = \frac{1}{2}$$

Thus the vertex is at $(h,k) = (\frac{1}{2}, -4)$, the axis is $x = \frac{1}{2}$, and the parabola opens downward since $a < 0$.

PROGRESS CHECK

Determine the vertex, axis, and direction of the graph of the parabola $3y^2 + 6y - 12x + 7 = 0$

Answer

vertex $\left(\frac{1}{3}, -1\right)$, axis $y = -1$, opens to the right

EXAMPLE 2

Locate the vertex and axis of symmetry of each of the given parabolas. Sketch the graph.

(a) $x^2 + 2x - 2y - 3 = 0$ (b) $y^2 - 4y + x + 1 = 0$

Solution

(a)   We complete the square in x;

$$x^2 + 2x = 2y + 3$$
$$x^2 + 2x + 1 = 2y + 3 + 1$$
$$(x + 1)^2 = 2(y + 2)$$

The vertex of the parabola is at $(-1,-2)$; the axis is $x = -1$. See Figure 8a.

(b)   We complete the square in y:

$$y^2 - 4y = -x - 1$$
$$y^2 - 4y + 4 = -x - 1 + 4$$
$$(y - 2)^2 = -(x - 3)$$

The vertex of the parabola is at $(3,2)$; the axis is $y = 2$. See Figure 8b.

PROGRESS CHECK

Write the equation of the parabola in standard form. Locate the vertex and the axis, and sketch the graph.

(a) $y^2 - 2y - 2x - 5 = 0$ (b) $x^2 - 2x + 2y - 1 = 0$

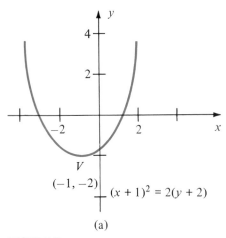

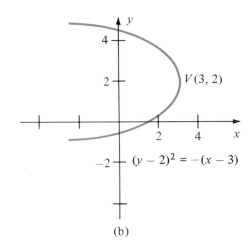

(a)   (b)

FIGURE 8

Answers
(a) $(y - 1)^2 = 2(x + 3)$; *vertex* $(-3,1)$; *axis* $y = 1$. *See Figure 9a.*
(b) $(x - 1)^2 = -2(y - 1)$; *vertex* $(1,1)$; *axis* $x = 1$. *See Figure 9b.*

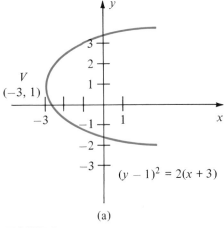

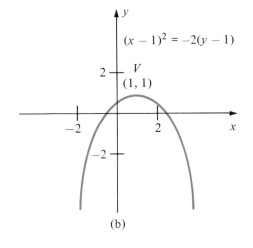

(a)   (b)

FIGURE 9

EXERCISE SET 5.3

Sketch the graph of the given equation.

1. $x^2 = 4y$ 2. $x^2 = -4y$ 3. $y^2 = 2x$

4. $y^2 = -\dfrac{3}{2}x$ 5. $x^2 + 5y = 0$ 6. $2y^2 - 3x = 0$

Determine the vertex, axis, and direction of each given parabola. Sketch the graph.

7. $(x - 2)^2 = 2(y + 1)$ 8. $(y - 2)^2 = -2(x + 1)$

9. $(x + 4)^2 = -\dfrac{1}{2}(y + 2)$ 10. $x^2 = \dfrac{1}{2}(y - 3)$

11. $y^2 = -2(x + 1)$ 12. $(y - 1)^2 = -3(x - 2)$

Determine the vertex, axis, and direction of each given parabola.

13. $x^2 - 2x - 3y + 7 = 0$ 14. $x^2 + 4x + 2y - 2 = 0$

15. $y^2 - 8y + 2x + 12 = 0$ 16. $y^2 + 6y - 3x + 12 = 0$

17. $x^2 - x + 3y + 1 = 0$ 18. $y^2 + 2y - 4x - 3 = 0$

19. $y^2 - 10y - 3x + 24 = 0$ 20. $x^2 + 2x - 5y - 19 = 0$

21. $x^2 - 3x - 3y + 1 = 0$ 22. $y^2 + 4y + x + 3 = 0$

23. $y^2 + 6y + \dfrac{1}{2}x + 7 = 0$ 24. $x^2 + 2x - 3y + 19 = 0$

25. $x^2 + 2x + 2y + 3 = 0$ 26. $y^2 - 6y + 2x + 17 = 0$

5.4
THE ELLIPSE AND HYPERBOLA

The geometric definition of an ellipse is as follows.

An **ellipse** is the set of all points the sum of whose distances from two fixed points is a constant.

The two fixed points are called the **foci** of the ellipse. The ellipse is in standard position if the two fixed points are on either the x-axis or the y-axis and are equidistant from the origin. Thus, if F_1 and F_2 are the foci of the ellipse in Figure 10 and P and Q are points on the ellipse, then $\overline{F_1P} + \overline{F_2P} = c$ and $\overline{F_1Q} + \overline{F_2Q} = c$, where c is a constant.

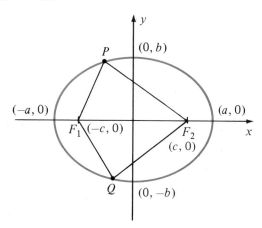

FIGURE 10

The equation of an ellipse in standard position can be shown to be as follows (see Exercise 35).

$$\frac{x^2}{a^2} + \frac{y^2}{b^2} = 1$$

This is called the **standard form of the equation of an ellipse**. Note that the equation indicates that the graph will be symmetric with respect to the x-axis, the y-axis, and the origin.

If we let $x = 0$ in the standard form, we find $y = \pm b$; if we let $y = 0$, we find $x = \pm a$. Thus, the ellipse whose equation is

$$\frac{x^2}{a^2} + \frac{y^2}{b^2} = 1$$

has intercepts $(\pm a, 0)$ and $(0, \pm b)$.

EXAMPLE 1
Find the intercepts and sketch the graph of the ellipse whose equation is

$$\frac{x^2}{16} + \frac{y^2}{9} = 1$$

Solution
The intercepts are found by first setting $x = 0$ and solving for y, and then by setting $y = 0$ and solving for x. Thus, the intercepts are $(\pm 4, 0)$ and $(0, \pm 3)$. The graph is then easily sketched (Figure 11).

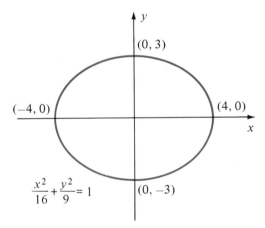

FIGURE 11

EXAMPLE 2
Write the equation of the ellipse in standard form and determine the intercepts.

(a) $4x^2 + 3y^2 = 12$ (b) $9x^2 + y^2 = 10$

Solution
(a)   Dividing by 12 to make the right-hand side equal to 1, we have

$$\frac{x^2}{3} + \frac{y^2}{4} = 1$$

The x-intercepts are $(\pm\sqrt{3}, 0)$; the y-intercepts are $(0, \pm 2)$.

(b)   Dividing by 10 we have

$$\frac{9x^2}{10} + \frac{y^2}{10} = 1$$

But this is *not* standard form. However, if we write

$$\frac{9x^2}{10} \quad \text{as} \quad \frac{x^2}{\frac{10}{9}}$$

then

$$\frac{x^2}{\frac{10}{9}} + \frac{y^2}{10} = 1$$

is the standard form of an ellipse. The intercepts are

$$\left(\frac{\pm\sqrt{10}}{3}, 0\right) \text{ and } (0, \pm\sqrt{10})$$

PROGRESS CHECK

Write the equation of each ellipse in standard form and determine the intercepts.

(a) $2x^2 + 3y^2 = 6$ (b) $3x^2 + y^2 = 5$

Answers

(a) $\frac{x^2}{3} + \frac{y^2}{2} = 1$ $(\pm\sqrt{3}, 0)$, $(0, \pm\sqrt{2})$

(b) $\frac{x^2}{\frac{5}{3}} + \frac{y^2}{5} = 1$ $\left(\frac{\pm\sqrt{15}}{3}, 0\right)$, $(0, \pm\sqrt{5})$

The hyperbola is the last of the conic sections we will study.

> A **hyperbola** is the set of all points the difference of whose distances from two fixed points is a constant.

The two fixed points are called the **foci** of the hyperbola. The hyperbola is in standard position if the two fixed points are on either the x-axis or the y-axis and are equidistant from the origin. The equations of the hyperbolas in standard position can be shown to be as follows (see Exercise 36).

$$\frac{x^2}{a^2} - \frac{y^2}{b^2} = 1 \qquad \text{Foci on the } x\text{-axis} \qquad (1)$$

$$\frac{y^2}{a^2} - \frac{x^2}{b^2} = 1 \qquad \text{Foci on the } y\text{-axis} \qquad (2)$$

These equations are called the **standard forms of the equations of a hyperbola**. These equations indicate that the graphs are symmetric with respect to the x-axis, the y-axis, and the origin.

Letting $y = 0$, we see that the x-intercepts of the graph of Equation (1) are $x = \pm a$. Letting $x = 0$, we find there are no y-intercepts since the equation $y^2 = -b^2$ has no real roots (Figure 12a). Similarly, the graph of Equation (2) has y-intercepts of $\pm a$ and no x-intercepts (Figure 12b).

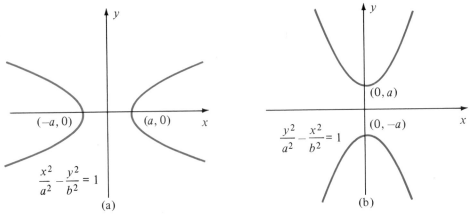

FIGURE 12

EXAMPLE 3
Find the intercepts and sketch the graph of each equation.

(a) $\dfrac{x^2}{9} - \dfrac{y^2}{4} = 1$ (b) $\dfrac{y^2}{4} - \dfrac{x^2}{3} = 1$

Solution
(a)   When $y = 0$, we have $x^2 = 9$ or $x = \pm 3$. The intercepts are $(3,0)$ and $(-3,0)$. With the assistance of a few plotted points, we can sketch the graph (Figure 13a)
(b)   When $x = 0$, we have $y^2 = 4$ or $y = \pm 2$. The intercepts are $(0,2)$ and $(0,-2)$. Plotting a few points, we can sketch the graph (Figure 13b).

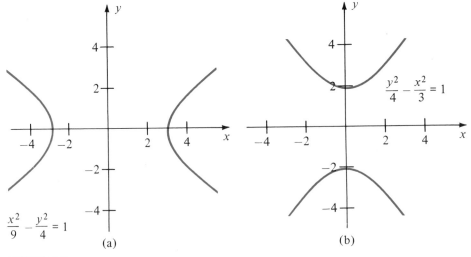

FIGURE 13

EXAMPLE 4
Write the equation of the hyperbola $9x^2 - 5y^2 = 10$ in standard form and determine the intercepts.

Solution
Dividing by 10 we have

$$\frac{9x^2}{10} - \frac{y^2}{2} = 1$$

Rewriting the equation in standard form, we have

$$\frac{x^2}{\frac{10}{9}} - \frac{y^2}{2} = 1$$

The x-intercepts are

$$\left(\frac{\pm\sqrt{10}}{3}, 0\right)$$

There are no y-intercepts.

PROGRESS CHECK
Write the equation of the hyperbola in standard form and determine the intercepts.

(a) $2x^2 - 5y^2 = 6$ (b) $4y^2 - x^2 = 5$

Answers

(a) $\frac{x^2}{3} - \frac{y^2}{\frac{6}{5}} = 1;$ $(\pm\sqrt{3}, 0)$ (b) $\frac{y^2}{\frac{5}{4}} - \frac{x^2}{5} = 1;$ $\left(0, \frac{\pm\sqrt{5}}{2}\right)$

There is a way of sketching the graph of a hyperbola without having to plot any auxiliary points on the curve itself. Given the equation of the hyperbola

$$\frac{x^2}{a^2} - \frac{y^2}{b^2} = 1$$

in standard form, we plot the four points $(a, \pm b)$, $(-a, \pm b)$ as in Figure 14

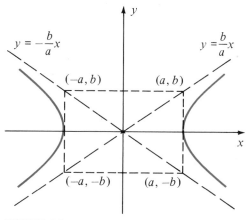

FIGURE 14

and draw the diagonals of the rectangle formed by the four points. The hyperbola opens from the intercepts $(\pm a, 0)$ and *approaches the lines formed by the diagonals of the rectangle.* We call these lines the **asymptotes** of the hyperbola. Since one asymptote passes through the points $(0,0)$ and (a,b), its equation is

$$y = \frac{b}{a} x$$

Similarly, the equation of the other asymptote is found to be

$$y = -\frac{b}{a} x$$

Of course, a similar discussion çan be carried out for the standard form

$$\frac{y^2}{a^2} - \frac{x^2}{b^2} = 1$$

In this case, the four points $(\pm b, \pm a)$ determine the rectangle, and the equations of the asymptotes are

$$y = \pm \frac{a}{b} x$$

EXAMPLE 5
Using asymptotes, sketch the graph of the equation

$$\frac{y^2}{4} - \frac{x^2}{9} = 1$$

Solution
The points $(\pm 3, \pm 2)$ form the vertices of the rectangle. See Figure 15. Using the fact that $(0, \pm 2)$ are intercepts, we can sketch the graph opening from these points and approaching the asymptotes.

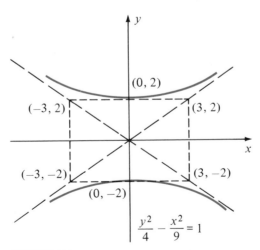

FIGURE 15

EXERCISE SET 5.4

Find the intercepts and sketch the graph of the ellipse.

1. $\dfrac{x^2}{25} + \dfrac{y^2}{4} = 1$

2. $\dfrac{x^2}{4} + \dfrac{y^2}{16} = 1$

3. $\dfrac{x^2}{8} + \dfrac{y^2}{4} = 1$

4. $\dfrac{x^2}{12} + \dfrac{y^2}{18} = 1$

5. $\dfrac{x^2}{16} + \dfrac{y^2}{25} = 1$

6. $\dfrac{x^2}{1} + \dfrac{y^2}{3} = 1$

Write the equation of the ellipse in standard form and determine its intercepts.

7. $4x^2 + 9y^2 = 36$ 8. $16x^2 + 9y^2 = 144$

9. $4x^2 + 16y^2 = 16$ 10. $25x^2 + 4y^2 = 100$

11. $4x^2 + 16y^2 = 4$ 12. $8x^2 + 4y^2 = 32$

13. $8x^2 + 6y^2 = 24$ 14. $5x^2 + 6y^2 = 50$

15. $36x^2 + 8y^2 = 9$ 16. $5x^2 + 4y^2 = 45$

Find the intercepts and sketch the graph of the hyperbola.

17. $\dfrac{x^2}{25} - \dfrac{y^2}{16} = -1$

18. $\dfrac{y^2}{9} - \dfrac{x^2}{4} = 1$

19. $\dfrac{x^2}{36} - \dfrac{y^2}{1} = 1$

20. $\dfrac{y^2}{49} - \dfrac{x^2}{25} = 1$

21. $\dfrac{x^2}{6} - \dfrac{y^2}{8} = 1$

22. $\dfrac{y^2}{8} - \dfrac{x^2}{10} = -1$

Write the equation of the hyperbola in standard form and determine its intercepts.

23. $16x^2 - y^2 = 64$ 24. $4x^2 - 25y^2 = 100$

25. $4y^2 - 4x^2 = 1$ 26. $2x^2 - 3y^2 = 6$

27. $4x^2 - 5y^2 = 20$ 28. $25y^2 - 16x^2 = 400$

In Exercises 29–34, use the asymptotes and intercepts of the hyperbola to sketch its graph.

29. $16x^2 - 9y^2 = 144$ 30. $16y^2 - 25x^2 = 400$

31. $9y^2 - 9x^2 = 1$ 32. $25x^2 - 9y^2 = 225$

33. $\dfrac{x^2}{25} - \dfrac{y^2}{36} = 1$ 34. $y^2 - 4x^2 = 4$

35. Derive the standard form of the equation of the ellipse from the geometric definition of an ellipse. (*Hint:* In Figure 10, let $P(x,y)$ be any point on the ellipse and let $F_1(-c,0)$ and $F_2(c,0)$ be the foci. Note that the point $B(a,0)$ lies on the ellipse and that $\overline{BF_1} + \overline{BF_2} = 2a$. Thus, the sum of the distances $\overline{PF_1} + \overline{PF_2}$ must also equal $2a$. Use the distance formula, simplify, and substitute $b^2 = a^2 - c^2$.)

36. Derive the standard form of the equation of the hyperbola from the geometric definition of a hyperbola. (*Hint:* Proceed in a manner similar to that of Exercise 35.)

5.5
IDENTIFYING THE CONIC SECTIONS

Each of the conic sections we have studied in this chapter has one or more axes of symmetry. We studied the circle and parabola when their axes of symmetry are the coordinate axes or lines parallel to them. Although our study of the ellipse and hyperbola was restricted to those that have the coordinate axes as their axes of symmetry, the same method of completing the square allows us to transform the **general equation of a conic section**

$$Ax^2 + Cy^2 + Dx + Ey + F = 0$$

into standard form. This transformation is very helpful in sketching the graph of the conic. It is easy, also, to identify the conic section from the general equation (see Table 2).

TABLE 2

$Ax^2 + Cy^2 + Dx$ $+ Ey + F = 0$	Conic Section	Remarks
$A = 0$ or $C = 0$	Parabola	Second degree in one variable, first degree in the other.
$A = C \, (\neq 0)$	Circle	Coefficients A and C are the same. *Caution:* Complete the square to obtain the standard form and check that radius $r > 0$.
$A \neq C$ $AC > 0$	Ellipse	A and C are unequal but have the same sign. *Caution:* Complete the square and check that the right-hand side is a positive constant.
$A \neq C$ $AC < 0$	Hyperbola	A and C have opposite signs.

EXAMPLE 1

Identify the conic section.

(a) $3x^2 + 3y^2 - 2y = 4$

Since the coefficients of x^2 and y^2 are the same, the graph will be a circle if the standard form yields $r > 0$. Completing the square, we have

$$3x^2 + 3\left(y - \frac{1}{3}\right)^2 = \frac{13}{3}$$

which is the equation of a circle.

(b) $3x^2 - 9y^2 + 2x - 4y = 7$

Since the coefficients of x^2 and y^2 are of opposite sign, the graph is a hyperbola.

(c) $2x^2 + 3y^2 - 2x + 6y + \frac{1}{2} = 0$

The coefficients of x^2 and y^2 are unequal but of like sign. We must complete the square in both x and y to obtain standard form.

$$2(x^2 - x) + 3(y^2 + 2y) = -\frac{1}{2}$$

$$2\left(x^2 - x + \frac{1}{4}\right) + 3(y^2 + 2y + 1) = -\frac{1}{2} + \frac{1}{2} + 3$$

$$2\left(x - \frac{1}{2}\right)^2 + 3(y + 1)^2 = 3$$

Since the right-hand side is positive, the graph is an ellipse.

(d) $3y^2 - 4x + 17y = -10$

The graph is a parabola since the equation is of the second degree in y and of first degree in x.

PROGRESS CHECK

Identify the conic section.

(a) $\dfrac{x^2}{5} - 3y^2 - 2x + 2y - 4 = 0$

(b) $x^2 - 2y - 3x = 2$

(c) $x^2 + y^2 - 4x - 6y = -11$

(d) $4x^2 + 3y^2 + 6x - 10 = 0$

Answers
(a) *hyperbola*   (b) *parabola*   (c) *circle*   (d) *ellipse*

A summary of the characteristics of the conic sections is given in Table 3.

TABLE 3

Curve and Standard Equation	Characteristics	Example
Circle $(x - h)^2 + (y - k)^2 = r^2$	Center: (h, k) Radius: r	$(x-2)^2 + (y + 4)^2 = 25$ Center: $(2,-4)$ Radius: 5
Parabola $(x - h)^2 = a(y - k)$ or	Vertex: (h, k) Axis: $x = h$ $a > 0$: Opens up $a < 0$: Opens down	$(x + 1)^2 = 2(y - 3)$ Vertex: $(-1,3)$ Axis: $x = -1$ Opens up
$(y - k)^2 = a(x - h)$	Vertex: (h, k) Axis: $y = k$ $a < 0$: Opens left $a > 0$: Opens right	$(y + 4)^2 = -3(x + 5)$ Vertex: $(-5, -4)$ Axis: $y = -4$ Opens left
Ellipse $\dfrac{x^2}{a^2} + \dfrac{y^2}{b^2} = 1$	Intercepts: $(\pm a,0)$, $(0,\pm b)$	$\dfrac{x^2}{4} + \dfrac{y^2}{6} = 1$ Intercepts: $(\pm 2,0)$, $(0,\pm \sqrt{6})$
Hyperbola $\dfrac{x^2}{a^2} - \dfrac{y^2}{b^2} = 1$ or	Intercepts: $(\pm a,0)$ Asymptotes: $y = \pm \dfrac{b}{a}x$ Opens to left and right	$\dfrac{x^2}{4} - \dfrac{y^2}{9} = 1$ Intercepts: $(\pm 2,0)$ Asymptotes: $y = \pm \dfrac{3}{2}x$ Opens to left and right
$\dfrac{y^2}{a^2} - \dfrac{x^2}{b^2} = 1$	Intercepts: $(0,\pm a)$ Asymptotes: $y = \pm \dfrac{a}{b}x$ Opens up and down	$\dfrac{y^2}{9} - \dfrac{x^2}{4} = 1$ Intercepts: $(0,\pm 3)$ Asymptotes: $y = \pm \dfrac{3}{2}x$ Opens up and down

EXERCISE SET 5.5

Identify the conic section.

1. $2x^2 + y - x + 3 = 0$
2. $4y^2 - x^2 + 2x - 3y + 5 = 0$
3. $4x^2 + 4y^2 - 2x + 3y - 4 = 0$
4. $3x^2 + 6y^2 - 2x + 8 = 0$
5. $36x^2 - 4y^2 + x - y + 2 = 0$
6. $x^2 + y^2 - 6x + 4y + 13 = 0$
7. $16x^2 + 4y^2 - 2y + 3 = 0$
8. $2y^2 - 3x + y + 4 = 0$
9. $x^2 + y^2 - 4x - 2y + 8 = 0$
10. $x^2 + y^2 - 2x - 2y + 6 = 0$
11. $4x^2 + 9y^2 - x + 2 = 0$
12. $3x^2 + 3y^2 - 3x + y = 0$
13. $4x^2 - 9y^2 + 2x + y + 3 = 0$
14. $x^2 + y^2 + 6x - 2y + 10 = 0$
15. $x^2 + y^2 - 4x + 4 = 0$
16. $4x^2 + y^2 = 32$

TERMS AND SYMBOLS

analytic geometry (p. 195)	parabola (p. 202)	standard form of the equation of an ellipse (p. 207)
midpoint formula (p. 197)	focus (p. 202)	
	directrix (p. 202)	hyperbola (p. 209)
conic sections (p. 199)	axis of a parabola (p. 203)	foci of a hyperbola (p. 209)
circle (p. 199)	vertex (p. 203)	standard forms of the equations of a hyperbola (p. 209)
center of a circle (p. 199)	standard forms of the equations of a parabola (p. 204)	
radius of a circle (p. 199)		asymptotes of a hyperbola (p. 212)
	ellipse (p. 207)	
standard form of the equation of a circle (p. 200)	foci of an ellipse (p. 207)	general equation of the conic sections (p. 213)
general form of the equation of a circle (p. 201)		

KEY IDEAS FOR REVIEW

■ The midpoint of the line segment joining the points $P_1(x_1,y_1)$ and $P_2(x_2,y_2)$ has coordinates

$$\left(\frac{x_1 + x_2}{2}, \frac{y_1 + y_2}{2} \right)$$

■ Theorems from plane geometry can be proven using the methods of analytic geometry. In general, place the given geometric figure in a convenient position relative to the origin and axes. The distance formula, the midpoint formula, and the computation of slope are the tools to apply in proving a theorem.

■ The conic sections represent the possible intersections of a plane and a cone. The conic sections are the circle, parabola, ellipse, and hyperbola. (In special cases these may reduce to a point, a line, or two lines.)

■ Each conic section has a geometric definition which can be used to derive a second-degree equation in two variables whose graph corresponds to the conic.

■ A second-degree equation in x and y can be converted from general form to standard form by completing the square in each variable. It is much simpler to sketch the graph of an equation when it is written in standard form.

■ It is often possible to distinguish the various conic sections even when the equation is given in general form.

REVIEW EXERCISES

In Exercises 1–3 find the midpoint of the line segment whose endpoints are given.

1. $(-5,4)$ $(3,-6)$ 2. $(-2,0), (-3,5)$ 3. $(2,-7), (-3,-2)$

4. Find the coordinates of the point P_2 if $(2,2)$ are the coordinates of the midpoint of the line segment joining $P_1(-6,-3)$ and P_2.

5. Use the distance formula to show that $P_1(-1,2)$, $P_2(4,3)$, $P_3(1,-1)$, and $P_4(-4,-2)$ are the coordinates of a parallelogram.

6. Show that the points $A(-8,4)$, $B(5,3)$, and $C(2,-2)$ are the vertices of a right triangle.

7. Find an equation of the perpendicular bisector of the line segment joining the points $A(-4,-3)$ and $B(1,3)$. (The perpendicular bisector passes through the midpoint of AB and is perpendicular to AB.)

8. Write an equation of the circle whose center is $(-5,2)$ and that has a radius of 4.

9. Write an equation of the circle whose center is $(-3,-3)$ and that has a radius of 2.

In Exercises 10–15 determine the center and radius of the circle with the given equation.

10. $(x - 2)^2 + (y + 3)^2 = 9$

11. $\left(x + \frac{1}{2}\right)^2 + (y - 4)^2 = \frac{1}{9}$

12. $x^2 + y^2 + 4x - 6y = -10$

13. $2x^2 + 2y^2 - 4x + 4y = -3$

14. $x^2 + y^2 - 6y + 3 = 0$

15. $x^2 + y^2 - 2x - 2y = 8$

In Exercises 16–17 determine the vertex and axis of each given parabola. Sketch the graph.

16. $(y + 5)^2 = 4\left(x - \frac{3}{2}\right)$

17. $(x - 1)^2 = 2 - y$

In Exercises 18–23 determine the vertex, axis, and direction of each given parabola.

18. $y^2 + 3x + 9 = 0$

19. $y^2 + 4y + x + 2 = 0$

20. $2x^2 - 12x - y + 16 = 0$

21. $x^2 + 4x + 2y + 5 = 0$

22. $y^2 - 2y - 4x + 1 = 0$

23. $x^2 + 6x + 4y + 9 = 0$

In Exercises 24–29 write the given equation in standard form and determine the intercepts.

24. $9x^2 - 4y^2 = 36$ 25. $9x^2 + y^2 = 9$ 26. $5x^2 + 7y^2 = 35$

27. $9x^2 - 16y^2 = 144$ 28. $3x^2 + 4y^2 = 9$ 29. $3y^2 - 5x^2 = 20$

In Exercises 30–31 use the intercepts and asymptotes of the hyperbola to sketch the graph.

30. $4x^2 - 4y^2 = 1$

31. $9y^2 - 4x^2 = 36$

In Exercises 32–35 identify the conic section.

32. $2y^2 + 6y - 3x + 2 = 0$

33. $6x^2 - 7y^2 - 5x + 6y = 0$

34. $2x^2 + y^2 + 12x - 2y + 17 = 0$

35. $9x^2 + 4y^2 = -36$

PROGRESS TEST 5A

1. Find the midpoint of the line segment whose endpoints are $(2,4)$ and $(-2,4)$.

2. Find the coordinates of the point P if $(-3,3)$ are the coordinates of the midpoint of the line segment joining P and $Q(-5,4)$.

3. By using the slope of a line, show that the points $A(-3,-1)$, $B(-5,4)$, $C(2,6)$, and $D(4,1)$ determine a parallelogram.

4. Write an equation of the circle of radius 6 whose center is at $(2,-3)$.

In Problems 5–6 determine the center and radius of the circle.

5. $x^2 + y^2 - 2x + 4y = -1$ 6. $x^2 - 4x + y^2 = 1$

In Problems 7–8 determine the vertex and axis of the parabola. Sketch the graph.

7. $x^2 + 6x + 2y + 7 = 0$ 8. $y^2 - 4x - 4y + 8 = 0$

In Problems 9–10 determine the vertex, axis, and direction of the parabola.

9. $x^2 - 6x + 2y + 5 = 0$ 10. $y^2 + 8y - x + 14 = 0$

In Problems 11–13 write the given equation in standard form and determine the intercepts.

11. $x^2 + 4y^2 = 4$ 12. $4y^2 - 9x^2 = 36$ 13. $4x^2 - 4y^2 = 1$

14. Use the intercepts and asymptotes of the hyperbola $9x^2 - y^2 = 9$ to sketch its graph.

In Problems 15–16 identify the conic section.

15. $x^2 + y^2 + 2x - 2y - 2 = 0$ 16. $x^2 + 9y^2 - 4x + 6y + 4 = 0$

PROGRESS TEST 5B

1. Find the midpoint of the line segment whose endpoints are $(-5,-3)$ and $(4,1)$.

2. Find the coordinates of the point A if $(-2, -1/2)$ are the coordinates of the midpoint of the line segment joining A and $B(3,-2)$.

3. Show that the diagonals of the quadrilateral whose vertices are $P(-3,1)$, $Q(-1,4)$, $R(5,0)$, and $S(3,-3)$ are equal.

4. Write an equation of the circle of radius 5 whose center is at $(-2,-5)$.

In Problems 5–6 determine the center and radius of the circle.

5. $x^2 + y^2 + 6x - 4y = -4$ 6. $4x^2 + 4y^2 - 4x - 8y = 35$

In Problems 7–8 determine the vertex and axis of the parabola. Sketch the graph.

7. $2y^2 - 8y - x + 30 = 0$ 8. $9x^2 + 18y - 6x + 7 = 0$

In Problems 9–10 determine the vertex, axis, and direction of the parabola.

9. $y^2 - 4y - 3x + 1 = 0$ 10. $4x^2 - 4x - 8y - 23 = 0$

In Problems 11–13 write the given equation in standard form and determine the intercepts.

11. $5x^2 + 9y^2 = 25$ 12. $7x^2 + 6y^2 = 21$ 13. $y^2 - 3x^2 = 9$

14. Use the intercepts and asymptotes of the hyperbola $4y^2 - x^2 = 1$ to sketch its graph.

In Problems 15–16 identify the conic section.

15. $5y^2 - 4x^2 - 6x + 2 = 0$ 16. $3x^2 - 5x + 6y = 3$

CHAPTER SIX
SYSTEMS OF EQUATIONS AND INEQUALITIES

Many problems in business and engineering require the solution of systems of equations or inequalities. In fact, systems of linear equations and inequalities occur with such frequency that mathematicians and computer scientists have devoted considerable energy to devising methods for their solution. With the aid of large-scale computers it is possible to solve systems involving thousands of equations or inequalities, a task that previous generations would not have dared tackle.

In this chapter we will study some of the basic methods for solving systems of linear equations. We will investigate both Gaussian elimination and Gauss-Jordan elimination, methods that can be extended to systems of linear equations of any size. We will also demonstrate that similar techniques can be applied to systems of equations that include at least one second-degree equation. Although algebraic means for solving systems of linear inequalities are beyond the scope of this book, we will illustrate a graphical technique.

6.1
SYSTEMS OF LINEAR EQUATIONS

A pile of nine coins consists of nickels and quarters. If the total value of the coins is $1.25, how many of each type of coin are there?

This type of word problem was handled in earlier chapters by using one variable. Now we tackle this problem by using two variables. If we let

$$x = \text{the number of nickels}$$

and

$$y = \text{the number of quarters}$$

then the requirements can be expressed as

$$x + y = 9$$
$$5x + 25y = 125$$

We call this a **system of linear equations** or simply a **linear system,** and we seek values of x and y that satisfy *both* equations. A pair of values $x = r$, $y = s$ that satisfies both equations is called a **solution** of the system, and a system that has a solution is said to be **consistent.** Thus,

$$x = 5, \quad y = 4$$

is a solution because substituting in the equations of the system gives

$$5 + 4 = 9$$
$$5(5) + 25(4) = 125$$

However,

$$x = 6, \quad y = 3$$

is not a solution, since

$$6 + 3 = 9$$

but

$$5(6) + 25(3) \neq 125$$

We will devote most of this chapter to methods of solving linear systems.

SOLVING BY GRAPHING

We know that the graph of a linear equation is a straight line. Returning to the linear system

$$x + y = 9$$
$$5x + 25y = 125$$

we can graph both lines on the same set of coordinate axes, as in Figure 1. The coordinates of every point on line L_1 satisfy the equation $x + y = 9$, while the coordinates of every point on line L_2 satisfy the equation $5x + 25y = 125$. Therefore the coordinates $(5,4)$ of the point where the lines

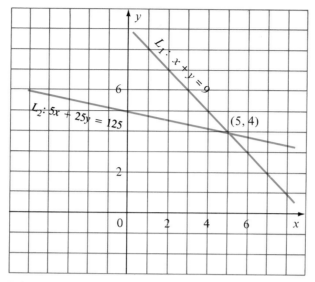

FIGURE 1

intersect must satisfy *both* equations, and hence these coordinates are a solution of the linear system. In general,

> If the straight lines obtained from a system of linear equations intersect, then the coordinates of the point of intersection are a solution of the system.

PROGRESS CHECK

Solve the linear system by graphing, and verify the solution algebraically.

$$2x - 3y = 12$$
$$3x - 2y = 13$$

Answer
x = 3,   y = −2

When we graph two linear equations on the same set of coordinate axes, there are three possibilities.

(a) The two lines intersect in a point (Figure 2a). The system is consistent and has a unique solution, the point of intersection.

(b) The two lines are parallel (Figure 2b). Since the lines do not intersect, the linear system has no solution and is said to be **inconsistent.**

(c) The equations are different forms of the same line (Figure 2c). The system is consistent and has an infinite number of solutions, namely, any point on the line.

The method of graphing has severe limitations since the accuracy of the solution depends on the accuracy of the graph. The algebraic methods that follow avoid this limitation.

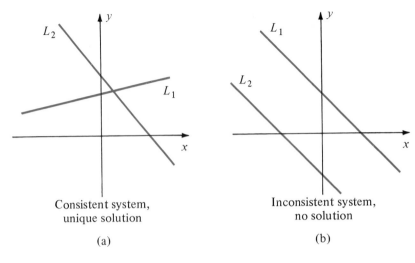

Consistent system,
unique solution

(a)

Inconsistent system,
no solution

(b)

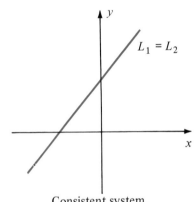

Consistent system,
infinite number of solutions

(c)

FIGURE 2

SOLVING BY SUBSTITUTION

We demonstrate the method of substitution by an example.

EXAMPLE 1
Solve the system by substitution.

$$x - 2y = -10$$
$$2x + 3y = 8$$

Solution
It is easy to solve the first equation for x.

$$x = 2y - 10$$

We now *substitute* this expression for x in the second equation.

$$2x + 3y = 8$$
$$2(2y - 10) + 3y = 8$$

This last equation is a linear equation in one variable, which we know how to solve.

$$4y - 20 + 3y = 8$$

$$7y = 28$$

$$y = 4$$

We can now substitute $y = 4$ in either of the original equations to find x.

$$x - 2y = 10$$

$$x - 2(4) = -10$$

$$x = -2$$

You can verify that $x = -2$, $y = 4$ is a solution of the linear system.

PROGRESS CHECK
Solve by substitution.

(a) $2x - y = 10$
     $3x - 2y = 14$

(b) $3x - 2y = -19$
     $2x + 3y = -4$

Answers
(a) $x = 6$, $y = 2$

(b) $x = -5$, $y = 2$

The following example illustrates the method of substitution applied to linear systems that represent parallel lines or that reduce to the same line.

EXAMPLE 2
Solve by substitution.

(a) $x - 3y = 5$
     $2x - 6y = 20$

(b) $x + 4y = 10$
     $-2x - 8y = -20$

Solution
We apply the method of substitution to each system.

Parallel Lines	Same Line
$x - 3y = 5$	$x + 4y = 10$
$2x - 6y = 20$	$-2x - 8y = -20$
Solving the first equation for x,	Solving the first equation for x,
$x = 5 + 3y$	$x = 10 - 4y$
and substituting in the second equation,	and substituting in the second equation,
$2(5 + 3y) - 6y = 20$	$-2(10 - 4y) - 8y = -20$
$10 + 6y - 6y = 20$	$-20 + 8y - 8y = -20$
$10 = 20$	$-20 = -20$
Impossible!	Always true!
The system is satisified by *no* values of x and y.	All solutions of one equation will also satisfy the other equation.

The results illustrated by Example 2 can be summarized as follows.

> If the method of substitution results in an impossible equation, then there is no solution of the system of linear equations, and the system is inconsistent.
>
> If the method of substitution results in an identity, then there are an infinite number of solutions and the original equations are really different forms of the same equation.

PROGRESS CHECK
Solve by substitution.

(a) $3x - y = 7$

$-9x + 3y = -22$

(b) $-5x + 2y = -4$

$\dfrac{5}{2}x - y = 2$

Answers
(a) *no solution*   (b) *any point on the line* $-5x + 2y = -4$

WARNING
The expression for x or y obtained from an equation *must not be substituted into the same equation*. From the first equation of the linear system

$$x + 2y = -1$$
$$3x + y = 2$$

we obtain

$$x = -1 - 2y$$

Subsituting *(incorrectly)* into the same equation would result in

$$(-1 - 2y) + 2y = -1$$
$$-1 = -1$$

The substitution $x = -1 - 2y$ must be made in the *second* equation.

EXERCISE SET 6.1
Find an approximate solution to the given linear system by graphing.

1. $x + y = 1$
    $x - y = 3$

2. $x - y = 1$
    $x + y = 5$

3. $3x - y = 4$
    $6x - 2y = -8$

4. $x + 2y = 8$
    $3x - 4y = 4$

5. $x + 3y = -2$
    $3x - 5y = 8$

6. $2x - y = -1$
    $3x - y = -1$

7. $2x + 5y = -6$
    $4x + 10y = -12$

8. $3x + 2y = 10$
    $-9x - 6y = 8$

Solve the given linear system by the method of substitution and check your

11. $x + 3y = 3$
    $x - 5y = -1$

12. $x - 4y = -7$
    $2x - 8y = -4$

13. $x + 6y = 4$
    $3x - 4y = 1$

14. $3x - y = -9$
    $2x - y = -7$

15. $3x + 3y = 9$
    $2x + 2y = -6$

16. $2x + 3y = -2$
    $-3x - 5y = 4$

17. $3x - y = 18$
    $\dfrac{3}{2} x - \dfrac{1}{2} y = 9$

18. $2x + y = 6$
    $x + \dfrac{1}{2} y = 3$

6.2
SOLVING BY ELIMINATION

When we solve a system of linear equations by graphing, we must estimate the coordinates of the point of intersection. If we require the answers to be accurate to, say, five decimal places, it is clear that graphing won't suffice. The method of substitution provides us with exact answers but suffers from the disadvantage that it is difficult to program for use in a digital computer.

The **method of elimination** overcomes these difficulties. The strategy of the method is to obtain an equation in just one variable that is easily solved. The procedure is illustrated in the following example.

EXAMPLE 1
Solve by elimination.

$$3x + y = 7$$
$$2x - 4y = 14$$

Solution

Solving by Elimination	Example
Step 1. Multiply each equation by a constant so that the coefficients of either x or y will differ only in sign.	*Step 1.* Multiply the first equation by 4 and the second equation by 1 so that the coefficients of y will be 4 and -4. $12x + 4y = 28$ $2x - 4y = 14$
Step 2. Add the equations. The resulting equation will contain (at most) one variable.	*Step 2.* $12x + 4y = 28$ $\underline{2x - 4y = 14}$ $14x \qquad = 42$
Step 3. Solve the resulting linear equation in one variable.	*Step 3.* $14x = 42$ $x = 3$

Step 4. Substitute in either of the *original* equations to solve for the second variable.	Step 4. Substituting $x = 3$ in the first equation of the original system, $$3x + y = 7$$ $$3(3) + y = 7$$ $$y = -2$$
Step 5. Check in both equations.	Step 5. $$\begin{array}{cc} 3x + y = 7 & 2x - 4y = 14 \\ 3(3) + (-2) \stackrel{?}{=} 7 & 2(3) - 4(-2) \stackrel{?}{=} 14 \\ 7 \stackrel{\checkmark}{=} 7 & 14 \stackrel{\checkmark}{=} 14 \end{array}$$

Note that in Step 2 we have "eliminated" y, which is why we call this the method of elimination.

PROGRESS CHECK

Solve by elimination.

(a) $2x - y = 2$
      $-4x - 2y = 8$

(b) $5x + 2y = 5$
      $-2x + 3y = -21$

Answers

(a) $x = -\dfrac{1}{2}, \quad y = -3$

(b) $x = 3, \quad y = -5$

In the first section of this chapter we saw that the graphs of the equations

$$x - 3y = 5$$

$$2x - 6y = 20$$

are parallel lines and that the system of equations has no solution. If we attempt to solve this system by elimination, we can multiply the first equation by -2 and then add to eliminate x.

$$\begin{array}{l} -2x + 6y = -10 \\ \underline{2x - 6y = 20} \\ 0x + 0y = 10 \qquad \text{Impossible.} \end{array}$$

We see that the elimination method signals us that there is no solution by yielding an impossible equation.

If the elimination method results in an equation of the form

$$0x + 0y = c$$

where $c \neq 0$, then there is no solution of the original system of linear equations, and the system is inconsistent.

We also saw in the first section of this chapter that the system of equations

$$x + 4y = 10$$

$$-2x - 8y = -20$$

consists of two forms of the same equation. We can eliminate x by multiplying the first equation by 2 and adding.

$$\begin{array}{r} 2x + 8y = 20 \\ -2x - 8y = -20 \\ \hline 0x + 0y = 0 \end{array} \qquad \text{Always true.}$$

The elimination method has signaled us that any point on the line is a solution.

> If the elimination method results in an equation of the form
>
> $$0x + 0y = 0$$
>
> then the equations in the original system are two forms of the same equation.

   You may have noticed that the special cases are signaled in essentially the same way by both the method of elimination and the method of substitution. The case of parallel lines (no solution) is signaled by an impossible equation; the case of the same line is signaled by an identity.

EXAMPLE 2
Solve by elimination.

(a) $6x - 2y = 7$

   $3x - y = 16$

(b) $5x + 6y = 4$

   $-10x - 12y = -8$

Solution
(a)   If we multiply the second equation by -2, we have

$$\begin{array}{r} 6x - 2y = 7 \\ -6x + 2y = -32 \\ \hline 0x + 0y = 25 \end{array}$$

   We conclude that the system of equations represents a pair of parallel lines and that the system is inconsistent.

(b)   Multiplying the first equation by 2, we have

$$\begin{array}{r} 10x + 12y = 8 \\ -10x - 12y = -8 \\ \hline 0x + 0y = 0 \end{array}$$

We conclude that the equations represent the same line and that the solution set consists of all points on the line $5x + 6y = 4$.

PROGRESS CHECK
Solve by elimination.

(a) $x - y = 2$

   $3x - 3y = -6$

(b) $4x + 6y = 3$

   $-2x - 3y = -\dfrac{3}{2}$

Answers

(a) no solution   *(b)   all points on the line $4x + 6y = 3$*

EXERCISE SET 6.2

In Exercises 1–14 solve by elimination and check the answer.

1. $x + y = -1$
    $x - y = 3$

2. $x - 2y = 8$
    $2x + y = 1$

3. $x + 4y = -1$
    $2x - 4y = 4$

4. $2x - 2y = 4$
    $x - y = 8$

5. $x + 2y = 6$
    $2x + 4y = 12$

6. $x - 2y = 4$
    $2x + y = 3$

7. $x + 3y = 2$
    $3x - 5y = -6$

8. $x + 2y = 0$
    $5x - y = 22$

9. $2x + 2y = 6$
    $3x + 3y = 6$

10. $2x + y = 2$
    $3x - y = 8$

11. $x + 2y = 1$
    $5x + 2y = 13$

12. $x - 4y = -7$
    $2x + 3y = -8$

13. $x - 3y = 9$
    $x + 5y = 11$

14. $2x - 3y = 8$
    $4x - 6y = 16$

In Exercises 15–20 use a pair of linear equations to solve the given problem.

15. A pile of 34 coins worth $4.10 consists of nickels and quarters. Find the number of each type of coin.

16. Car A can travel 20 kilometers per hour faster than car B. If car A travels 240 kilometers in the same time that car B travels 200 kilometers, what is the speed of each car?

17. How many pounds of nuts worth $2.10 per pound and how many pounds of raisins worth $0.90 per pound must be mixed to obtain a mixture of two pounds that is worth $1.62 per pound?

18. A part of $8000 was invested at an annual interest of 7% and the remainder at 8%. If the total interest received at the end of one year is $590, how much was invested at each rate?

19. The owner of a service station sold 1325 gallons of gasoline and collected 200 ration tickets. If Type-A ration tickets are used to purchase 10 gallons of gasoline and Type-B are used to purchase 1 gallon of gasoline, how many of each type of ration ticket did the station collect?

20. A bank is paying 12% annual interest on one-year certificates, and treasury notes are paying 10% annual interest. An investor received $620 interest at the end of one year by investing a total of $6000. How much was invested at each rate?

6.3
APPLICATIONS

In the first section of this chapter we saw that many of the word problems that we previously solved by using one variable can be recast as a system of linear equations. There are, in addition, many word problems that are difficult to handle with one variable but are easily formulated by using two variables.

EXAMPLE 1

If 3 sulfa pills and 4 penicillin pills cost 69 cents, while 5 sulfa pills and 2 penicillin pills cost 73 cents, what is the cost of each type of pill?

Solution

Using two variables, we let

$$x = \text{the cost of each sulfa pill}$$

$$y = \text{the cost of each penicillin pill}$$

Then

$$3x + 4y = 69$$

$$5x + 2y = 73$$

Multiplying the second equation by -2 and adding to eliminate y,

$$
\begin{array}{rcl}
3x + 4y & = & 69 \\
-10x - 4y & = & -146 \\
\hline
7x & = & -77 \\
\end{array}
$$

$$\boxed{x = 11}$$

Substituting in the first equation,

$$3(11) + 4y = 69$$

$$4y = 36$$

$$\boxed{y = 9}$$

Thus, each sulfa pill costs 11 cents and each penicillin pill costs 9 cents. (Could you have set up this problem using only one variable? Unlikely!)

EXAMPLE 2

Swimming downstream, a swimmer can cover 2 kilometers in 15 minutes. The return trip upstream requires 20 minutes. What is the rate of the swimmer and of the current in km per hour? (The rate of the swimmer is the speed at which he would swim if there were no current.)

Solution

Let

$$x = \text{the rate of the swimmer (in km per hour)}$$

$$y = \text{the rate of the current (in km per hour)}$$

When swimming downstream, the rate of the current is added to the rate of the swimmer so that $x + y$ is the rate downstream. Similarly, $x - y$ is the rate while swimming upstream. We display the information we have, expressing time in hours.

	Rate	$\times$	Time	$=$	Distance
Downstream	$x + y$		$\dfrac{1}{4}$		$\dfrac{1}{4}(x + y)$
Upstream	$x - y$		$\dfrac{1}{3}$		$\dfrac{1}{3}(x - y)$

Since distance upstream = distance downstream = 2 kilometers,

$$\frac{1}{4}(x + y) = 2$$

$$\frac{1}{3}(x - y) = 2$$

or, equivalently,

$$x + y = 8$$
$$x - y = 6$$

Solving, we have

$$x = 7 \qquad \text{rate of the swimmer}$$
$$y = 1 \qquad \text{rate of the current}$$

Thus, the rate of the swimmer is 7 kilometers per hour and the rate of the current is 1 kilometer per hour. (The student is urged to verify the solution.)

EXAMPLE 3

The sum of a two-digit number and its units digit is 64, and the sum of the number and its tens digit is 62. Find the number.

Solution

The basic idea in solving digit problems is to note that if we let

$$t = \text{tens digit}$$

and

$$u = \text{units digit}$$

then

$$10t + u = \text{the two-digit number}$$

Then "the sum of a two-digit number and its units digit is 64" translates into

$$(10t + u) + u = 64 \quad \text{or} \quad 10t + 2u = 64$$

Also, "the sum of the number and its tens digit is 62" becomes

$$(10t + u) + t = 62 \quad \text{or} \quad 11t + u = 62$$

Solving, we find that $t = 5$, $u = 7$ (verify) and the number we seek is 57.

APPLICATIONS IN BUSINESS AND ECONOMICS: BREAK-EVEN ANALYSIS

An important problem faced by a manufacturer is that of determining the **level of production,** that is, the number of units of the product to be manufactured during a given time period such as a day, a week, or a month. Suppose that

$$C = 400 + 2x \tag{1}$$

is the total cost (in thousands of dollars) of producing x units of the product and that

$$R = 4x \tag{2}$$

is the total revenue (in thousands of dollars) when x units of the product are sold. This would happen, for example, if after setting up production at a cost of $400,000 there is an additional cost of $2000 to make each unit (Equation (1)) and a revenue of $4000 is earned from the sale of each unit (Equation (2)). If all units that are manufactured are sold, then the total profit P is the difference between total revenue and total cost.

$$\begin{aligned} P &= R - C \\ &= 4x - (400 + 2x) \\ &= 2x - 400 \end{aligned}$$

The value of x for which $R = C$, so that the profit is zero, is called the **break-even point.** When that many units of the product have been produced and sold the manufacturer neither makes money nor loses money. To find the break-even point we set $R = C$. Using Equations (1) and (2) we obtain

$$400 + 2x = 4x$$

$$x = 200$$

Thus, the break-even point is 200 units.

The break-even point can also be obtained graphically as follows. Observe that Equations (1) and (2) are linear equations and therefore equations of straight lines. The break-even point is the x-coordinate of the point where the two lines intersect. Figure 3 shows the lines and their point of

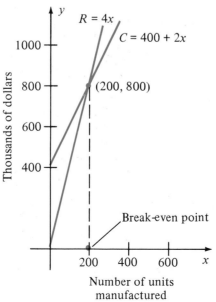

FIGURE 3

intersection (200, 800). When 200 units of the product are made, the cost ($800,000) is exactly equal to the revenue, and the profit is $0. If $x > 200$, then $R > C$, so that the manufacturer is making a profit. If $x < 200$, $R < C$ and the manufacturer is losing money.

PROGRESS CHECK

A producer of photographic developer finds that the total weekly cost of producing x liters of developer is given (in dollars) by $C = 550 + 0.40x$. The manufacturer sells the product at $0.50 per liter.
(a)   What is the total revenue received when x liters of developer are sold?
(b)   Find the break-even point graphically.
(c)   What is the total revenue received at the break-even point?

Answers
(a) $R = 0.50x$ (b)   *5500 liters*   (c)   *$2750*

APPLICATIONS IN BUSINESS AND ECONOMICS: SUPPLY AND DEMAND

A manufacturer of a product is free to set any price p (in dollars) for each unit of the product. Of course, if the price is too high, not enough people will buy the product; if the price is too low, so many people will rush to buy the product that the producer will not be able to satisfy demand. Thus, in setting price, the manufacturer must take into consideration the demand for the product.

Let S be the number of units that the manufacturer is willing to supply at the price p; S is called the **supply.** Generally, the value of S will increase as p increases; that is, the manufacturer is willing to supply more of the product as the price p increases. Let D be the number of units of the product that consumers are willing to buy at the price p; D is called the **demand.** Generally, the value of D will decrease as p increases; that is, consumers are willing to buy fewer units of the product as the price rises. For example, suppose that S and D are given by

$$S = 2p + 3 \tag{3}$$

$$D = -p + 12 \tag{4}$$

Equations (3) and (4) are linear equations, so they are equations of straight lines (see Figure 4). The price at which supply S and demand D are equal is called the **equilibrium price.** At this price, every unit that is supplied is purchased. Thus there is neither a surplus nor a shortage. In Figure 4 the equilibrium price is $p = 3$. At this price, the number of units supplied equals the number of units demanded and is given by substituting in Equation (3): $S = 2(3) + 3 = 9$. This value can also be obtained by finding the ordinate at the point of intersection in Figure 4.

If we are in an economic system in which there is pure competition, then the law of supply and demand states that the selling price of a product will be its equilibrium price. That is, if the selling price were higher than the equilibrium price, then the consumers' reduced demand would leave the manufacturer with an unsold surplus. This would force the manufacturer to reduce the selling price. If the selling price is below the equilibrium

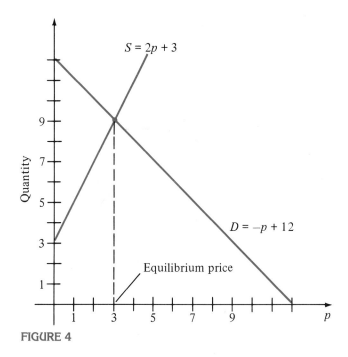

FIGURE 4

price, then the increased demand would cause a shortage of the product. This would lead the manufacturer to raise the selling price. Of course, in actual practice, the marketplace does not operate under pure competition. Also, deeper mathematical analysis of economic systems requires the use of more sophisticated equations.

EXAMPLE 4
Suppose that the supply and demand for ball-point pens are given by

$$S = p + 5$$
$$D = -p + 7$$

(a)   Find the equilibrium price.
(b)   Find the number of pens sold at that price.

Solution
(a)   Figure 5 illustrates the graphical solution. Thus, the equilibrium price is $p = 1$. Algebraic methods will, of course, yield the same solution.
(b)   When $p = 1$, the number of pens sold is $S = 1 + 5 = 6$, the value of the ordinate at the point of intersection.

PROGRESS CHECK
Suppose that the supply and demand for radios are given by

$$S = 3p + 120$$
$$D = -p + 200$$

(a)   Find the equilibrium price.
(b)   Find the number of radios sold at this price.

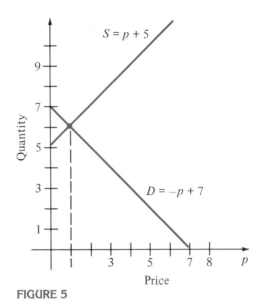

FIGURE 5

Answers
(a)   *20*   *(b)*   *180*

EXERCISE SET 6.3

1. A pile of 40 coins consists of nickels and dimes. If the value of the coins is $2.75, how many of each type of coin are there?
2. An automatic vending machine in the post office, which charges no more than a clerk, provides a packet of 27 ten-cent and twenty-cent stamps worth $3.00. How many of each type of stamp are there?
3. A photography store sells sampler A, consisting of six rolls of color film and four rolls of black and white film for $21.00. It also sells sampler B, consisting of four rolls of color film and six rolls of black and white film for $19.00. What is the cost per roll of each type of film?
4. A hardware store sells power pack A, consisting of four D cells and two C cells for $1.70, and power pack B, consisting of six D cells and four C cells for $2.80. What is the price of each cell?
5. A fund manager invested $6000 in two types of bonds, A and B. Bond A is safer than bond B and pays annual interest of 8 percent, while bond B pays annual interest of 10 percent. If the total annual return of both investments is $520, how much was invested in each type of bond?
6. A trash removal company carries waste material in sealed containers weighing 4 and 3 kilograms, respectively. On a certain trip there are a total of 30 containers weighing 100 kilograms. How many of each type of container are there?
7. A paper firm makes rolls of paper 12″ wide and 15″ wide by cutting a sheet that is 180″ wide. Suppose that a total of 14 rolls of paper are to be cut without any waste. How many of each type of roll will be made?
8. An animal-feed producer mixes two types of grain, A and B. Each unit of grain A contains 2 grams of fat and 80 calories, and each unit of grain B contains 3 grams of fat and 60 calories. If the producer wants the final product to provide 18 grams of fat and 480 calories, how much of each type of grain should be used?

9. A supermarket mixes coffee which sells for $1.20 per pound with coffee selling for $1.80 per pound to obtain 24 pounds of coffee selling for $1.60 per pound. How much of each type of coffee should be used?

10. An airplane flying against the wind covers a distance of 3000 kilometers in 6 hours. The return trip with the aid of the wind takes 5 hours. What is the speed of the airplane in still air and what is the speed of the wind?

11. A cyclist who travels against the wind can cover a distance of 45 miles in 4 hours. The return trip with the aid of the wind takes 3 hours. What is the speed of the cyclist in still air and what is the speed of the wind?

12. The sum of a two-digit number and its units digit is 20 and the sum of the number and its tens digit is 16. Find the number.

13. The sum of the digits of a two-digit number is 7. If the digits are reversed, the resulting number exceeds the given number by 9. Find the number.

14. The sum of three times the tens digit and the units digit of a two-digit number is 14, and the sum of the tens digit and twice the units digit is 18. Find the number.

15. A health food shop mixes nuts and raisins into a snack pack. How many pounds of nuts, selling for $2 per pound, and how many pounds of raisins, selling for $1.50 per pound, must be mixed to produce a 50-pound mixture selling for $1.80 per pound?

16. A movie theatre charges $3 admission for an adult and $1.50 for a child. If 600 tickets were sold and the total revenue received was $1350, how many tickets of each type were sold?

17. A moped dealer selling a model A and a model B moped has $18,000 in inventory. The profit on selling a model A moped is 12%, while the profit on a model B moped is 18%. If the profit on the entire stock would be 16%, how much was invested in each type of model?

18. The cost of sending a telegram is determined as follows. There is a flat charge for the first 10 words and a uniform rate for each additional word. Suppose that an 18-word telegram costs $1.94 and a 22-word telegram costs $2.16. Find the cost of the first 10 words and the rate for each additional word.

19. A certain epidemic disease is treated by a combination of the drugs Epiline I and Epiline II. Suppose that each unit of Epiline I contains 1 milligram of factor X and 2 milligrams of factor Y, while each unit of Epiline II contains 2 milligrams of factor X and 3 milligrams of factor Y. Successful treatment of the disease calls for 13 milligrams of factor X and 22 milligrams of factor Y. How many milligrams of Epiline I and Epiline II should be administered to a patient?

20. (**Break-even analysis**) An animal feed manufacturer finds that the weekly cost of making x kilograms of feed is given (in dollars) by $C = 2000 + 0.50x$, and the revenue received from selling the feed is given by $R = 0.75x$.
    (a) Find the break-even point graphically.
    (b) What is the total revenue at the break-even point?

21. (**Break-even analysis**) A small manufacturer of a new solar device finds that the annual cost of making x units is given (in dollars) by $C = 24,000 + 55x$. Each device sells for $95.
    (a) What is the total revenue received when x devices are sold?
    (b) Find the break-even point graphically.
    (c) What is the total revenue received at the break-even point?

22. (**Supply and demand**) A manufacturer of calculators finds that the supply and demand are given by

$$S = 0.5p + 0.5$$
$$D = -2p + 8$$

    (a) Find the equilibrium price.

    (b) What is the number of calculators sold at this price?

23. **(Supply and demand)** A manufacturer of mopeds finds that the supply and demand are given by

$$S = 2p + 10$$

$$D = -p + 22$$

    (a) Find the equilibrium price.

    (b) What is the number of mopeds sold at this price?

6.4
SYSTEMS OF LINEAR EQUATIONS IN THREE UNKNOWNS

The method of substitution and the method of elimination can both be applied to systems of linear equations in three unknowns and, more generally, to systems of linear equations in any number of unknowns. There is yet another method known as **Gaussian elimination** that is ideally suited for computers and which we will now apply to solving linear systems in three unknowns.

The objective of Gaussian elimination is to transform a given linear system into triangular form such as

$$3x - y + 3z = -11$$

$$2y + z = 2$$

$$2z = -4$$

A linear system is in **triangular form** when the only nonzero coefficient of x appears in the first equation, the only nonzero coefficients of y appear in the first and second equations, and so on.

Note that when a linear system is in triangular form, the last equation immediately yields the value of an unknown. In our example, we see that

$$2z = -4$$

$$z = -2$$

Substituting $z = -2$ in the second equation yields

$$2y + (-2) = 2$$

$$y = 2$$

Finally, substituting $z = -2$ and $y = 2$ in the first equation yields

$$3x - (2) + 3(-2) = -11$$

$$3x = -3$$

$$x = -1$$

This process of **back-substitution** thus allows us to solve a linear system quickly when it is in triangular form.

The challenge, then, is to find a means of transforming a linear system into triangular form. We now offer (without proof) a list of operations that transform a system of linear equations into an equivalent system.

(a) Interchange any two equations.

(b) Multiply an equation by a nonzero constant.

(c) Replace an equation by the sum of itself plus a constant times another equation.

Using these operations we can now demonstrate the method of Gaussian elimination.

EXAMPLE 1
Solve the linear system.

$$2y - z = -5$$
$$x - 2y + 2z = 9$$
$$2x + y - z = -2$$

Solution

Gaussian Elimination	
Step 1. (a) If necessary, interchange equations to obtain a nonzero coefficient for x in the first equation.	*Step 1.* (a) Interchanging the first two equations yields $$x - 2y + 2z = 9$$ $$2y - z = -5$$ $$2x + y - z = -2$$
(b) Replace the second equation by the sum of itself plus an appropriate multiple of the first equation which will result in a zero coefficient for x.	(b) The coefficient of x in the second equation is already 0.
(c) Replace the third equation by the sum of itself plus an appropriate multiple of the first equation which will result in a zero coefficient for x.	(c) Replace the third equation by the sum of itself plus -2 times the first equation. $$x - 2y + 2z = 9$$ $$2y - z = -5$$ $$5y - 5z = -20$$
Step 2. Apply the procedures of Step 1 to the second and third equations.	*Step 2.* Replace the third equation by $\frac{5}{2}$ times the second equation. $$x - 2y + 2z = 9$$ $$2y - z = -5$$ $$-\frac{5}{2}z = -\frac{15}{2}$$
Step 3. The system is now in triangular form. The solution is obtained by back-substitution.	*Step 3.* From the third equation, $$-\frac{5}{2}z = -\frac{15}{2}$$ $$\boxed{z = 3}$$

	Substituting this value of z in the second equation,$$2y - (3) = -5$$ $$\boxed{y = -1}$$ Substituting for y and for z in the first equation,$$x - 2(-1) + 2(3) = 9$$ $$x + 8 = 9$$ $$\boxed{x = 1}$$ The solution is $x = 1$, $y = -1$, $z = 3$.

PROGRESS CHECK

Solve by Gaussian elimination.

(a) $2x - 4y + 2z = 1$
      $3x + y + 3z = 5$
      $x - y - 2z = -8$

(b) $-2x + 3y - 12z = -17$
      $3x - y - 15z = 11$
      $-x + 5y + 3z = -9$

Answers

(a) $x = -\dfrac{3}{2}$, $y = \dfrac{1}{2}$, $z = 3$

(b) $x = 5$, $y = -1$, $z = \dfrac{1}{3}$

The graph of a linear equation in three unknowns is a plane in three-dimensional space. A system of three linear equations in three unknowns corresponds to three planes (Figure 6). If the planes intersect in a point P (Figure 6a), then the coordinates of the point P are a solution of the system and can be found by Gaussian elimination. Interestingly, the cases of no solution and of an infinite number of solutions are signaled in essentially the same way as we described in Section 2 of this chapter (see Figure 6b and 6c).

If Gaussian elimination results in an equation of the form

(a) $0x + 0y + 0z = c$, $c \neq 0$

then the system is inconsistent.

(b) $0x + 0y + 0z = 0$

then the system is consistent and has an infinite number of solutions.

EXAMPLE 2

Solve the linear system.

$$x - 2y + 2z = -4$$
$$x + y - 7z = 8$$
$$-x - 4y + 16z = -20$$

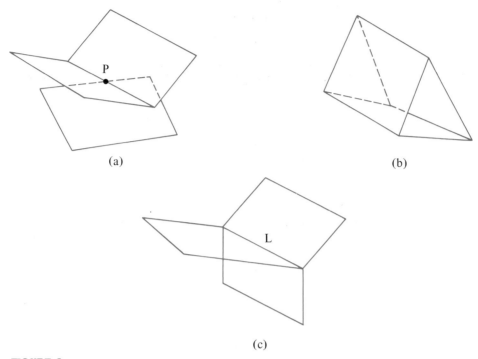

(a)

(b)

(c)

FIGURE 6

Solution

Replacing the second equation by itself minus the first equation, and replacing the third equation by itself plus the first equation, we have

$$x - 2y + 2z = -4$$
$$3y - 9z = 12$$
$$-6y + 18z = -24$$

Replacing the third equation of the last system by itself plus 2 times the second equation results in the system

$$x - 2y + 2z = -4$$
$$3y - 9z = 12$$
$$0x + 0y + 0z = 0$$

in which the last equation indicates that the system is consistent and has an infinite number of solutions. If we solve the second equation of the last system for y we have

$$y = 3z + 4$$

Then solving the first equation for x, we have

$$x = 2y - 2z - 4$$
$$= 2(3z + 4) - 2z - 4 \qquad \text{Substituting for } y$$
$$= 4z + 4$$

The equations

$$x = 4z + 4$$
$$y = 3z + 4$$

yield a solution of the original system for every real value of z. For example, if $z = 0$, then $x = 4$, $y = 4$, $z = 0$ satisfies the original system; if $z = -2$, then $x = -4$, $y = -2$, $z = -2$ is another solution.

PROGRESS CHECK
Verify that the linear system

(a)
$$x - 2y + z = 3$$
$$2x + y - 2z = -1$$
$$-x - 8y + 7z = 5$$

is inconsistent.

(b)
$$2x + y + 2z = 1$$
$$x - 4y + 7z = -4$$
$$x - y + 3z = -1$$

has an infinite number of solutions.

EXERCISE SET 6.4
Solve by Gaussian elimination. Indicate if the system is inconsistent or has an infinite number of solutions.

1.
$$x + 2y + 3z = -6$$
$$2x - 3y - 4z = 15$$
$$3x + 4y + 5z = -8$$

2.
$$2x + 3y + 4z = -12$$
$$x - 2y + z = -5$$
$$3x + y + 2z = 1$$

3.
$$x + y + z = 1$$
$$x + y - 2z = 3$$
$$2x + y + z = 2$$

4.
$$2x - y + z = 3$$
$$x - 3y + z = 4$$
$$-5x - 2z = -5$$

5.
$$x + y + z = 2$$
$$x - y + 2z = 3$$
$$3x + 5y + 2z = 6$$

6.
$$x + y + z = 0$$
$$x + y = 3$$
$$y + z = 1$$

7.
$$x + 2y + z = 7$$
$$x + 2y + 3z = 11$$
$$2x + y + 4z = 12$$

8.
$$4x + 2y - z = 5$$
$$3x + 3y + 6z = 1$$
$$5x + y - 8z = 8$$

9.
$$x + y + z = 2$$
$$x + 2y + z = 3$$
$$x + y - z = 2$$

10.
$$x + y - z = 2$$
$$x + 2y + z = 3$$
$$x + y + 4z = 3$$

11.
$$2x + y + 3z = 8$$
$$-x + y + z = 10$$
$$x + y + z = 12$$

12.
$$2x - 3z = 4$$
$$x + 4y - 5z = -6$$
$$3x + 4y - z = -2$$

13.
$$x + 3y + 7z = 1$$
$$3x - y - 5z = 9$$
$$2x + y + z = 4$$

14.
$$2x - y + z = 2$$
$$3x + y + 2z = 3$$
$$x + y - z = -1$$

15.
$$x - 2y + 3z = -2$$
$$x - 5y + 9z = 4$$
$$2x - y = 6$$

16.
$$x + 2y - 2z = 8$$
$$5y - z = 6$$
$$-2x + y + 3z = -2$$

17. $x - 2y + z = -5$
      $2x \quad\quad + z = -10$
      $\quad\quad y - z = 15$

18. $\quad\quad 2y - 3z = \quad 4$
      $x \quad\quad + 2z = -2$
      $x - 8y + 14z = -18$

19.   A special low-calorie diet consists of dishes A, B and C. Each unit of A has 2 grams of fat, 1 gram of carbohydrate, and 3 grams of protein. Each unit of B has 1 gram of fat, 2 grams of carbohydrate, and 1 gram of protein. Each unit of C has 1 gram of fat, 2 grams of carbohydrate, and 3 grams of protein. The diet must provide exactly 10 grams of fat, 14 grams of carbohydrate, and 18 grams of protein. How much of each dish should be used?

20.   A furniture manufacturer makes chairs, coffee tables, and dining-room tables. Each chair requires 2 minutes of sanding, 2 minutes of staining, and 4 minutes of varnishing. Each coffee table requires 5 minutes of sanding, 4 minutes of staining, and 3 minutes of varnishing. Each dining room table requires 5 minutes of sanding, 4 minutes of staining, and 6 minutes of varnishing. The sanding benches, staining benches, and varnishing benches are available 6, 5, and 6 hours per day, respectively. How many of each type of furniture can be made if all facilities are used to capacity?

21.   A manufacturer produces 12″, 16″, and 19″ television sets that require assembly, testing, and packing. The 12″ sets each require 45 minutes to assemble, 30 minutes to test, and 10 minutes to package. The 16″ sets each require 1 hour to assemble, 45 minutes to test, and 15 minutes to package. The 19″ sets each require 1½ hours to assemble, 1 hour to test, and 15 minutes to package. If the assembly line operates for 17¾ hours per day, the test facility is used for 12½ hours per day, and the packing equipment is used for 3¾ hours per day, how many of each type of set can be produced?

6.5
SYSTEMS INVOLVING NONLINEAR EQUATIONS

The algebraic methods of substitution and elimination can be applied to systems of equations that involve at least one second-degree equation. It is also a good idea to consider the graph of each equation since this indicates the maximum number of points of intersection or solutions of the system. Here are some examples.

EXAMPLE 1
Solve the system of equations.

$$x^2 + y^2 = 25$$
$$x + y = -1$$

Solution
From the second equation we have

$$y = -1 - x$$

Substituting for y in the first equation,

$$x^2 + (-1 - x)^2 = 25$$
$$x^2 + 1 + 2x + x^2 = 25$$

$$2x^2 + 2x - 24 = 0$$

$$x^2 + x - 12 = 0$$

$$(x + 4)(x - 3) = 0$$

which yields $x = -4$ and $x = 3$. Substituting these values for x in the equation $x + y = -1$ we obtain the corresponding values of y.

$x = -4$: $-4 + y = -1$ $x = 3$: $3 + y = -1$

$y = 3$ $y = -4$

Thus, $x = -4$, $y = 3$ and $x = 3$, $y = -4$ are solutions of the system of equations. (Verify.)

Note that the equations represent a circle and a line; the algebraic solution tells us that they intersect in two points. See Figure 7.

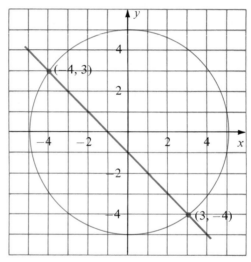

FIGURE 7

PROGRESS CHECK
Solve the system of equations.

(a) $x^2 + 3y^2 = 12$ (b) $x^2 + y^2 = 34$

$x + 3y = 6$ $x - y = 2$

Answers
(a)   x = 3,   y = 1;   x = 0,   y = 2
(b)   x = -3,   y = -5;   x = 5,   y = 3

EXAMPLE 2
Solve the system.

$$3x^2 + 8y^2 = 21$$

$$x^2 + 4y^2 = 10$$

Solution

We can employ the method of elimination to obtain an equation that has just one variable. If we multiply the second equation by -3 and add the result to the first equation we have

$$3x^2 + 8y^2 = 21$$
$$\underline{-3x^2 - 12y^2 = -30}$$
$$-4y^2 = -9$$

$$y^2 = \frac{9}{4}$$

$$y = \pm\frac{3}{2}$$

We can now substitute $+\frac{3}{2}$ and $-\frac{3}{2}$ for y in either of the original equations. Using the second equation,

$$x^2 + 4y^2 = 10 \qquad\qquad\qquad x^2 + 4y^2 = 10$$
$$x^2 + 4\left(\frac{3}{2}\right)^2 = 10 \qquad\qquad x^2 + 4\left(-\frac{3}{2}\right)^2 = 10$$
$$x^2 + 9 = 10 \qquad\qquad\qquad x^2 + 9 = 10$$
$$x^2 = 1 \qquad\qquad\qquad\qquad x^2 = 1$$
$$x = \pm 1 \qquad\qquad\qquad\qquad x = \pm 1$$

We then have four solutions: $x = 1$, $y = \frac{3}{2}$; $x = -1$, $y = \frac{3}{2}$; $x = 1$, $y = -\frac{3}{2}$; $x = -1$, $y = -\frac{3}{2}$.

Note that the equations represent two ellipses; the algebraic solution tells us that the ellipses intersect in four points. The student is urged to sketch the graphs.

PROGRESS CHECK

Solve the system.

$$2x^2 + y^2 = 5$$
$$2x^2 - 3y^2 = 3$$

Answer

$$x = \frac{3}{2}, \quad y = \frac{\sqrt{2}}{2}; \qquad x = -\frac{3}{2}, \quad y = \frac{\sqrt{2}}{2};$$

$$x = \frac{3}{2}, \quad y = -\frac{\sqrt{2}}{2}; \qquad x = -\frac{3}{2}, \quad y = -\frac{\sqrt{2}}{2}$$

EXAMPLE 3

Solve the system.

$$4x - y = 7$$
$$x^2 - y = 3$$

Solution

By subtracting the first equation from the second equation we obtain

$$x^2 - 4x = -4$$
$$x^2 - 4x + 4 = 0$$
$$(x - 2)(x - 2) = 0$$

which yields $x = 2$. Substituting $x = 2$ in the first equation,

$$4(2) - y = 7$$
$$y = 1$$

Thus, $x = 2$, $y = 1$ is a solution of the system. Note that the equations represent a line and a parabola. Since our algebraic techniques yield just one solution and the line is not the axis of the parabola, the line is tangent to the parabola at the point (2,1).

PROGRESS CHECK

Find the real solutions of the system.

$$x^2 - 4x + y^2 - 4y = 1$$
$$x^2 - 4x \qquad + y = -5$$

Answer
$x = 2$, $y = -1$ (*The parabola is tangent to the circle.*)

EXERCISE SET 6.5

Find all solutions of each of the following systems.

1. $x^2 + y^2 = 13$
    $2x - y = 4$

2. $x^2 + 4y^2 = 32$
    $x + 2y = 0$

3. $y^2 - x = 0$
    $y - 4x = -3$

4. $xy = -4$
    $4x - y = 8$

5. $x^2 - 2x + y^2 = 3$
    $2x + y = 4$

6. $4x^2 + y^2 = 4$
    $x - y = 3$

7. $xy = 1$
    $x - y + 1 = 0$

8. $x^2 - y^2 = 3$
    $x^2 + y^2 = 5$

9. $4x^2 + 9y^2 = 72$
    $4x - 3y^2 = 0$

10. $x^2 + y^2 + 2y = 9$
    $y - 2x = 4$

11. $2y^2 - x^2 = -1$
    $4y^2 + x^2 = 25$

12. $x^2 + 4y^2 = 25$
    $4x^2 + y^2 = 25$

13. $25y^2 - 16x^2 = 400$
    $9y^2 - 4x^2 = 36$

14. $16y^2 + 5x^2 = 26$
    $25y^2 - 4x^2 = 17$

15. $y^2 - 8x^2 = 9$
    $y^2 + 3x^2 = 31$

16. $4y^2 + 3x^2 = 24$
    $3y^2 - 2x^2 = 35$

17. $x^2 - 3xy - 2y^2 - 2 = 0$
    $x - y - 2 = 0$

18. $3x^2 + 8y^2 = 21$
    $x^2 + 4y^2 = 10$

19. The sum of the squares of the sides of a rectangle is 100 square meters. If the area of the rectangle is 48 square meters, find the length of each side of the rectangle.
20. Find the dimensions of a rectangle with an area of 30 square feet and perimeter of 22 feet.
21. Find two numbers such that their product is 20 and their sum is 9.
22. Find two numbers such that the sum of their squares is 65 and their sum is 11.

6.6
SYSTEMS OF LINEAR INEQUALITIES IN TWO VARIABLES

When we draw the graph of a linear equation, say

$$y = 2x - 1$$

we can readily see that the graph of the line divides the plane into two regions, which we call **half-planes** (see Figure 8).

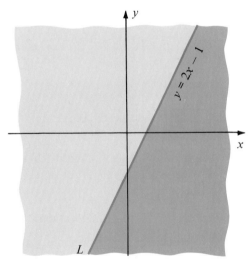

FIGURE 8

If, in the equation $y = 2x - 1$, we replace the equals sign by any of the symbols $<$, $>$, $\leq$, or $\geq$, we have a **linear inequality in two variables.** By the **graph of a linear inequality** such as

$$y < 2x - 1$$

we mean the set of all points whose coordinates satisfy the inequality. Thus, the point $(4,2)$ lies on the graph of $y < 2x - 1$ since

$$2 < (2)(4) - 1 = 7$$

shows that $x = 4$, $y = 2$ satisfy the inequality. However, the point $(1,5)$ does *not* lie on the graph of $y < 2x - 1$ since

$$5 < (2)(1) - 1 = 1$$

is not true. Since the coordinates of every point on the line L in Figure 8 satisfy the *equation* $y = 2x - 1$, we readily see that the coordinates of those points in the half-plane below the line must satisfy the *inequality* $y < 2x - 1$. Similarly, the coordinates of those points in the half-plane above the line must satisfy the *inequality* $y > 2x - 1$. This suggests that the graph of a linear inequality in two variables is a half-plane and leads to a straightforward method for graphing linear inequalities.

Graphing Linear Inequalities	Example: $y \leq x - 1$
Step 1. Replace the inequality sign by an equal sign and plot the line. (a) If the inequality is $\leq$ or $\geq$, plot a solid line (points on the line will satisfy the inequality). (b) If the inequality is $<$ or $>$, plot a dashed line (points on the line will not satisfy the inequality).	*Step 1.* $y = x - 1$ 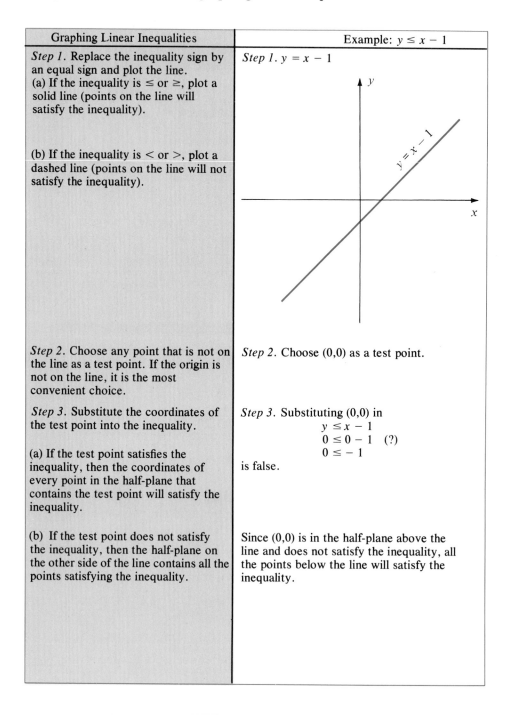
Step 2. Choose any point that is not on the line as a test point. If the origin is not on the line, it is the most convenient choice.	*Step 2.* Choose (0,0) as a test point.
Step 3. Substitute the coordinates of the test point into the inequality. (a) If the test point satisfies the inequality, then the coordinates of every point in the half-plane that contains the test point will satisfy the inequality.	*Step 3.* Substituting (0,0) in $$y \leq x - 1$$ $$0 \leq 0 - 1 \quad (?)$$ $$0 \leq -1$$ is false.
(b) If the test point does not satisfy the inequality, then the half-plane on the other side of the line contains all the points satisfying the inequality.	Since (0,0) is in the half-plane above the line and does not satisfy the inequality, all the points below the line will satisfy the inequality.

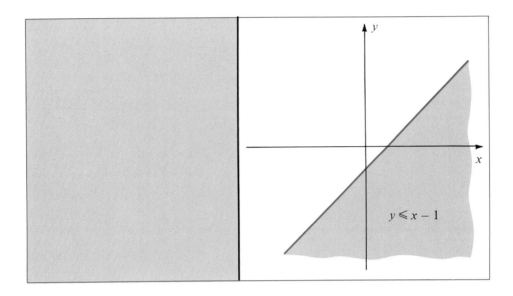

EXAMPLE 1
Graph $2x - 3y > 6$.

Solution
We first graph the line $2x - 3y = 6$. We draw a dashed or broken line to indicate that $2x - 3y = 6$ is not part of the graph (see Figure 9). Since $(0,0)$ is not on the line, we can use it as a test point.

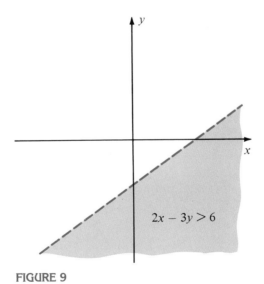

FIGURE 9

is false. Since (0,0) is in the half-plane above the line, the graph consists of the half-plane below the line.

PROGRESS CHECK
Graph the inequalities.

(a) $y \leq 2x + 1$ (b) $y + 3x > -2$ (c) $y \geq -x + 1$

Answers

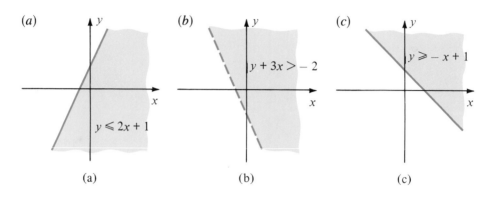

(a)   (b)   (c)

EXAMPLE 2
Graph the inequalities.
(a) $y > x$ (b) $2x \geq 5$

Solution
(a)   Since the origin lies on the line $y = x$, we choose another test point, say $(0,1)$ above the line. Since $(0,1)$ does satisfy the inequality, the graph of the inequality is the half-plane above the line. See Figure 10a.

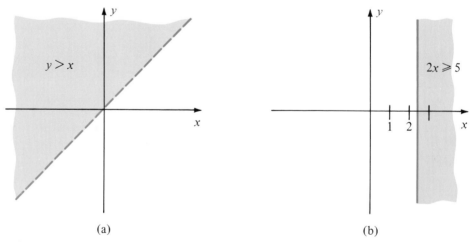

(a)   (b)

FIGURE 10

(b)   The graph of $2x = 5$ is a vertical line and the graph of $2x \geq 5$ is the half-plane to the right of the line and also the line itself. See Figure 10b.

PROGRESS CHECK

Graph the inequalities.

(a) $2y \geq 7$ (b) $x < -2$ (c) $1 \leq y < 3$

Answers

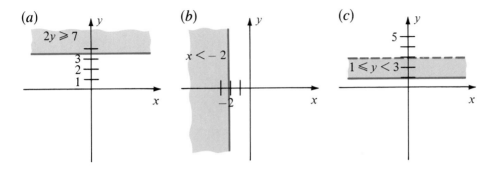

We may also consider **systems of linear inequalities** in two variables, x and y. Examples of such systems are

$$2x - 3y > 6 \qquad\qquad\qquad 2x - 5y \leq 12$$
$$x + 2y < 2 \qquad\qquad\qquad 2x + y \leq 18$$
$$x \geq 0$$
$$y \geq 0$$

The **solution of a system of linear inequalities** consists of all ordered pairs (a,b) such that the substitution $x = a$, $y = b$ satisifies *all* of the inequalities. Thus, the ordered pair $(2,1)$ is a solution of the system

$$2x - 3y \leq 2$$
$$x + y \leq 6$$

since the substitution $x = 2$, $y = 1$ satisfies both inequalities.

$$(2)(2) - (3)(1) = 1 \leq 2$$
$$2 + 1 = 3 \leq 6$$

We can graph the solution set of a system of linear inequalities by graphing the solution set of each inequality and marking that portion of the graph that satisfies *all* of the inequalities.

EXAMPLE 3

Graph the solution set of the system.

$$2x - 3y \leq 2$$
$$x + y \leq 6$$

Solution

In Figure 11 we have graphed the solution set of each of the inequalities. The

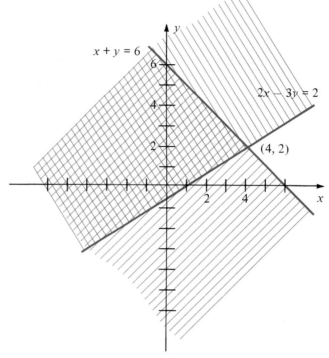

FIGURE 11

cross-hatched region indicates those points that satisfy both inequalities and is therefore the solution set of the system of inequalities.

EXAMPLE 4
Graph the solution set of the system.

$$x + \ y < 2$$
$$2x + 3y \geq 9$$
$$x \geq 1$$

Solution
See Figure 12. Since there are no points satisfying *all* of the inequalities, we conclude that the system is inconsistent and has no solutions.

PROGRESS CHECK
Graph the solution set of the given system.

(a) $x + \ y \geq 3$
$x + 2y < 8$

(b) $2x + y \leq 4$
$x + y \leq 3$
$x \geq 0$
$y \geq 0$

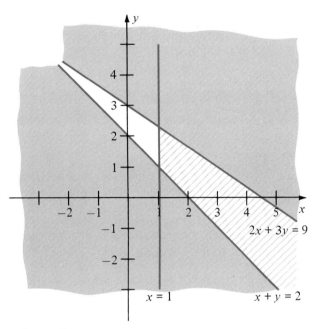

FIGURE 12

Answer
(a)   (b)

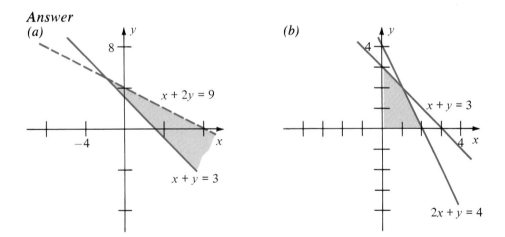

EXAMPLE 5

A dietician at a university is planning a menu for a meal to consist of two primary foods, A and B, whose nutritional contents are shown in the table. The dietician insists that the meal provide at most 12 units of fat, at least 2 units of carbohy-

	Nutritional Content in Units per Gram		
	Fat	Carbohydrate	Protein
A	2	2	0
B	3	1	1

drate, and at least 1 unit of protein. If x and y represent the number of grams of food types A and B, respectively, write a system of linear inequalities expressing the restrictions. Graph the solution set.

Solution
The number of units of fat contained in the meal is $2x + 3y$ so that x and y must satisfy the inequality

$$2x + 3y \leq 12 \qquad \text{fat requirement}$$

Similarly, the requirements for carbohydrate and protein result in the inequalities

$$2x + y \geq 2 \qquad \text{carbohydrate requirement}$$

$$y \geq 1 \qquad \text{protein requirement}$$

Of course, we must also have $x \geq 0$ since negative quantities of food type A would make no sense. The system of linear inequalities is then

$$2x + 3y \leq 12$$

$$2x + y \geq 2$$

$$x \geq 0$$

$$y \geq 1$$

and the graph is shown in Figure 13.

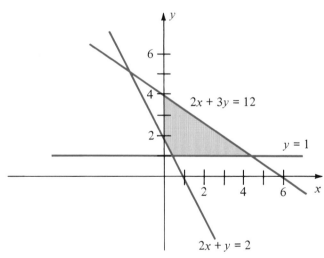

FIGURE 13

EXERCISE SET 6.6
Graph the solution set of the given inequality in the following exercises.

1. $y \leq x + 2$ 2. $y \geq x + 3$
3. $y > x - 4$ 4. $y < x - 5$
5. $y \leq 4 - x$ 6. $y \geq 2 - x$
7. $y > x$ 8. $y \leq 2x$
9. $3x - 5y > 15$ 10. $2y - 3x < 12$
11. $x \leq 4$ 12. $3x > -2$

13. $y > -3$ 14. $5y \leq 25$

15. $x > 0$ 16. $y < 0$

17. $-2 \leq x \leq 3$ 18. $-6 < y < -2$

19. A steel producer makes two types of steel, regular and special. A ton of regular steel requires 2 hours in the open-hearth furnace and a ton of special steel requires 5 hours. Let x and y denote the number of tons of regular and special steel, respectively, made per day. If the open-hearth furnace is available at most 15 hours per day, write an inequality that must be satisfied by x and y. Graph this inequality.

20. A patient is placed on a diet which restricts caloric intake to 1500 calories per day. The patient plans to eat x ounces of cheese, y slices of bread, and z apples on the first day of the diet. If cheese contains 100 calories per ounce, bread 110 calories per slice, and apples 80 calories each, write an inequality that must be satisfied by x, y, and z.

Graph the solution set of the system of linear inequalities.

21. $2x - y \leq 3$ 22. $x - y \leq 4$
    $3x - y \leq 2$ $2x + y \geq 6$

23. $3x - y \geq -7$ 24. $3x - 2y > 1$
    $3x + y \leq -2$ $2x + 3y \leq 18$

25. $3x - 2y \geq -4$ 26. $2x - y \geq -3$
    $2x - y \leq 5$ $x + y \leq 5$
    $y \geq 1$ $x \geq 1$

27. $2x - y \leq 5$ 28. $-x + 3y \leq 2$
    $x + 2y \geq 1$ $4x + 3y \leq 18$
    $x \geq 0$ $x \geq 0$
    $y \geq 0$ $y \geq 0$

29. $3x + y \leq 6$ 30. $x - y \geq -2$
    $x - 2y \leq -1$ $x + y \geq -5$
    $x \geq 2$ $y \geq 0$

31. $3x - 2y \leq -6$ 32. $2x + 3y \geq 18$
    $8x + 3y \leq 24$ $x + 3y \geq 12$
    $5x + 4y \geq 20$ $4x + 3y \geq 24$
    $x \geq 0$ $x \geq 0$
    $y \geq 0$ $y \geq 0$

33. A farmer has 10 quarts of milk and 15 quarts of cream which he will use to make ice cream and yogurt. Each quart of ice cream requires 0.4 quart of milk and 0.2 quart of cream, and each quart of yogurt requires 0.2 quart of milk and 0.4 quart of cream. Graph the set of points representing the possible production of ice cream and of yogurt.

34. A coffee packer uses Jamaican and Colombian coffee to prepare a mild blend and a strong blend. Each pound of mild blend contains $\frac{1}{2}$ pound of Jamaican coffee and $\frac{1}{2}$ pound of Colombian coffee, and each pound of the strong blend requires $\frac{1}{4}$ pound of Jamaican coffee and $\frac{3}{4}$ pound of Colombian coffee. The packer has available 100 pounds of Jamaican coffee and 125 pounds of Colombian coffee. Graph the set of points representing the possible production of the two blends.

35. A trust fund of $100,000 that has been established to provide university scholarships must adhere to certain restrictions.
    (a) No more than half of the fund may be invested in common stocks.
    (b) No more than $35,000 may be invested in preferred stocks.
    (c) No more than $60,000 may be invested in all types of stocks.
    (d) The amount invested in common stocks cannot be more than twice the amount invested in preferred stocks.
    Graph the solution set representing the possible investments in common and preferred stocks.

36. An institution serves a luncheon consisting of two dishes, A and B, whose nutritional content in units per gram is given in the accompanying table.

	Fat	Carbohydrate	Protein
A	1	1	2
B	2	1	6

The luncheon is to provide no more than 10 grams of fat, 7 grams of carbohydrate, and at least 6 grams of protein. Graph the solution set of possible quantities of dishes A and B.

TERMS AND SYMBOLS

system of linear equations (p. 220)	**level of production** (p. 230)	**back-substitution** (p. 237)
linear system (p. 220)	**revenue and cost** (p. 231)	**nonlinear system** (p. 241)
solution of a system of linear equations (p. 220)	**break-event point** (p. 231)	**half-plane** (p. 245)
consistent system (p. 220)	**supply and demand** (p. 232)	**linear inequality in two variables** (p. 245)
inconsistent system (p. 221)	**equilibrium price** (p. 232)	**graph of a linear inequality** (p. 245)
method of substitution (p. 222)	**Gaussian elimination** (p. 236)	**systems of linear inequalities** (p. 249)
method of elimination (p. 225)	**triangular form** (p. 236)	**solution of a system of linear inequalities** (p. 249)

KEY IDEAS FOR REVIEW

■ The graph of a pair of linear equations in two variables is two straight lines which may either (a) intersect in a point, (b) be parallel, or (c) be the same line. If the two straight lines intersect, the coordinates of the point of intersection are a solution of the system of linear equations. If the lines do not intersect, the system is inconsistent.

■ The method of substitution involves solving an equation for one variable and substituting the result into another equation.

■ The method of elimination involves multiplying an equation by a nonzero constant so that when it is added to a second equation, a variable drops out.

■ When using any method of solution, it is possible to detect the special cases when lines are parallel or reduce to the same line.

■ It is often easier and more natural to set up word problems using two or more variables.

■ Gaussian elimination is a systematic way of transforming a linear system to triangular form. A linear system in triangular form is easily solved by back-substitution.

■ Many of the methods used in solving linear systems are applicable in solving nonlinear systems. Nonlinear systems often have more than one solution.

■ The solution of a system of linear inequalities can be found graphically as the region satisfying all of the inequalities.

REVIEW EXERCISES

In Exercises 1–2 solve the given linear system by graphing.

1. $2x + 3y = 2$
    $4x + 5y = 3$

2. $4x - y = 4$
    $2x + y = -1$

In Exercises 3–8 solve the given linear system by the method of substitution.

3. $-x + 6y = -11$
    $2x + 5y = 5$

4. $2x - 4y = -14$
    $-x - 6y = -5$

5. $2x - y = 0$
    $x - 3y = \frac{7}{4}$

6. $-x + 2y = 4$
    $2x - 2y = -7$

7. $3x - 2y = 4$
    $2x + y = -2$

8. $2x + 3y = -7$
    $-x + 2y = 7$

In Exercises 9–14 solve the given linear system by the method of elimination.

9. $x + 4y = 17$
    $2x - 3y = -21$

10. $5x - 2y = 14$
     $-x - 3y = 4$

11. $-3x + y = -13$
     $2x - 3y = 11$

12. $7x - 2y = -20$
     $3x - y = -9$

13. $2x + 3y = -1$
     $-3x + 4y = -\frac{11}{4}$

14. $\frac{1}{3}x + \frac{1}{2}y = -1$
     $\frac{1}{2}x - \frac{1}{4}y = \frac{5}{2}$

15. The sum of a two-digit number and its tens digit is 49. If we reverse the digits of the number, the resulting number is 9 more than the original number. Find the number.

16. The sum of the digits of a two-digit number is 9. The sum of the number and its units digit is 74. Find the number.

17. Five pounds of hamburger and 4 pounds of steak cost $22, and 3 pounds of hamburger and 7 pounds of steak cost $28.15. Find the cost per pound of hamburger and of steak.

18. An airplane flying with a tail wind can complete a journey of 3500 kilometers in 5 hours. Flying the reverse direction, the plane completes the same trip in 7 hours. What is the speed of the plane in still air?

19. A manufacturer of faucets finds that the supply S and demand D are related to price p as follows.

$$S = 3p + 2$$
$$D = -2p + 17$$

Find the equilibrium price and the number of faucets sold at that price.

20. An auto repair shop finds that the monthly expenditures (in dollars) is given by $C = 4025 + 9x$ where x is the total number of hours worked by all employees. If the revenue received (in dollars) is given by $R = 16x$, find the break-even point in the number of work hours, and the total revenue received at that point.

In Exercises 21–24 use Gaussian elimination to solve the given linear systems.

21. $-3x - y + z = 12$
$2x + 5y - 2z = -9$
$-x + 4y + 2z = 15$

22. $3x + 2y - z = -8$
$2x + 3z = 5$
$x - 4y = -4$

23. $5x - y + 2z = 10$
$-2x + 3y - z = -7$
$3x + 2z = 7$

24. $x + 4y = 4$
$-x + 3z = -4$
$2x + 2y - z = \dfrac{41}{6}$

In Exercises 25–28 solve by any method.

25. $2x + 3y = 6$
$3x - y = -13$

26. $x + 2y = 0$
$-x + 4y = 5$

27. $2x + 3y - z = -4$
$x - 2y + 2z = -6$
$2x - 3z = 5$

28. $2x + 2y - 3z = -4$
$3y - z = -4$
$4x - y + z = 4$

In Exercises 29–33 find all real solutions of the given nonlinear system.

29. $x^2 + y^2 = 25$
$x + 3y = 5$

30. $y^2 = 2x - 1$
$x - y = 2$

31. $x^2 - 4y^2 = 9$
$y - 2x = 0$

32. $x^2 + y^2 = 9$
$y = x^2 + 3$

33. $y^2 = 4x$
$y^2 + x - 2y = 12$

In Exercises 34–35 use the graphs of the equations to find approximate solutions to the given systems.

34. $y^2 = x - 1$
$x + y = 7$

35. $x^2 + y^2 = 16$
$x^2 + y = -4$

In Exercises 36–41 graph the solution set of the linear inequality or system of linear inequalities.

36. $x - 2y \le 5$

37. $2x + y > 4$

38. $2x + 3y \le 2$
$x - y \ge 1$

39. $x - 2y \ge 4$
$2x - y \le 2$

40. $2x + 3y \le 6$
$x \ge 0$
$y \ge 1$

41. $2x + y \le 4$
$2x - y \le 3$
$x \ge 0$
$y \ge 0$

PROGRESS TEST 6A

1. Solve the linear system by graphing.
$$3x - y = -17$$
$$x + 2y = -1$$

In Problems 2–3 solve the given linear system by the method of substitution.

2. $2x + y = 4$
    $3x - 2y = -15$

3. $2x - 5y = 8$
    $3x + 2y = 12$

In Problems 4–5 solve the given linear system by the method of elimination.

4. $x - 2y = 7$
    $3x + 4y = -9$

5. $\frac{1}{4}x + \frac{1}{2}y = 0$
    $3x - 2y = 8$

6. The sum of the digits of a two-digit number is 11. If the sum of the number and its tens digit is 41, find the number.

7. An elegant men's shop is having a post-Christmas sale. All shirts are reduced to one low price and all ties are reduced to an even lower price. A customer purchases 3 ties and 7 shirts, paying $135. Another customer selects 5 ties and 3 shirts, paying $95. What is the sale price of each tie and of each shirt?

8. A school cafeteria manager finds that the weekly cost of operation is $1375 plus $1.25 for every meal served. If the average meal produces a revenue of $2.50, find the number of meals served that results in zero profit and zero loss.

9. Solve by Gaussian elimination.
$$3x + 2y - z = -4$$
$$x - y + 3z = 12$$
$$2x - y - 2z = -20$$

Solve Problems 10–11 by any method.

10. $-3x + 2y = -1$
     $6x - y = -1$

11. $3x + y - 2z = 8$
     $3y - 4z = 14$
     $3x + \frac{1}{2}y + z = 1$

In Problems 12–13 find all real solutions of the given nonlinear system.

12. $x^2 + y^2 = 25$
     $4x^2 - y^2 = 20$

13. $y^2 = 5x$
     $y^2 - x^2 = 6$

In Problems 14–15 graph the solution set of the system of linear inequalities.

14. $x - 2y \le 1$
     $3x + 2y \ge 4$

15. $2x + y \le 10$
     $-x + 3y \le 12$
     $x \ge 0$
     $y \ge 0$

PROGRESS TEST 6B

1. Solve the linear system by graphing.

$$x + 3y = -1$$
$$2x - 3y = -8$$

In Problems 2–3 solve the given linear system by the method of substitution.

2. $3x + y = 1$

$$x - \frac{1}{3}y = 1$$

3. $2x - 3y = 1$

$$3x - 2y = 1$$

In Problems 4–5 solve the given linear system by the method of elimination.

4. $-2x + 4y = 5$
$$-x + 3y = 2$$

5. $2x - 3y = 11$
$$3x + 5y = -12$$

6. The sum of the digits of a two-digit number is 14. The difference between the number and that obtained by reversing the digits of the number is 18. Find the number.

7. A motorboat can travel 60 kilometers downstream in 3 hours, and the return trip requires 4 hours. What is the rate of the current?

8. Suppose that supply and demand for a particular tennis racquet is related to price p by

$$S = 5p + 1$$
$$D = -2p + 43$$

Find the equilibrium price and the number of racquets sold at this price.

9. Solve by Gaussian elimination.

$$x \qquad + 2z = 7$$
$$3y + 4z = -10$$
$$-2x + y - 2z = -14$$

In Problems 10–11 solve by any method.

10. $x - 2y = 1$
$$3x + 2y = 1$$

11. $3x + y - 7z = -4$
$$2x - 2y - z = 9$$
$$-2x + y + 3z = -4$$

In Problems 12–13 find all real solutions of the given nonlinear system.

12. $x^2 + 3y^2 = 12$
$$x + 3y = 6$$

13. $x^2 - y^2 = 9$
$$x^2 + y^2 = 41$$

In Problems 14–15 graph the solution set of the system of linear inequalities.

14. $2x - 3y \geq 6$
$$3x + y \leq 3$$

15. $2x + y \geq 4$
$$2x - 5y \leq 5$$
$$y \geq 1$$

CHAPTER SEVEN
MATRICES AND DETERMINANTS

The material on matrices and determinants presented in this chapter serves as an introduction to linear algebra, a mathematical subject that is used in the natural sciences, business and economics, and the social sciences. Computers have played an important role in expanding the use of matrix techniques to a wide variety of practical problems since the matrix methods may require millions of numerical computations.

Our study of matrices and determinants will focus on their application to the problem of solving systems of linear equations. We will see that the method of Gaussian elimination studied in the previous chapter can be neatly implemented using matrices. We will show that matrix notation provides a convenient means for writing linear systems and that the inverse of a matrix enables us to solve such a system. Determinants will also provide us with an additional technique, known as Cramer's rule, for the solution of certain linear systems.

It should be emphasized that this material is a very brief introduction to matrices and determinants. Their properties and applications are both extensive and important.

7.1
MATRICES AND LINEAR SYSTEMS

In the previous chapter we studied methods for solving a system of linear equations such as the following:

$$2x + 3y = -7$$

$$3x - y = 17$$

This system can be displayed by a **matrix,** which is simply a rectangular array of mn real numbers arranged in m horizontal rows and n vertical columns. The numbers are called the **entries** or **elements** of the matrix and are enclosed within brackets. Thus,

$$A = \begin{bmatrix} 2 & 3 & -7 \\ 3 & -1 & 17 \end{bmatrix} \begin{matrix} \leftarrow \\ \leftarrow \end{matrix} \text{rows}$$

$$\underset{\text{columns}}{\uparrow \qquad \uparrow \qquad \uparrow}$$

is a matrix consisting of 2 rows and 3 columns, whose entries are obtained from the two given equations. In general, a matrix of m rows and n columns is said to be of **dimension m by n**, written $m \times n$. The matrix A is seen to be of dimension 2×3. If the number of rows and columns of a matrix are both equal to n, the matrix is called a **square matrix** of **order** n.

EXAMPLE 1

(a) $A = \begin{bmatrix} -1 & 4 \\ 0.1 & -2 \end{bmatrix}$ is a 2×2 matrix. Since matrix A has two rows and two columns, it is a square matrix of order 2.

(b) $B = \begin{bmatrix} 4 & -5 \\ -2 & 1 \\ 3 & 0 \end{bmatrix}$ has three rows and two columns and is a 3×2 matrix.

(c) $C = \begin{bmatrix} -8 & 6 & 1 \end{bmatrix}$ is a 1×3 matrix and is called a **row matrix** since it has precisely one row.

(d) $D = \begin{bmatrix} 2 \\ -4 \end{bmatrix}$ is a 2×1 matrix and is called a **column matrix** since it has precisely one column.

There is a convenient way of denoting a general $m \times n$ matrix, using "double subscripts."

$$A = \begin{bmatrix} a_{11} & a_{12} & \cdots & a_{1j} & \cdots & a_{1n} \\ a_{21} & a_{22} & \cdots & a_{2j} & \cdots & a_{2n} \\ \cdot & \cdot & & \cdot & & \cdot \\ \cdot & \cdot & & \cdot & & \cdot \\ \cdot & \cdot & & \cdot & & \cdot \\ a_{i1} & a_{i2} & \cdots & a_{ij} & \cdots & a_{in} \\ \cdot & \cdot & & \cdot & & \cdot \\ \cdot & \cdot & & \cdot & & \cdot \\ \cdot & \cdot & & \cdot & & \cdot \\ a_{m1} & a_{m2} & \cdots & a_{mj} & \cdots & a_{mn} \end{bmatrix} \begin{matrix} \leftarrow \text{first row} \\ \leftarrow \text{second row} \\ \\ \\ \\ \leftarrow i\text{th row} \\ \\ \\ \\ \leftarrow m\text{th row} \end{matrix}$$

$$\underset{\substack{\text{first} \\ \text{column}}}{\uparrow} \qquad \underset{\substack{\text{second} \\ \text{column}}}{\uparrow} \qquad \underset{\substack{j\text{th} \\ \text{column}}}{\uparrow} \qquad \underset{\substack{n\text{th} \\ \text{column}}}{\uparrow}$$

Thus, a_{ij} is the entry in the ith row and jth column of the matrix A. It is customary to write $A = [a_{ij}]$ to indicate that a_{ij} is the entry in row i and column j of matrix A.

EXAMPLE 2
Let

$$A = \begin{bmatrix} 3 & -2 & 4 & 5 \\ 9 & 1 & 2 & 0 \\ -3 & 2 & -4 & 8 \end{bmatrix}$$

Matrix A is of dimension 3×4. The element a_{12} is found in the first row and second column and is seen to be -2. Similarly, we see that $a_{31} = -3$, $a_{33} = -4$, and $a_{34} = 8$.

PROGRESS CHECK
Let

$$B = \begin{bmatrix} 4 & 8 & 1 \\ 2 & -5 & 3 \\ -8 & 6 & -4 \\ 0 & 1 & -1 \end{bmatrix}$$

Find (a) b_{11} (b) b_{23} (c) b_{31} (d) b_{42}

Answers
(a) *4*   (b) *3*   (c) *-8*   (d) *1*

If we begin with the system of linear equations

$$2x + 3y = -7$$
$$3x - y = 17$$

the matrix

$$\begin{bmatrix} 2 & 3 \\ 3 & -1 \end{bmatrix}$$

in which the first column is formed from the coefficients of x and the second column is formed from the coefficients of y is called the **coefficient matrix**. The matrix

$$\begin{bmatrix} 2 & 3 & | & -7 \\ 3 & -1 & | & 17 \end{bmatrix}$$

which includes the column consisting of the right-hand sides of the equations separated by a dashed line, is called the **augmented matrix.**

EXAMPLE 3
Write a system of linear equations that corresponds to the augmented matrix.

$$\begin{bmatrix} -5 & 2 & -1 & | & 15 \\ 0 & -2 & 1 & | & -7 \\ \frac{1}{2} & 1 & -1 & | & 3 \end{bmatrix}$$

Solution

We attach the unknown x to the first column, the unknown y to the second column, and the unknown z to the third column. The resulting system is

$$-5x + 2y - z = 15$$
$$- 2y + z = -7$$
$$\frac{1}{2}x + y - z = 3$$

Now that we have seen how a matrix can be used to represent a system of linear equations, we next proceed to show how routine operations on that matrix can yield the solution of the system. These "matrix methods" are simply a clever streamlining of the methods studied in the previous chapter.

In Section 4 of the previous chapter we used three elementary operations to transform a system of linear equations into triangular form. When applying the same procedures to a matrix, we speak of rows, columns, and elements instead of equations, variables, and coefficients. The three elementary operations of Section 4 of the previous chapter which yield an equivalent system now become the **elementary row operations.**

The following elementary row operations transform an augmented matrix into an equivalent system.
(1) Interchange any two rows.
(2) Multiply each element of any row by a constant $k(\neq 0)$.
(3) Replace each element of a given row by the sum of itself plus k times the corresponding element of any other row.

The method of Gaussian elimination introduced in Section 4 of the previous chapter can now be restated in terms of matrices. By use of elementary row operations we seek to transform an augmented matrix into a matrix for which $a_{ij} = 0$ when $i > j$. The resulting matrix will have the following appearance for a system of three linear equations in three unknowns.

$$\begin{bmatrix} * & * & * & | & * \\ 0 & * & * & | & * \\ 0 & 0 & * & | & * \end{bmatrix}$$

Since this matrix represents a linear system in triangular form, back-substitution will provide a solution of the original system. We will illustrate the process with an example.

EXAMPLE 4

Solve the system.

$$x - y + 4z = 4$$
$$2x + 2y - z = 2$$
$$3x - 2y + 3z = -3$$

Solution
We describe and illustrate the steps of the procedure.

Gaussian Elimination	
Step 1. Form the augmented matrix.	*Step 1.* The augmented matrix is $$\begin{bmatrix} 1 & -1 & 4 & \vert & 4 \\ 2 & 2 & -1 & \vert & 2 \\ 3 & -2 & 3 & \vert & -3 \end{bmatrix}$$
Step 2. If necessary, interchange rows to make sure that a_{11}, the first element of the first row, is nonzero. We call a_{11} the **pivot element** and row 1 the **pivot row.**	*Step 2.* We see that $a_{11} = 1 \neq 0$. The pivot element is a_{11} and is shown in color.
Step 3. Arrange to have 0 as the first element of every row below row 1. This is done by replacing row 2, row 3, and so on, by the sum of itself and an appropriate multiple of row 1.	*Step 3.* To make $a_{21} = 0$, replace row 2 by the sum of itself and (-2) times row 1; to make $a_{31} = 0$, replace row 3 by the sum of itself and (-3) times row 1. $$\begin{bmatrix} 1 & -1 & 4 & \vert & 4 \\ 0 & 4 & -9 & \vert & -6 \\ 0 & 1 & -9 & \vert & -15 \end{bmatrix}$$
Step 4. Repeat the process defined by Steps 2 and 3, allowing row 2, row 3, and so on, to play the role of the first row. Thus row 2, row 3, and so on, serve as the pivot rows.	*Step 4.* Since $a_{22} = 4 \neq 0$, it will serve as the next pivot element and is shown in color. To make $a_{32} = 0$, replace row 3 by the sum of itself and $(-\frac{1}{4})$ times row 2. $$\begin{bmatrix} 1 & -1 & 4 & \vert & 4 \\ 0 & 4 & -9 & \vert & -6 \\ 0 & 0 & -\frac{27}{4} & \vert & -\frac{27}{4} \end{bmatrix}$$
Step 5. The corresponding linear system is in triangular form. Solve by back-substitution.	*Step 5.* The third row of the final matrix yields $$-\frac{27}{4}z = -\frac{27}{2}$$ $$\boxed{z = 2}$$ Substituting $z = 2$, we obtain from the second row of the final matrix $$4y - 9z = -6$$ $$4y - 9(2) = -6$$ $$\boxed{y = 3}$$ Substituting $y = 3$, $z = 2$, we obtain from the first row of the final matrix $$x - y + 4z = 4$$ $$x - 3 + 4(2) = 4$$ $$\boxed{x = -1}$$ The solution is $x = -1$, $y = 3$, $z = 2$.

PROGRESS CHECK

Solve the linear system by matrix methods.

$$2x + 4y - z = 0$$
$$x - 2y - 2z = 2$$
$$-5x - 8y + 3z = -2$$

Answer
$x = 6, \quad y = -2, \quad z = 4$

Note that we have described the process of Gaussian elimination in a manner that will apply to any augmented matrix that is $n \times (n + 1)$; that is, Gaussian elimination may be used on any system of n linear equations in n unknowns that has a unique solution.

It is also permissible to perform elementary row operations in clever ways to simplify the arithmetic. For instance, you may wish to interchange rows, or to multiply a row by a constant to obtain a pivot element equal to 1. We will illustrate these ideas with an example.

EXAMPLE 5

Solve by matrix methods.

$$2y + 3z \qquad = 4$$
$$4x + y + 8z + 15w = -14$$
$$x - y + 2z \qquad = 9$$
$$-x - 2y - 3z - 6w = 10$$

Solution

We begin with the augmented matrix and perform a sequence of elementary row operations. The pivot element is shown in color.

$$\begin{bmatrix} 0 & 2 & 3 & 0 & | & 4 \\ 4 & 1 & 8 & 15 & | & -14 \\ 1 & -1 & 2 & 0 & | & 9 \\ -1 & -2 & -3 & -6 & | & 10 \end{bmatrix}$$
Augmented matrix
Note that $a_{11} = 0$.

$$\begin{bmatrix} 1 & -1 & 2 & 0 & | & 9 \\ 4 & 1 & 8 & 15 & | & -14 \\ 0 & 2 & 3 & 0 & | & 4 \\ -1 & -2 & -3 & -6 & | & 10 \end{bmatrix}$$
Interchanged rows 1 and 3 so that $a_{11} = 1$.

$$\begin{bmatrix} 1 & -1 & 2 & 0 & | & 9 \\ 0 & 5 & 0 & 15 & | & -50 \\ 0 & 2 & 3 & 0 & | & 4 \\ 0 & -3 & -1 & -6 & | & 19 \end{bmatrix}$$
To make $a_{21} = 0$, replaced row 2 by the sum of itself and (-4) times row 1.
To make $a_{41} = 0$, replaced row 4 by the sum of itself and row 1.

$$\begin{bmatrix} 1 & -1 & 2 & 0 & | & 9 \\ 0 & 1 & 0 & 3 & | & -10 \\ 0 & 2 & 3 & 0 & | & 4 \\ 0 & -3 & -1 & -6 & | & 19 \end{bmatrix}$$
Multiplied row 2 by $\frac{1}{5}$ so that $a_{22} = 1$.

$$\begin{bmatrix} 1 & -1 & 2 & 0 & | & 9 \\ 0 & 1 & 0 & 3 & | & -10 \\ 0 & 0 & 3 & -6 & | & 24 \\ 0 & 0 & -1 & 3 & | & -11 \end{bmatrix}$$

To make $a_{32} = 0$, replaced row 3 by the sum of itself and (-2) times row 2.
To make $a_{42} = 0$, replaced row 4 by the sum of itself and 3 times row 2.

$$\begin{bmatrix} 1 & -1 & 2 & 0 & | & 9 \\ 0 & 1 & 0 & 3 & | & -10 \\ 0 & 0 & -1 & 3 & | & -11 \\ 0 & 0 & 3 & -6 & | & 24 \end{bmatrix}$$

Interchanged rows 3 and 4 so that the next pivot will be $a_{33} = -1$.

$$\begin{bmatrix} 1 & -1 & 2 & 0 & | & 9 \\ 0 & 1 & 0 & 3 & | & -10 \\ 0 & 0 & -1 & 3 & | & -11 \\ 0 & 0 & 0 & 3 & | & -9 \end{bmatrix}$$

To make $a_{43} = 0$, replaced row 4 by the sum of itself and 3 times row 3.

The last row of the matrix indicates that

$$3w = -9$$

$$\boxed{w = -3}$$

The remaining variables are found by back-substitution.

Third row of final matrix	Second row of final matrix	First row of final matrix
$-z + 3w = -11$	$y + 3w = -10$	$x - y + 2z = 9$
$-z + 3(-3) = -11$	$y + 3(-3) = -10$	$x - (-1) + 2(2) = 9$
$\boxed{z = 2}$	$\boxed{y = -1}$	$\boxed{x = 4}$

The solution is $x = 4$, $y = -1$, $z = 2$, $w = -3$

There is an important variant of Gaussian elimination known as **Gauss–Jordan elimination.** The objective is to transform a linear system into a form that yields a solution without back-substitution. For a 3×3 system that has a unique solution, the final matrix and equivalent linear system will look like this.

$$\begin{bmatrix} 1 & 0 & 0 & | & c_1 \\ 0 & 1 & 0 & | & c_2 \\ 0 & 0 & 1 & | & c_3 \end{bmatrix}, \qquad \begin{aligned} x + 0y + 0z &= c_1 \\ 0x + y + 0z &= c_2 \\ 0x + 0y + z &= c_3 \end{aligned}$$

The solution is then seen to be $x = c_1$, $y = c_2$, and $z = c_3$.

The execution of the Gauss–Jordan method is essentially the same as that of Gaussian elimination except that

(a) the pivot elements are always required to be equal to 1, and

(b) all elements in a column (other than the pivot element) are forced to be 0.

These objectives are accomplished by the use of elementary row operations as illustrated in the following example.

EXAMPLE 6

Solve the linear system by the Gauss–Jordan method.

$$x - 3y + 2z = 12$$
$$2x + y - 4z = -1$$
$$x + 3y - 2z = -8$$

Solution

We begin with the augmented matrix. At each stage, the pivot element is shown in color and is used to force all elements in that column (other than the pivot element itself) to be zero.

$$\begin{bmatrix} 1 & -3 & 2 & | & 12 \\ 2 & 1 & -4 & | & -1 \\ 1 & 3 & -2 & | & -8 \end{bmatrix}$$ Pivot element is a_{11}.

$$\begin{bmatrix} 1 & -3 & 2 & | & 12 \\ 0 & 7 & -8 & | & -25 \\ 0 & 6 & -4 & | & -20 \end{bmatrix}$$ To make $a_{21} = 0$, replaced row 2 by the sum of itself and (-2) times row 1.
To make $a_{31} = 0$, replaced row 3 by the sum of itself and (-1) times row 1.

$$\begin{bmatrix} 1 & -3 & 2 & | & 12 \\ 0 & 1 & -4 & | & -5 \\ 0 & 6 & -4 & | & -20 \end{bmatrix}$$ Replaced row 2 by the sum of itself and (-1) times row 3 to yield the next pivot, $a_{22} = 1$.

$$\begin{bmatrix} 1 & 0 & -10 & | & -3 \\ 0 & 1 & -4 & | & -5 \\ 0 & 0 & 20 & | & 10 \end{bmatrix}$$ To make $a_{12} = 0$, replaced row 1 by the sum of itself and 3 times row 2.
To make $a_{32} = 0$, replaced row 3 by the sum of itself and (-6) times row 2.

$$\begin{bmatrix} 1 & 0 & -10 & | & -3 \\ 0 & 1 & -4 & | & -5 \\ 0 & 0 & 1 & | & \frac{1}{2} \end{bmatrix}$$ Multiplied row 3 by $\frac{1}{20}$ so that $a_{33} = 1$.

$$\begin{bmatrix} 1 & 0 & 0 & | & 2 \\ 0 & 1 & 0 & | & -3 \\ 0 & 0 & 1 & | & \frac{1}{2} \end{bmatrix}$$ To make $a_{13} = 0$, replaced row 1 by the sum of itself and 10 times row 3.
To make $a_{23} = 0$, replaced row 2 by the sum of itself and 4 times row 3.

We can see the solution directly from the final matrix: $x = 2$, $y = -3$, and $z = \frac{1}{2}$.

EXERCISE SET 7.1

In Exercises 1–6 state the dimension of each matrix.

1. $\begin{bmatrix} 3 & -1 \\ 2 & 4 \end{bmatrix}$

   2. $\begin{bmatrix} 1 & 2 & 3 & -1 \end{bmatrix}$

   3. $\begin{bmatrix} 4 & 2 & 3 \\ 5 & -1 & 4 \\ 2 & 3 & 6 \\ -8 & -1 & 2 \end{bmatrix}$

4. $\begin{bmatrix} -1 \\ 3 \\ 2 \end{bmatrix}$

   5. $\begin{bmatrix} 4 & 2 & 1 \\ 3 & 1 & 5 \\ -4 & -2 & 3 \end{bmatrix}$

   6. $\begin{bmatrix} 3 & -1 & 2 & 6 \\ 2 & 8 & 4 & 1 \end{bmatrix}$

7. Given

$$A = \begin{bmatrix} 3 & -4 & -2 & 5 \\ 8 & 7 & 6 & 2 \\ 1 & 0 & 9 & -3 \end{bmatrix}$$

   find   (a) a_{12} (b) a_{22} (c) a_{23} (d) a_{34}

8. Given

$$B = \begin{bmatrix} -5 & 6 & 8 \\ 4 & 1 & 3 \\ 0 & 2 & -6 \\ -3 & 9 & 7 \end{bmatrix}$$

   find   (a) b_{13} (b) b_{21} (c) b_{33} (d) b_{42}

In Exercises 9–12 write the coefficient matrix and the augmented matrix for each given linear system.

9. $3x - 2y = 12$
   $5x + y = -8$

10. $3x - 4y = 15$
    $4x - 3y = 12$

11. $\frac{1}{2}x + y + z = 4$

    $2x - y - 4z = 6$
    $4x + 2y - 3z = 8$

12. $2x + 3y - 4z = 10$

    $-3x + y = 12$
    $5x - 2y + z = -8$

In Exercises 13–16 write the linear system whose augmented matrix is given.

13. $\left[\begin{array}{cc|c} \frac{3}{2} & 6 & -1 \\ 4 & 5 & 3 \end{array}\right]$

14. $\left[\begin{array}{cc|c} 4 & 0 & 2 \\ -7 & 8 & 3 \end{array}\right]$

15. $\left[\begin{array}{ccc|c} 1 & 1 & 3 & -4 \\ -3 & 4 & 0 & 8 \\ 2 & 0 & 7 & 6 \end{array}\right]$

16. $\left[\begin{array}{ccc|c} 4 & 8 & 3 & 12 \\ 1 & -5 & 3 & -14 \\ 0 & 2 & 7 & 18 \end{array}\right]$

In Exercises 17–20 the augmented matrix corresponding to a linear system has been transformed to the given matrix by elementary row operations. Find a solution of the original linear system.

17. $\left[\begin{array}{ccc|c} 1 & 2 & 0 & 3 \\ 0 & 1 & -2 & 4 \\ 0 & 0 & 1 & 2 \end{array}\right]$

18. $\left[\begin{array}{ccc|c} 1 & 0 & 2 & -1 \\ 0 & 1 & 3 & 2 \\ 0 & 0 & 1 & 5 \end{array}\right]$

19. $\begin{bmatrix} 1 & -2 & 1 & | & 3 \\ 0 & 1 & 3 & | & 2 \\ 0 & 0 & 1 & | & -4 \end{bmatrix}$ 20. $\begin{bmatrix} 1 & -4 & 2 & | & -4 \\ 0 & 1 & 3 & | & -2 \\ 0 & 0 & 1 & | & 5 \end{bmatrix}$

In Exercises 21–30 solve the given linear system by applying Gaussian elimination on the augmented matrix.

21. $\quad x - 2y = -4$
$\quad\quad 2x + 3y = 13$

22. $\quad 2x + y = -1$
$\quad\quad 3x - y = -7$

23. $\quad x + y + z = 4$
$\quad\quad 2x - y + 2z = 11$
$\quad\quad x + 2y + z = 3$

24. $\quad x - y + z = -5$
$\quad\quad 3x + y + 2z = -5$
$\quad\quad 2x - y - z = -2$

25. $\quad 2x + y - z = 9$
$\quad\quad x - 2y + 2z = -3$
$\quad\quad 3x + 3y + 4z = 11$

26. $\quad 2x + y - z = -2$
$\quad\quad -2x - 2y + 3z = 2$
$\quad\quad 3x + y - z = -4$

27. $\quad -x - y + 2z = 9$
$\quad\quad x + 2y - 2z = -7$
$\quad\quad 2x - y + z = -9$

28. $\quad 4x + y - z = -1$
$\quad\quad x - y + 2z = 3$
$\quad\quad -x + 2y - z = 0$

29. $\quad x + y - z + 2w = 0$
$\quad\quad 2x + y - w = -2$
$\quad\quad 3x + 2z = -3$
$\quad\quad -x + 2y + 3w = 1$

30. $\quad 2x + y - 3w = -7$
$\quad\quad 3x + 2z + w = 0$
$\quad\quad -x + 2y + 3w = 10$
$\quad\quad -2x - 3y + 2z - w = 7$

31–40.   Solve the linear systems of Exercises 21–30 by Gauss–Jordan elimination applied to the augmented matrix.

7.2
MATRIX OPERATIONS AND APPLICATIONS (Optional)

After defining a new type of mathematical entity, it is useful to define operations using this entity. It is common practice to begin with a definition of equality, in this case the **equality of matrices.** We say two matrices are equal if they are of the same dimension and their corresponding elements are equal.

EXAMPLE 1
Solve for all unknowns.

$$\begin{bmatrix} -2 & 2x & 9 \\ y-1 & 3 & -4s \end{bmatrix} = \begin{bmatrix} z & 6 & 9 \\ -4 & r & 7 \end{bmatrix}$$

Solution
Equating corresponding elements, we must have

$$-2 = z \quad\quad \text{or} \quad z = -2$$
$$2x = 6 \quad\quad \text{or} \quad x = 3$$

$$y - 1 = -4 \qquad \text{or} \qquad y = -3$$
$$3 = r \qquad \text{or} \qquad r = 3$$
$$-4s = 7 \qquad \text{or} \qquad s = -\frac{7}{4}$$

Matrix addition can be performed only when the matrices are of the same order.

> The sum of two $m \times n$ matrices A and B is the $m \times n$ matrix obtained by adding the corresponding elements of A and B.

EXAMPLE 2
Given the following matrices,

$$A = \begin{bmatrix} 2 & -3 & 4 \end{bmatrix} \qquad B = \begin{bmatrix} 5 & 3 & 2 \end{bmatrix}$$

$$C = \begin{bmatrix} 1 & 6 & -1 \\ -2 & 4 & 5 \end{bmatrix} \qquad D = \begin{bmatrix} 16 & 2 & 9 \\ 4 & -7 & -1 \end{bmatrix}$$

find (if possible): (a) $A + B$ (b) $A + D$ (c) $C + D$

Solution
(a) Since A and B are both 1×3 matrices, they can be added, giving

$$A + B = \begin{bmatrix} 2 + 5 & -3 + 3 & 4 + 2 \end{bmatrix} = \begin{bmatrix} 7 & 0 & 6 \end{bmatrix}$$

(b) Matrices A and D are not of the same dimension and cannot be added.
(c) C and D are both 2×3 matrices. Thus,

$$C + D = \begin{bmatrix} 1 + 16 & 6 + 2 & -1 + 9 \\ -2 + 4 & 4 + (-7) & 5 + (-1) \end{bmatrix} = \begin{bmatrix} 17 & 8 & 8 \\ 2 & -3 & 4 \end{bmatrix}$$

Matrices are a natural way of writing the information displayed in a table. For example, Table 1 displays the current inventory of the Quality

TABLE 1

TV Sets	Boston	Miami	Chicago
17"	140	84	25
19"	62	17	48

TV Company at its various outlets. The same data is displayed by the matrix

$$S = \begin{bmatrix} 140 & 84 & 25 \\ 62 & 17 & 48 \end{bmatrix}$$

where we understand the columns to represent the cities and the rows to represent the sizes of the television sets. If the matrix

$$M = \begin{bmatrix} 30 & 46 & 15 \\ 50 & 25 & 60 \end{bmatrix}$$

specifies the number of sets of each size received at each outlet the following month, then the matrix

$$T = S + M = \begin{bmatrix} 170 & 130 & 40 \\ 112 & 42 & 108 \end{bmatrix}$$

gives the revised inventory.

Suppose the salespeople at each outlet are told that half of the revised inventory is to be placed on sale. To determine the number of sets of each size to be placed on sale, we need to multiply each element of the matrix T by 0.5. When working with matrices we call a real number such as 0.5 a **scalar** and define **scalar multiplication** as follows.

To multiply a matrix A by a scalar c, multiply each element of A by c.

EXAMPLE 3
The matrix Q

$$Q = \begin{array}{cccc} & \text{Regular} & \text{Unleaded} & \text{Premium} \\ & \begin{bmatrix} 130 & 250 & 60 \\ 110 & 180 & 40 \end{bmatrix} & & \begin{array}{l} \text{City A} \\ \text{City B} \end{array} \end{array}$$

shows the quantity (in thousands of gallons) of the principal types of gasolines stored by a refiner at two different locations. It is decided to increase the quantity of each type of gasoline stored at each site by 10%. Use scalar multiplication to determine the desired inventory levels.

Solution
To increase each entry of matrix Q by 10% we compute the scalar product $1.1Q$.

$$1.1Q = 1.1 \begin{bmatrix} 130 & 250 & 60 \\ 110 & 180 & 40 \end{bmatrix}$$

$$= \begin{bmatrix} 1.1(130) & 1.1(250) & 1.1(60) \\ 1.1(110) & 1.1(180) & 1.1(40) \end{bmatrix} = \begin{bmatrix} 143 & 275 & 66 \\ 121 & 198 & 44 \end{bmatrix}$$

We denote $A + (-1)B$ by $A - B$ and refer to this as the **difference** between A and B. The difference, then, is found by subtracting corresponding entries.

EXAMPLE 4
Using the matrices C and D of Example 2, find $C - D$.

Solution
By definition,

$$C - D = \begin{bmatrix} 1 - 16 & 6 - 2 & -1 - 9 \\ -2 - 4 & 4 - (-7) & 5 - (-1) \end{bmatrix} = \begin{bmatrix} -15 & 4 & -10 \\ -6 & 11 & 6 \end{bmatrix}$$

We will use the Quality TV Company again, this time to help us arrive at a definition of matrix multiplication. Suppose S

$$S = \begin{matrix} \text{Boston} & \text{Miami} & \text{Chicago} \\ \begin{bmatrix} 60 & 85 & 70 \\ 40 & 100 & 20 \end{bmatrix} & \begin{matrix} 17'' \\ 19'' \end{matrix} \end{matrix}$$

is a matrix representing the supply of television sets at the end of the year. Further, suppose the cost of each 17″ set is \$80 and the cost of each 19″ set is \$125. To find the total cost of the inventory at each outlet, we need to multiply the number of 17″ sets by \$80, the number of 19″ sets by \$125, and sum the two products. If we let

$$C = [80 \quad 125]$$

be the cost matrix, we seek to define the product

$$[80 \quad 125] \begin{bmatrix} 60 & 85 & 70 \\ 40 & 100 & 20 \end{bmatrix}$$

so that the result will be a matrix displaying the total cost at each outlet. To find the total cost at the Boston outlet, we need to calculate

$$(80)(60) + (125)(40) = 9800$$

$$[80 \quad 125] \begin{bmatrix} 60 & 85 & 70 \\ 40 & 100 & 20 \end{bmatrix}$$

At the Miami outlet, the total cost is

$$(80)(85) + (125)(100) = 19{,}300$$

$$[80 \quad 125] \begin{bmatrix} 60 & 85 & 70 \\ 40 & 100 & 20 \end{bmatrix}$$

At the Chicago outlet, the total cost is

$$(80)(70) + (125)(20) = 8100$$

$$[80 \quad 125] \begin{bmatrix} 60 & 85 & 70 \\ 40 & 100 & 20 \end{bmatrix}$$

The total cost at each outlet can then be displayed by the 1×3 matrix

$$[9800 \quad 19{,}300 \quad 8100]$$

which is the product of C and S. Thus,

$$CS = [80 \quad 125] \begin{bmatrix} 60 & 85 & 70 \\ 40 & 100 & 20 \end{bmatrix}$$

$$= [(80)(60) + (125)(40) \quad (80)(85) + (125)(100) \quad (80)(70) + (125)(20)]$$

$$= [9800 \quad 19{,}300 \quad 8100]$$

Our example illustrates the process for multiplying a matrix by a row matrix. If the matrix C had more than one row, we would repeat the process using each row of C. Here is an example.

EXAMPLE 5
Find the product AB if

$$A = \begin{bmatrix} 2 & 1 \\ 3 & 5 \end{bmatrix} \qquad B = \begin{bmatrix} 4 & -6 & -2 & 4 \\ 2 & 0 & 1 & -5 \end{bmatrix}$$

Solution

$$AB = \begin{bmatrix} (2)(4) + (1)(2) & (2)(-6) + (1)(0) & (2)(-2) + (1)(1) & (2)(4) + (1)(-5) \\ (3)(4) + (5)(2) & (3)(-6) + (5)(0) & (3)(-2) + (5)(1) & (3)(4) + (5)(-5) \end{bmatrix}$$

$$= \begin{bmatrix} 10 & -12 & -3 & 3 \\ 22 & -18 & -1 & -13 \end{bmatrix}$$

PROGRESS CHECK
Find the product AB if

$$A = \begin{bmatrix} -2 & -1 & 2 \\ 4 & 3 & 1 \end{bmatrix} \qquad B = \begin{bmatrix} 5 & -4 \\ 3 & 1 \\ -1 & 0 \end{bmatrix}$$

Answer

$$AB = \begin{bmatrix} -15 & 7 \\ 28 & -13 \end{bmatrix}$$

It is important to note that the product AB of an $m \times n$ matrix A and an $n \times r$ matrix B exists only when the number of columns of A equals the number of rows of B (see Figure 1). The product AB will then be of dimension $m \times r$.

EXAMPLE 6
Given the matrices

$$A = \begin{bmatrix} 1 & -1 \\ 2 & 3 \end{bmatrix} \qquad B = \begin{bmatrix} 5 & -3 \\ -2 & 2 \end{bmatrix} \qquad C = \begin{bmatrix} 3 & -1 & -2 \\ 1 & 0 & 4 \end{bmatrix} \qquad D = \begin{bmatrix} 1 \\ 2 \\ 3 \end{bmatrix}$$

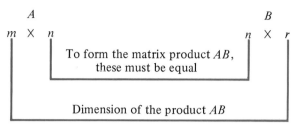

FIGURE 1

(a)   Show that $AB \neq BA$.
(b)   Determine the dimension of AC.

Solution

(a) $AB = \begin{bmatrix} (1)(5) + (-1)(-2) & (1)(-3) + (-1)(2) \\ (2)(5) + (3)(-2) & (2)(-3) + (3)(2) \end{bmatrix} = \begin{bmatrix} 7 & -5 \\ 4 & 0 \end{bmatrix}$

$BA = \begin{bmatrix} (5)(1) + (-3)(2) & (5)(-1) + (-3)(3) \\ (-2)(1) + (2)(2) & (-2)(-1) + (2)(3) \end{bmatrix} = \begin{bmatrix} -1 & -14 \\ 2 & 8 \end{bmatrix}$

Since the corresponding elements of AB and BA are not equal, $AB \neq BA$.

(b)   The product of a 2×2 matrix and a 2×3 matrix is a 2×3 matrix.

PROGRESS CHECK
If possible, find the dimension of CD and of CB using the matrices of Example 6.

Answer
2×1; not defined

We saw in Example 6 that $AB \neq BA$, that is, the commutative law does not hold for matrix multiplication. However, the associative law $A(BC) = (AB)C$ does hold when the dimensions of A, B, and C permit us to find the necessary products.

PROGRESS CHECK
Verify that $A(BC) = (AB)C$ for the matrices A, B, and C of Example 6.

Matrix multiplication provides a convenient shorthand means of writing a linear system. For example, the linear system

$$2x - y - 2z = 3$$
$$3x + 2y + z = -1$$
$$x + y - 3z = 14$$

can be expressed as

$$AX = B$$

where

$$A = \begin{bmatrix} 2 & -1 & -2 \\ 3 & 2 & 1 \\ 1 & 1 & -3 \end{bmatrix} \qquad X = \begin{bmatrix} x \\ y \\ z \end{bmatrix} \qquad B = \begin{bmatrix} 3 \\ -1 \\ 14 \end{bmatrix}$$

To verify this, simply form the matrix product AX and then apply the definition of matrix equality to the matrix equation $AX = B$.

EXAMPLE 7
Write out the linear system $AX = B$ if

$$A = \begin{bmatrix} -2 & 3 \\ 1 & 4 \end{bmatrix} \qquad X = \begin{bmatrix} x \\ y \end{bmatrix} \qquad B = \begin{bmatrix} 16 \\ -3 \end{bmatrix}$$

Solution
Equating corresponding elements of the matrix equation $AX = B$ yields

$$-2x + 3y = 16$$
$$x + 4y = -3$$

EXERCISE SET 7.2
1. For what values of a, b, c, and d are the matrices A and B equal?

$$A = \begin{bmatrix} a & b \\ 6 & -2 \end{bmatrix} \qquad B = \begin{bmatrix} 3 & -4 \\ c & d \end{bmatrix}$$

2. For what values of a, b, c, and d are the matrices A and B equal?

$$A = \begin{bmatrix} a+b & 2c \\ a & c-d \end{bmatrix} \qquad B = \begin{bmatrix} -1 & 6 \\ 5 & 10 \end{bmatrix}$$

In Exercises 3–18 the following matrices are given.

$$A = \begin{bmatrix} 2 & 3 & 1 \\ -3 & 4 & 1 \end{bmatrix} \qquad B = \begin{bmatrix} 2 & -1 \\ 3 & 2 \\ 4 & 1 \end{bmatrix} \qquad C = \begin{bmatrix} 1 & 2 & 3 \\ 4 & -1 & 2 \\ 3 & 2 & 5 \end{bmatrix}$$

$$D = \begin{bmatrix} -3 & 2 \\ 4 & 1 \end{bmatrix} \qquad E = \begin{bmatrix} 1 & -3 & 2 \\ 3 & 2 & 4 \\ 1 & 1 & 2 \end{bmatrix} \qquad F = \begin{bmatrix} 1 & 3 \\ -2 & 4 \end{bmatrix}$$

$$G = \begin{bmatrix} -2 & 4 & 2 \\ 1 & 0 & 3 \end{bmatrix}$$

If possible, compute the indicated matrix.

3. $C + E$	4. $C - E$	5. $2A + 3G$
6. $3G - 4A$	7. $A + F$	8. $2B - D$
9. AB	10. BA	11. $CB + D$
12. $EB - FA$	13. $DF + AB$	14. $AC + 2DG$
15. $DA + EB$	16. $FG + B$	17. $2GE - 3A$
18. $AB + FG$		

19. If $A = \begin{bmatrix} -2 & 3 \\ 2 & -3 \end{bmatrix}$, $B = \begin{bmatrix} -1 & 3 \\ 2 & 0 \end{bmatrix}$, $C = \begin{bmatrix} -4 & -3 \\ 0 & -4 \end{bmatrix}$, show that $AB = AC$.

20. If $A = \begin{bmatrix} 1 & 2 \\ 3 & 2 \end{bmatrix}$ and $B = \begin{bmatrix} 2 & -1 \\ -3 & 4 \end{bmatrix}$, show that $AB \neq BA$.

21. If $A = \begin{bmatrix} -2 & 3 \\ 2 & -3 \end{bmatrix}$ and $B = \begin{bmatrix} 3 & 6 \\ 2 & 4 \end{bmatrix}$, show that $AB = \begin{bmatrix} 0 & 0 \\ 0 & 0 \end{bmatrix}$.

22. If $A = \begin{bmatrix} 0 & 1 \\ 1 & 0 \end{bmatrix}$, show that $A \cdot A = \begin{bmatrix} 1 & 0 \\ 0 & 1 \end{bmatrix}$.

23. If $I = \begin{bmatrix} 1 & 0 & 0 \\ 0 & 1 & 0 \\ 0 & 0 & 1 \end{bmatrix}$ and $A = \begin{bmatrix} a_{11} & a_{12} & a_{13} \\ a_{21} & a_{22} & a_{23} \\ a_{31} & a_{32} & a_{33} \end{bmatrix}$,

show that $AI = A$ and $IA = A$.

24. Pesticides are sprayed on plants to eliminate harmful insects. However, some of the pesticide is absorbed by the plant, and the pesticide is then absorbed by herbivores (plant-eating animals such as cows) when they eat the plants that have been sprayed. Suppose that we have three pesticides and four plants and that the amounts of pesticide absorbed by the different plants are given by the matrix

$$
\begin{array}{cccc}
\text{Plant 1} & \text{Plant 2} & \text{Plant 3} & \text{Plant 4}
\end{array}
$$

$$
A = \begin{bmatrix} 3 & 2 & 4 & 3 \\ 6 & 5 & 2 & 4 \\ 4 & 3 & 1 & 5 \end{bmatrix} \quad \begin{array}{l} \text{Pesticide 1} \\ \text{Pesticide 2} \\ \text{Pesticide 3} \end{array}
$$

where a_{ij} denotes the amount of pesticide i in milligrams that has been absorbed by plant j. Thus, plant 4 has absorbed 5 mg of pesticide 3. Now suppose that we have three herbivores and that the number of plants eaten by these animals are given by the matrix

$$
\begin{array}{ccc}
\text{Herbivore 1} & \text{Herbivore 2} & \text{Herbivore 3}
\end{array}
$$

$$
B = \begin{bmatrix} 18 & 30 & 20 \\ 12 & 15 & 10 \\ 16 & 12 & 8 \\ 6 & 4 & 12 \end{bmatrix} \quad \begin{array}{l} \text{Plant 1} \\ \text{Plant 2} \\ \text{Plant 3} \\ \text{Plant 4} \end{array}
$$

How much of pesticide 2 has been absorbed by herbivore 3?

25. What does entry (2,3) in the matrix product AB of Exercise 24 represent?

In Exercises 26–29 indicate the matrices A, X, and B so that the matrix equation $AX = B$ is equivalent to the given linear system.

26. $\quad 7x - 2y = 6$
$\quad -2x + 3y = -2$

27. $\quad 3x + 4y = -3$
$\quad 3x - y = 5$

28. $5x + 2y - 3z = 4$

$\quad 2x - \dfrac{1}{2}y + z = 10$

$\quad x + y - 5z = -3$

29. $\quad 3x - y + 4z = 5$

$\quad 2x + 2y + \dfrac{3}{4}z = -1$

$\quad x - \dfrac{1}{4}y + z = \dfrac{1}{2}$

In Exercises 30–33 write out the linear system that is represented by the matrix equation $AX = B$.

30. $A = \begin{bmatrix} 2 & -1 \\ -3 & 4 \end{bmatrix}$ $X = \begin{bmatrix} x \\ y \end{bmatrix}$ $B = \begin{bmatrix} -2 \\ 10 \end{bmatrix}$

31. $A = \begin{bmatrix} 1 & -5 \\ 4 & 3 \end{bmatrix}$ $X = \begin{bmatrix} x_1 \\ x_2 \end{bmatrix}$ $B = \begin{bmatrix} 0 \\ 2 \end{bmatrix}$

32. $A = \begin{bmatrix} 1 & 7 & -2 \\ 3 & 6 & 1 \\ -4 & 2 & 0 \end{bmatrix}$ $X = \begin{bmatrix} x \\ y \\ z \end{bmatrix}$ $B = \begin{bmatrix} 3 \\ -3 \\ 2 \end{bmatrix}$

33. $A = \begin{bmatrix} 4 & 5 & -2 \\ 0 & 3 & -1 \\ 0 & 0 & 2 \end{bmatrix}$ $X = \begin{bmatrix} x_1 \\ x_2 \\ x_3 \end{bmatrix}$ $B = \begin{bmatrix} 2 \\ -5 \\ 4 \end{bmatrix}$

34. The $m \times n$ matrix all of whose elements are zero is called the **zero matrix** and is denoted by O. Show that $A + O = A$ for every $m \times n$ matrix A.

35. The square matrix of order n such that $a_{ii} = 1$ and $a_{ij} = 0$ when $i \neq j$ is called the **identity matrix** of order n and is denoted by I_n. (Note: The definition indicates that the diagonal elements are all equal to 1 and all elements off the diagonal are 0.) Show that $AI_n = I_n A$ for every square matrix A of order n.

36. The matrix B, each of whose entries is the negative of the corresponding entry of matrix A, is called the **additive inverse** of the matrix A. Show that $A + B = O$ where O is the zero matrix (see Exercise 34).

7.3
INVERSES OF MATRICES (Optional)

If $a \neq 0$, then the linear equation $ax = b$ can be solved easily by multiplying both sides by the reciprocal of a. Thus, we obtain $x = \dfrac{1}{a} \cdot b$. It would be nice if we could multiply both sides of the matrix equation $AX = B$ by the "reciprocal of A." Unfortunately, a matrix has *no* reciprocal. However, we shall discuss a notion that, for a square matrix, provides an analogue of the reciprocal of a real number, and will enable us to solve the linear system in a manner distinct from the Gauss–Jordan method discussed in Section 1 of this chapter.

In this section we confine our attention to square matrices. The $n \times n$ matrix

$$I_n = \begin{bmatrix} 1 & 0 & 0 \ldots 0 \\ 0 & 1 & 0 \ldots 0 \\ \cdot & \cdot & \cdot & \cdot \\ \cdot & \cdot & \cdot & \cdot \\ \cdot & \cdot & \cdot & \cdot \\ 0 & 0 & 0 \ldots 1 \end{bmatrix}$$

that has 1's on the main diagonal and 0's elsewhere is called the **identity matrix**. Examples of identity matrices are

$$I_2 = \begin{bmatrix} 1 & 0 \\ 0 & 1 \end{bmatrix} \quad I_3 = \begin{bmatrix} 1 & 0 & 0 \\ 0 & 1 & 0 \\ 0 & 0 & 1 \end{bmatrix} \quad I_4 = \begin{bmatrix} 1 & 0 & 0 & 0 \\ 0 & 1 & 0 & 0 \\ 0 & 0 & 1 & 0 \\ 0 & 0 & 0 & 1 \end{bmatrix}$$

If A is any $n \times n$ matrix we can show that

$$AI_n = I_n A = A$$

(see Exercise 35, Section 2). Thus, I_n is the matrix analogue of the real number 1.

An $n \times n$ matrix A is called **invertible** or **nonsingular** if we can find an $n \times n$ matrix B such that

$$AB = BA = I_n$$

The matrix B is called an **inverse** of A.

EXAMPLE 1
Let

$$A = \begin{bmatrix} 2 & 1 \\ 3 & 2 \end{bmatrix} \quad \text{and} \quad B = \begin{bmatrix} 2 & -1 \\ -3 & 2 \end{bmatrix}$$

Since

$$AB = BA = \begin{bmatrix} 1 & 0 \\ 0 & 1 \end{bmatrix} \quad \text{(Verify this.)}$$

we conclude that A is an invertible matrix and that B is an inverse of A. Of course, if B is an inverse of A, then A is an inverse of B.

If an $n \times n$ matrix A has an inverse, then it can be shown that it can only have one inverse. We denote the inverse of A by A^{-1}. Thus, we have

$$AA^{-1} = I_n \quad \text{and} \quad A^{-1}A = I_n$$

Note that the products AA^{-1} and $A^{-1}A$ yield the *identity matrix*, and that the products $a \cdot \dfrac{1}{a}$ and $\dfrac{1}{a} \cdot a$ yield the *identity element*. For this reason, A^{-1} may be thought of as the matrix analogue of the reciprocal $\dfrac{1}{a}$ of the real number a.

PROGRESS CHECK
Verify that the matrices

$$A = \begin{bmatrix} 4 & 5 \\ 2 & 2 \end{bmatrix} \quad \text{and} \quad B = \begin{bmatrix} -1 & \dfrac{5}{2} \\ 1 & -2 \end{bmatrix}$$

are inverses of each other.

WARNING

If $a \neq 0$ is a real number, then a^{-1} has the property that $aa^{-1} = a^{-1}a = 1$. Since $a^{-1} = \dfrac{1}{a}$, we may refer to a^{-1} as the inverse *or* reciprocal of a. However, the matrix A^{-1} is the inverse of the $n \times n$ matrix A since $AA^{-1} = A^{-1}A = I_n$ but cannot be referred to as the reciprocal of A since *matrix division is not defined.*

We now develop a practical method for finding the inverse of an invertible matrix. Suppose we want to find the inverse of the matrix

$$A = \begin{bmatrix} 1 & 3 \\ 2 & 5 \end{bmatrix}$$

Let the inverse be denoted by

$$B = \begin{bmatrix} b_1 & b_2 \\ b_3 & b_4 \end{bmatrix}$$

Then we must have

$$AB = I_2 \tag{1}$$

and

$$BA = I_2 \tag{2}$$

Equation (1) now becomes

$$\begin{bmatrix} 1 & 3 \\ 2 & 5 \end{bmatrix} \begin{bmatrix} b_1 & b_2 \\ b_3 & b_4 \end{bmatrix} = \begin{bmatrix} 1 & 0 \\ 0 & 1 \end{bmatrix}$$

or

$$\begin{bmatrix} b_1 + 3b_3 \\ 2b_1 + 5b_3 \end{bmatrix} \begin{bmatrix} b_2 + 3b_4 \\ 2b_2 + 5b_4 \end{bmatrix} = \begin{bmatrix} 1 & 0 \\ 0 & 1 \end{bmatrix}$$

Since two matrices are equal if and only if their corresponding entries are equal, we have

$$\begin{aligned} b_1 + 3b_3 &= 1 \\ 2b_1 + 5b_3 &= 0 \end{aligned} \tag{3}$$

and

$$\begin{aligned} b_2 + 3b_4 &= 0 \\ 2b_2 + 5b_4 &= 1 \end{aligned} \tag{4}$$

We solve the linear systems (3) and (4) by Gauss–Jordan elimination. We begin with the augmented matrices of the linear systems and perform a sequence of elementary row operations as follows.

$$
\begin{array}{cc}
(3) & (4) \\
\begin{bmatrix} 1 & 3 & | & 1 \\ 2 & 5 & | & 0 \end{bmatrix} &
\begin{bmatrix} 1 & 3 & | & 0 \\ 2 & 5 & | & 1 \end{bmatrix}
\end{array}
$$
Augmented matrices of (3) and (4).

$$
\begin{bmatrix} 1 & 3 & | & 1 \\ 0 & -1 & | & -2 \end{bmatrix} \qquad
\begin{bmatrix} 1 & 3 & | & 0 \\ 0 & -1 & | & 1 \end{bmatrix}
$$
To make $a_{21} = 0$, replaced row 2 by the sum of itself and -2 times row 1.

$$
\begin{bmatrix} 1 & 3 & | & 1 \\ 0 & 1 & | & 2 \end{bmatrix} \qquad
\begin{bmatrix} 1 & 3 & | & 0 \\ 0 & 1 & | & -1 \end{bmatrix}
$$
Multiplied row 2 by -1 to obtain $a_{22} = 1$.

$$
\begin{bmatrix} 1 & 0 & | & -5 \\ 0 & 1 & | & 2 \end{bmatrix} \qquad
\begin{bmatrix} 1 & 0 & | & 3 \\ 0 & 1 & | & -1 \end{bmatrix}
$$
To make $a_{12} = 0$, replaced row 1 by the sum of itself and -3 times row 2.

Thus, $b_1 = -5$ and $b_3 = 2$ is the solution of (3), and $b_2 = 3$ and $b_4 = -1$ is the solution of (4). We can check that

$$
B = \begin{bmatrix} -5 & 3 \\ 2 & -1 \end{bmatrix}
$$

also satisfies the requirement $BA = I_2$ of Equation (2).

Observe that the linear systems (3) and (4) have the same coefficient matrix and that an identical sequence of elementary row operations was performed in the Gauss–Jordan elimination. This suggests that we can solve the systems *at the same time*. We simply write the coefficient matrix A and next to it list the right-hand sides of (3) and (4) to obtain the matrix

$$
\begin{bmatrix} 1 & 3 & | & 1 & 0 \\ 2 & 5 & | & 0 & 1 \end{bmatrix} \tag{5}
$$

Note that the columns of right-hand sides to the right of the dashed line in (5) form the identity matrix I_2. Performing the same sequence of elementary row operations on matrix (5) that we did on matrices (3) and (4) yields

$$
\begin{bmatrix} 1 & 0 & | & -5 & 3 \\ 0 & 1 & | & 2 & -1 \end{bmatrix} \tag{6}
$$

Then A^{-1} is the matrix to the right of the dashed line in (6).

The procedure outlined for the 2×2 matrix A applies in general. Thus, we have the following method for finding the inverse of an invertible $n \times n$ matrix A.

Computing A^{-1}

Step 1. Form the $n \times 2n$ matrix $[A \mid I_n]$ by adjoining the identity matrix I_n to the given matrix A.

Step 2. Apply elementary row operations to the matrix $[A \mid I_n]$ to transform the matrix A to I_n.

Step 3. The final matrix is of the form $[I_n \mid B]$ where B is A^{-1}.

EXAMPLE 2
Find the inverse of

$$A = \begin{bmatrix} 1 & 2 & 3 \\ 2 & 5 & 7 \\ 1 & 1 & 1 \end{bmatrix}$$

Solution
We form the 3×6 matrix $[A \mid I_3]$ and transform it by elementary row operations to the form $[I_3 \mid A^{-1}]$. The pivot element at each stage is shown in color.

$$\begin{bmatrix} 1 & 2 & 3 & \vline & 1 & 0 & 0 \\ 2 & 5 & 7 & \vline & 0 & 1 & 0 \\ 1 & 1 & 1 & \vline & 0 & 0 & 1 \end{bmatrix}$$ Matrix A augmented by I_3.

$$\begin{bmatrix} 1 & 2 & 3 & \vline & 1 & 0 & 0 \\ 0 & 1 & 1 & \vline & -2 & 1 & 0 \\ 0 & -1 & -2 & \vline & -1 & 0 & 1 \end{bmatrix}$$ To make $a_{21} = 0$, replaced row 2 by the sum of itself and -2 times row 1.

To make $a_{31} = 0$, replaced row 3 by the sum of itself and -1 times row 1.

$$\begin{bmatrix} 1 & 0 & 1 & \vline & 5 & -2 & 0 \\ 0 & 1 & 1 & \vline & -2 & 1 & 0 \\ 0 & 0 & -1 & \vline & -3 & 1 & 1 \end{bmatrix}$$ To make $a_{12} = 0$, replaced row 1 by the sum of itself and -2 times row 2.

To make $a_{32} = 0$, replaced row 3 by the sum of itself and row 2.

$$\begin{bmatrix} 1 & 0 & 1 & \vline & 5 & -2 & 0 \\ 0 & 1 & 1 & \vline & -2 & 1 & 0 \\ 0 & 0 & 1 & \vline & 3 & -1 & -1 \end{bmatrix}$$ Multiplied row 3 by -1.

$$\begin{bmatrix} 1 & 0 & 0 & \vdots & 2 & -1 & 1 \\ 0 & 1 & 0 & \vdots & -5 & 2 & 1 \\ 0 & 0 & 1 & \vdots & 3 & -1 & -1 \end{bmatrix}$$

To make $a_{13} = 0$, replaced row 1 by the sum of itself and -1 times row 3.

To make $a_{23} = 0$, replaced row 2 by the sum of itself and -1 times row 3.

The final matrix is of the form $[I_3 \mid A^{-1}]$, that is,

$$A^{-1} = \begin{bmatrix} 2 & -1 & 1 \\ -5 & 2 & 1 \\ 3 & -1 & -1 \end{bmatrix}$$

We now have a practical method for finding the inverse of an invertible matrix, but we don't know whether a given square matrix *has* an inverse. It can be shown that if the preceding procedure is carried out with the matrix $[A \mid I_n]$ and we arrive at a point at which all possible candidates for the next pivot element are zero, then the matrix is not invertible and we may stop our calculations.

EXAMPLE 3
Find the inverse of

$$A = \begin{bmatrix} 1 & 2 & 6 \\ 0 & 0 & 2 \\ -3 & -6 & -9 \end{bmatrix}$$

Solution
We begin with $[A \mid I_3]$.

$$\begin{bmatrix} 1 & 2 & 6 & \vdots & 1 & 0 & 0 \\ 0 & 0 & 2 & \vdots & 0 & 1 & 0 \\ -3 & -6 & -9 & \vdots & 0 & 0 & 1 \end{bmatrix}$$

$$\begin{bmatrix} 1 & 2 & 6 & \vdots & 1 & 0 & 0 \\ 0 & 0 & 2 & \vdots & 0 & 1 & 0 \\ 0 & 0 & 9 & \vdots & 3 & 0 & 1 \end{bmatrix}$$

To make $a_{31} = 0$, replaced row 3 by the sum of itself and 3 times row 1.

Note that $a_{22} = a_{32} = 0$ in the last matrix. We cannot perform any elementary row operations upon rows 2 and 3 that will produce a nonzero pivot element for a_{22}. We conclude that the matrix A does not have an inverse.

PROGRESS CHECK
Show that the matrix A is not invertible.

$$A = \begin{bmatrix} 1 & 2 & -3 \\ 3 & 2 & 1 \\ 5 & 6 & -5 \end{bmatrix}$$

SOLVING LINEAR SYSTEMS

Consider a linear system of n equations in n unknowns.

$$
\begin{aligned}
a_{11}x_1 + a_{12}x_2 + \cdots + a_{1n}x_n &= b_1 \\
a_{12}x_2 + a_{22}x_2 + \cdots + a_{2n}x_n &= b_2 \\
&\ \ \vdots \\
a_{n1}x_1 + a_{n2}x_2 + \cdots + a_{nn}x_n &= b_n
\end{aligned}
\tag{7}
$$

As has already been pointed out in Section 2 of this chapter, we can write the linear system (7) in matrix form as

$$AX = B \tag{8}$$

where

$$
A = \begin{bmatrix}
a_{11} & a_{12}\dots a_{1n} \\
a_{21} & a_{22}\dots a_{2n} \\
\vdots & \vdots \quad \vdots \\
a_{n1} & a_{n2}\dots a_{nn}
\end{bmatrix}
\qquad
X = \begin{bmatrix}
x_1 \\ x_2 \\ \vdots \\ x_n
\end{bmatrix}
\qquad
B = \begin{bmatrix}
b_1 \\ b_2 \\ \vdots \\ b_n
\end{bmatrix}
$$

Suppose now that the coefficient matrix A is invertible so that we can compute A^{-1}. Multiplying both sides of (8) by A^{-1} we have

$$
\begin{aligned}
A^{-1}(AX) &= A^{-1}B \\
(A^{-1}A)X &= A^{-1}B &&\text{Associative law} \\
I_nX &= A^{-1}B &&A^{-1}A = I_n \\
X &= A^{-1}B &&I_nX = X
\end{aligned}
$$

Thus, we have the following result.

> If $AX = B$ is a linear system of n equations in n unknowns and if the coefficient matrix A is invertible, then the system has exactly one solution given by
>
> $$X = A^{-1}B$$

EXAMPLE 4

Solve the linear system by finding the inverse of the coefficient matrix.

$$
\begin{aligned}
x + 2y + 3z &= -3 \\
2x + 5y + 7z &= 4 \\
x + y + z &= 5
\end{aligned}
$$

Solution

The coefficient matrix

$$A = \begin{bmatrix} 1 & 2 & 3 \\ 2 & 5 & 7 \\ 1 & 1 & 1 \end{bmatrix}$$

is the matrix whose inverse was obtained in Example 2 as

$$A^{-1} = \begin{bmatrix} 2 & -1 & 1 \\ -5 & 2 & 1 \\ 3 & -1 & -1 \end{bmatrix}$$

Since

$$B = \begin{bmatrix} -3 \\ 4 \\ 5 \end{bmatrix}$$

we obtain the solution of the given system as

$$X = A^{-1}B = \begin{bmatrix} 2 & -1 & 1 \\ -5 & 2 & 1 \\ 3 & -1 & -1 \end{bmatrix} \begin{bmatrix} -3 \\ 4 \\ 5 \end{bmatrix} = \begin{bmatrix} -5 \\ 28 \\ -18 \end{bmatrix}$$

Thus $x = -5$, $y = 28$, $z = -18$.

PROGRESS CHECK

Solve the linear system by finding the inverse of the coefficient matrix.

$$\begin{aligned} x - 2y + z &= 1 \\ x + 3y + 2z &= 2 \\ -x \quad\quad + z &= -11 \end{aligned}$$

Answer
x = 7, y = 1, z = −4

The inverse of the coefficient matrix is especially useful when we need to solve a number of linear systems

$$AX = B_1, \quad AX = B_2, \quad \cdots, \quad AX = B_k$$

where the coefficient matrix is the same and the right-hand side changes.

EXAMPLE 5

A steel producer makes two types of steel, regular and special. A ton of regular steel requires 2 hours in the open-hearth furnace and 5 hours in the soaking pit; a ton of special steel requires 2 hours in the open-hearth furnace and 3 hours in the

soaking pit. How many tons of each type of steel can be manufactured daily if
(a)   the open-hearth furnace is available 8 hours per day and the soaking pit is
available 15 hours per day?
(b)   the open-hearth furnace is available 9 hours per day and the soaking pit is
available 15 hours per day?

Solution

Let $x =$ the number of tons of regular steel to be made

$y =$ the number of tons of special steel to be made

Then the total amount of time required in the open-hearth furnace is

$$2x + 2y$$

Similarly, the total amount of time required in the soaking pit is

$$5x + 3y$$

If we let b_1 and b_2 denote the number of hours that the open-hearth furnace and the
soaking pit, respectively, are available per day, then we have

$$2x + 2y = b_1$$
$$5x + 3y = b_2$$

or

$$\begin{bmatrix} 2 & 2 \\ 5 & 3 \end{bmatrix} \begin{bmatrix} x \\ y \end{bmatrix} = \begin{bmatrix} b_1 \\ b_2 \end{bmatrix}$$

Then

$$\begin{bmatrix} x \\ y \end{bmatrix} = \begin{bmatrix} 2 & 2 \\ 5 & 3 \end{bmatrix}^{-1} \begin{bmatrix} b_1 \\ b_2 \end{bmatrix}$$

We find (verify) the inverse of the coefficient matrix to be

$$\begin{bmatrix} 2 & 2 \\ 5 & 3 \end{bmatrix}^{-1} = \begin{bmatrix} -\frac{3}{4} & \frac{1}{2} \\ \frac{5}{4} & -\frac{1}{2} \end{bmatrix}$$

(a)   We are given $b_1 = 8$ and $b_2 = 15$. Then

$$\begin{bmatrix} x \\ y \end{bmatrix} = \begin{bmatrix} -\frac{3}{4} & \frac{1}{2} \\ \frac{5}{4} & -\frac{1}{2} \end{bmatrix} \begin{bmatrix} 8 \\ 15 \end{bmatrix} = \begin{bmatrix} \frac{3}{2} \\ \frac{5}{2} \end{bmatrix}$$

That is, $\frac{3}{2}$ tons of regular steel and $\frac{5}{2}$ tons of special steel can be manufactured
daily.

(b) We are given $b_1 = 9$ and $b_2 = 15$. Then

$$\begin{bmatrix} x \\ y \end{bmatrix} = \begin{bmatrix} -\frac{3}{4} & \frac{1}{2} \\ \frac{5}{4} & -\frac{1}{2} \end{bmatrix} \begin{bmatrix} 9 \\ 15 \end{bmatrix} = \begin{bmatrix} \frac{3}{4} \\ \frac{-15}{4} \end{bmatrix}$$

That is, $\frac{3}{4}$ tons of regular steel and $\frac{-15}{4}$ tons of special steel can be manufactured daily.

EXERCISE SET 7.3

In Exercises 1–4 determine whether the matrix B is the inverse of the matrix A.

1. $A = \begin{bmatrix} 2 & \frac{1}{2} \\ -1 & 3 \end{bmatrix}$ $B = \begin{bmatrix} 1 & -1 \\ -2 & 4 \end{bmatrix}$

2. $A = \begin{bmatrix} 3 & -1 \\ -2 & 2 \end{bmatrix}$ $B = \begin{bmatrix} \frac{1}{2} & \frac{1}{4} \\ \frac{1}{2} & \frac{3}{4} \end{bmatrix}$

3. $A = \begin{bmatrix} 1 & 2 & 2 \\ -1 & 3 & 0 \\ 0 & 2 & 1 \end{bmatrix}$ $B = \begin{bmatrix} 3 & 2 & -6 \\ 1 & 1 & -2 \\ -2 & -2 & 5 \end{bmatrix}$

4. $A = \begin{bmatrix} 1 & 0 & -2 \\ 2 & 1 & 3 \\ -4 & 1 & 2 \end{bmatrix}$ $B = \begin{bmatrix} 1 & 2 & -2 \\ -2 & -4 & 1 \\ 0 & 1 & -1 \end{bmatrix}$

In Exercises 5–10 find the inverse of the given matrix.

5. $\begin{bmatrix} -1 & 5 \\ 2 & -4 \end{bmatrix}$ 6. $\begin{bmatrix} 2 & 0 \\ -1 & -2 \end{bmatrix}$ 7. $\begin{bmatrix} -1 & 1 \\ -2 & 1 \end{bmatrix}$

8. $\begin{bmatrix} 2 & 1 & 0 \\ 1 & 1 & 0 \\ 1 & 1 & 1 \end{bmatrix}$ 9. $\begin{bmatrix} 1 & -2 & 3 \\ -1 & 3 & -4 \\ 0 & 5 & -4 \end{bmatrix}$ 10. $\begin{bmatrix} 1 & 1 & 0 \\ 1 & 0 & 0 \\ 1 & 2 & 2 \end{bmatrix}$

In Exercises 11–18 find the inverse, if possible.

11. $\begin{bmatrix} 1 & 3 \\ -1 & 4 \end{bmatrix}$ 12. $\begin{bmatrix} 6 & -4 \\ 9 & -6 \end{bmatrix}$

13. $\begin{bmatrix} 1 & 1 & 3 \\ 2 & -8 & -4 \\ -1 & 2 & 0 \end{bmatrix}$ 14. $\begin{bmatrix} 8 & 7 & -1 \\ -5 & -5 & 1 \\ -4 & -4 & 1 \end{bmatrix}$

15. $\begin{bmatrix} 2 & 0 \\ 0 & -3 \end{bmatrix}$ 16. $\begin{bmatrix} -1 & 0 & 0 \\ 0 & 4 & 0 \\ 0 & 0 & 2 \end{bmatrix}$

17. $\begin{bmatrix} 1 & 0 & -1 \\ 2 & 1 & 0 \\ 0 & 1 & 1 \end{bmatrix}$ 18. $\begin{bmatrix} 1 & 0 & -3 & 0 \\ 0 & 1 & 0 & 0 \\ -1 & 0 & 4 & 0 \\ 2 & 0 & -6 & 1 \end{bmatrix}$

In Exercises 19–24 solve the given linear system by finding the inverse of the coefficient matrix.

19. $2x + y = 5$
$\quad x - 3y = 6$

20. $2x - 3y = -5$
$\quad 3x + y = -13$

21. $3x + y - z = 2$
$\quad x - 2y \quad = 8$
$\quad\quad 3y + z = -8$

22. $3x + 2y - z = 10$
$\quad 2x - y + z = -1$
$\quad -x + y - 2z = 5$

23. $2x - y + 3z = -11$
$\quad 3x - y + z = -5$
$\quad x + y + z = -1$

24. $2x + 3y - 2z = 13$
$\quad 4x + 2y + z = 3$
$\quad\quad\quad y - z = 5$

25–34. Solve the linear systems of Section 1 of this chapter, Exercises 21–30 by finding the inverse of the coefficient matrix.

35. Solve the linear systems $AX = B_1$ and $AX = B_2$ given

$$A^{-1} = \begin{bmatrix} 3 & -2 & 4 \\ 2 & -1 & 0 \\ 0 & 4 & 1 \end{bmatrix} \quad B_1 = \begin{bmatrix} 1 \\ -1 \\ 5 \end{bmatrix} \quad B_2 = \begin{bmatrix} 4 \\ 3 \\ -2 \end{bmatrix}$$

36. Solve the linear systems $AX = B_1$ and $AX = B_2$ given

$$A^{-1} = \begin{bmatrix} 1 & 0 & -1 \\ 1 & 2 & 0 \\ -1 & -1 & 3 \end{bmatrix} \quad B_1 = \begin{bmatrix} 2 \\ -3 \\ 2 \end{bmatrix} \quad B_2 = \begin{bmatrix} 4 \\ -3 \\ -5 \end{bmatrix}$$

37. Show that the matrix

$$\begin{bmatrix} a & b & c \\ 0 & 0 & 0 \\ d & e & f \end{bmatrix}$$

is not invertible.

38. A trustee decides to invest \$30,000 in two mortgages which yield 10% and 15% per year, respectively. How should the \$30,000 be invested in the two mortgages if the total annual interest is to be

(a) \$3600   (b) \$4000   (c) \$5000

(*Hint:* Some of these investment objectives cannot be attained.)

7.4
DETERMINANTS AND CRAMER'S RULE

In this section we will define a determinant and will develop manipulative skills for evaluating determinants. We will then show that determinants have important applications and can be used to solve linear systems.

Associated with every square matrix A is a number called the **determinant of A,** denoted by $|A|$. If A is the 2×2 matrix

$$A = \begin{bmatrix} a_{11} & a_{12} \\ a_{21} & a_{22} \end{bmatrix}$$

then $|A|$ is said to be a **determinant of second order** and is defined by the rule

$$|A| = \begin{vmatrix} a_{11} & a_{12} \\ a_{21} & a_{22} \end{vmatrix} = a_{11}a_{22} - a_{21}a_{12}$$

EXAMPLE 1

Compute the real number represented by

$$\begin{vmatrix} 4 & -5 \\ 3 & -1 \end{vmatrix}$$

Solution

We apply the rule for a determinant of second order.

$$\begin{vmatrix} 4 & -5 \\ 3 & -1 \end{vmatrix} = (4)(-1) - (3)(-5) = 11$$

PROGRESS CHECK

Compute the real number represented by

(a) $\begin{vmatrix} -6 & 2 \\ -1 & -2 \end{vmatrix}$ (b) $\begin{vmatrix} \frac{1}{2} & \frac{1}{4} \\ -4 & -2 \end{vmatrix}$

Answer
(a) 14 *(b) 0*

 To simplify matters, when we want to compute the determinant of a matrix we will say "evaluate the determinant." This is not technically correct, however, since a determinant *is* a real number.
 The rule for evaluating a determinant of order 3 is

$$\begin{vmatrix} a_{11} & a_{12} & a_{13} \\ a_{21} & a_{22} & a_{23} \\ a_{31} & a_{32} & a_{33} \end{vmatrix} = \begin{matrix} a_{11}a_{22}a_{33} - a_{11}a_{32}a_{23} - a_{12}a_{21}a_{33} \\ + a_{12}a_{31}a_{23} + a_{13}a_{21}a_{32} - a_{13}a_{31}a_{22} \end{matrix}$$

The situation becomes even more cumbersome for determinants of higher order! Fortunately, we don't have to memorize this rule; instead, we shall see that it is possible to evaluate a determinant of order 3 by reducing the problem to that of evaluating a number of determinants of order 2.
 The **minor of an element** a_{ij} is the determinant of the matrix remaining after deleting the row and column in which the element a_{ij} appears. Given the matrix

$$\begin{bmatrix} 4 & 0 & -2 \\ 1 & -6 & 7 \\ -3 & 2 & 5 \end{bmatrix}$$

the minor of the element in row 2, column 3, is

$$
\begin{vmatrix} 4 & 0 & -2 \\ 1 & 6 & 7 \\ -3 & 2 & 5 \end{vmatrix} = \begin{vmatrix} 4 & 0 \\ -3 & 2 \end{vmatrix} = 8 - 0 = 8
$$

The **cofactor** of the element a_{ij} is the minor of the element a_{ij} multiplied by $(-1)^{i+j}$. Since $(-1)^{i+j}$ is $+1$ if $i + j$ is even and is -1 if $i + j$ is odd, we see that the cofactor is the minor with a sign attached. The cofactor attaches the sign to the minor according to this pattern.

$$
\begin{array}{ccccc}
+ & - & + & - & \cdots \\
- & + & - & + & \cdots \\
+ & - & + & - & \cdots \\
- & + & - & + & \cdots
\end{array}
$$

EXAMPLE 2
Find the cofactor of each element in the first row of the matrix.

$$
\begin{bmatrix} -2 & 0 & 12 \\ -4 & 5 & 3 \\ 7 & 8 & -6 \end{bmatrix}
$$

Solution
The cofactors are

$$
(-1)^{1+1} \begin{vmatrix} -2 & 0 & 12 \\ -4 & 5 & 3 \\ 7 & 8 & -6 \end{vmatrix} = \begin{vmatrix} 5 & 3 \\ 8 & -6 \end{vmatrix}
$$
$$
= -30 - 24 = -54
$$

$$
(-1)^{1+2} \begin{vmatrix} -2 & 0 & 12 \\ -4 & 5 & 3 \\ 7 & 8 & -6 \end{vmatrix} = - \begin{vmatrix} -4 & 3 \\ 7 & -6 \end{vmatrix}
$$
$$
= -(24 - 21) = -3
$$

$$
(-1)^{1+3} \begin{vmatrix} -2 & 0 & 12 \\ -4 & 5 & 3 \\ 7 & 8 & -6 \end{vmatrix} = \begin{vmatrix} -4 & 5 \\ 7 & 8 \end{vmatrix}
$$
$$
= -32 - 35 = -67
$$

PROGRESS CHECK

Find the cofactor of each element in the second column of the matrix.

$$\begin{bmatrix} 16 & -9 & 3 \\ -5 & 2 & 0 \\ -3 & 4 & -1 \end{bmatrix}$$

Answer

Cofactor of −9 is −5; cofactor of 2 is −7; cofactor of 4 is −15.

The cofactor is the key to the process of evaluating determinants of order 3 or higher.

To evaluate a determinant, form the sum of the products obtained by multiplying each element of any row or any column by its cofactor. This process is called **expansion by cofactors.**

Let's illustrate the process with an example.

EXAMPLE 3

Evaluate the determinant by cofactors.

$$\begin{vmatrix} -2 & 7 & 2 \\ 6 & -6 & 0 \\ 4 & 10 & -3 \end{vmatrix}$$

Solution

Expansion by Cofactors	
Step 1. Choose a row or column about which to expand. In general, a row or column containing zeros will simplify the work.	*Step 1.* We will expand about column 3.
Step 2. Expand about the cofactors of the chosen row or column by multiplying each element of the row or column by its cofactor.	*Step 2.* The expansion about column 3 is $$(2)(-1)^{1+3}\begin{vmatrix} 6 & -6 \\ 4 & 10 \end{vmatrix}$$ $$+(0)(-1)^{2+3}\begin{vmatrix} -2 & 7 \\ 4 & 10 \end{vmatrix}$$ $$+(-3)(-1)^{3+3}\begin{vmatrix} -2 & 7 \\ 6 & -6 \end{vmatrix}$$
Step 3. Evaluate the cofactors and form their sum.	*Step 3.* Using the rule for evaluating a determinant of order 2, we have $$(2)(1)[(6)(10) - (4)(-6)] + 0$$ $$+ (-3)(1)[(-2)(-6) - (6)(7)]$$ $$= 2(60 + 24) - 3(12 - 42)$$ $$= 258$$

Note that expansion by cofactors of *any row or any column* will produce the same result. This important property of determinants can be used to simplify the arithmetic. The best choice of a row or column about which to expand is one that has the most zero elements. The reason for this is that if an element is zero, the element times its cofactor will be zero, so we don't have to evaluate that cofactor.

PROGRESS CHECK

Evaluate the determinant of Example 3 by expanding about the second row.

Answer
258

EXAMPLE 4

Verify the rule for evaluating a determinant of order 3.

$$
\begin{vmatrix} a_{11} & a_{12} & a_{13} \\ a_{21} & a_{22} & a_{23} \\ a_{31} & a_{32} & a_{33} \end{vmatrix} = \begin{array}{l} a_{11}a_{22}a_{33} - a_{11}a_{32}a_{23} - a_{12}a_{21}a_{33} \\[2mm] + a_{12}a_{31}a_{23} + a_{13}a_{21}a_{32} - a_{13}a_{31}a_{22} \end{array}
$$

Solution

Expanding about the first row we have

$$
\begin{vmatrix} a_{11} & a_{12} & a_{13} \\ a_{21} & a_{22} & a_{23} \\ a_{31} & a_{32} & a_{33} \end{vmatrix} = a_{11} \begin{vmatrix} a_{22} & a_{23} \\ a_{32} & a_{33} \end{vmatrix} - a_{12} \begin{vmatrix} a_{21} & a_{23} \\ a_{31} & a_{33} \end{vmatrix} + a_{13} \begin{vmatrix} a_{21} & a_{22} \\ a_{31} & a_{32} \end{vmatrix}
$$

$$
= a_{11}(a_{22}a_{33} - a_{32}a_{23}) - a_{12}(a_{21}a_{33} - a_{31}a_{23}) + a_{13}(a_{21}a_{32} - a_{31}a_{22})
$$

$$
= a_{11}a_{22}a_{33} - a_{11}a_{32}a_{33} - a_{12}a_{21}a_{33} + a_{12}a_{31}a_{23} + a_{13}a_{21}a_{32} - a_{13}a_{31}a_{22}
$$

PROGRESS CHECK

Show that the determinant is equal to zero.

$$
\begin{vmatrix} a & b & c \\ a & b & c \\ d & e & f \end{vmatrix}
$$

The process of expanding by cofactors works for determinants of any order. If we apply the method to a determinant of order 4, we will produce determinants of order 3; applying the method again will result in determinants of order 2.

EXAMPLE 5

Evaluate the determinant.

$$\begin{vmatrix} -3 & 5 & 0 & -1 \\ 1 & 2 & 3 & -3 \\ 0 & 4 & -6 & 0 \\ 0 & -2 & 1 & 2 \end{vmatrix}$$

Solution
Expanding about the cofactors of the first column, we have

$$\begin{vmatrix} -3 & 5 & 0 & -1 \\ 1 & 2 & 3 & -3 \\ 0 & 4 & -6 & 0 \\ 0 & -2 & 1 & 2 \end{vmatrix} = -3 \begin{vmatrix} 2 & 3 & -3 \\ 4 & -6 & 0 \\ -2 & 1 & 2 \end{vmatrix} - 1 \begin{vmatrix} 5 & 0 & -1 \\ 4 & -6 & 0 \\ -2 & 1 & 2 \end{vmatrix}$$

Each determinant of order 3 can then be evaluated.

$$-3 \begin{vmatrix} 2 & 3 & -3 \\ 4 & -6 & 0 \\ -2 & 1 & 2 \end{vmatrix} = (-3)(-24) \qquad -1 \begin{vmatrix} 5 & 0 & -1 \\ 4 & -6 & 0 \\ -2 & 1 & 2 \end{vmatrix} = (-1)(-52)$$
$$= 72 \qquad\qquad\qquad\qquad = 52$$

The original determinant has the value $72 + 52 = 124$.

PROGRESS CHECK
Evaluate.

$$\begin{vmatrix} 0 & -1 & 0 & 2 \\ 3 & 0 & 4 & 0 \\ 0 & 5 & 0 & -3 \\ 1 & 0 & 1 & 0 \end{vmatrix}$$

Answer
7

CRAMER'S RULE
Determinants provide a convenient way of expressing formulas in many areas of mathematics, particularly in geometry. One of the best known uses of determinants is in solving systems of linear equations, a procedure known as **Cramer's rule.**

In the previous chapter we solved systems of linear equations by the method of elimination. Let's apply this method to the general system of two equations in two unknowns.

$$a_{11}x + a_{12}y = c_1 \qquad (1)$$
$$a_{21}x + a_{22}y = c_2 \qquad (2)$$

If we multiply Equation (1) by a_{22} and Equation (2) by $-a_{12}$ and add, we will eliminate y.

$$a_{11}a_{22}x + a_{12}a_{22}y = \quad c_1a_{22}$$

$$\underline{-a_{21}a_{12}x - a_{12}a_{22}y = -c_2a_{12}}$$

$$a_{11}a_{22}x - a_{21}a_{12}x = \quad c_1a_{22} - c_2a_{12}$$

Thus,

$$x(a_{11}a_{22} - a_{21}a_{12}) = c_1a_{22} - c_2a_{12}$$

or

$$x = \frac{c_1a_{22} - c_2a_{12}}{a_{11}a_{22} - a_{21}a_{12}}$$

Similarly, multiplying Equation (1) by a_{21} and Equation (2) by $-a_{11}$ and adding, we can eliminate x and solve for y.

$$y = \frac{c_2a_{11} - c_1a_{21}}{a_{11}a_{22} - a_{21}a_{12}}$$

The denominators in the expressions for x and y are identical and can be written as a determinant.

$$|D| = \begin{vmatrix} a_{11} & a_{12} \\ a_{21} & a_{22} \end{vmatrix}$$

If we apply this same idea to the numerators, we have

$$x = \frac{\begin{vmatrix} c_1 & a_{12} \\ c_2 & a_{22} \end{vmatrix}}{|D|} \qquad y = \frac{\begin{vmatrix} a_{11} & c_1 \\ a_{21} & c_2 \end{vmatrix}}{|D|} \qquad |D| \neq 0$$

What we have arrived at is Cramer's rule, which is a means of expressing the solution of a system of linear equations in determinant form.

The following example outlines the steps for using Cramer's rule.

EXAMPLE 6
Solve by Cramer's rule.

$$3x - y = 9$$

$$x + 2y = -4$$

Solution

Cramer's Rule	
Step 1. Form the determinant $|D|$ from the coefficients of x and y.	*Step 1.* $|D| = \begin{vmatrix} 3 & -1 \\ 1 & 2 \end{vmatrix}$

Step 2. The numerator for x is obtained from the determinant $\|D\|$ by replacing the column of coefficients of x by the column of right-hand sides of the equations.	*Step 2.* $$x = \frac{\begin{vmatrix} 9 & -1 \\ -4 & 2 \end{vmatrix}}{\|D\|}$$
Step 3. The numerator for y is obtained from the determinant $\|D\|$ by replacing the column of coefficients of y by the column of right-hand sides of the equations.	*Step 3.* $$y = \frac{\begin{vmatrix} 3 & 9 \\ 1 & -4 \end{vmatrix}}{\|D\|}$$
Step 4. Evaluate the determinants to obtain the solution. If $\|D\| = 0$, Cramer's rule cannot be used.	*Step 4.* $\|D\| = 6 + 1 = 7$ $x = \dfrac{18 - 4}{7} = \dfrac{14}{7} = 2$ $y = \dfrac{12 - 9}{7} = -\dfrac{21}{7} = -3$

PROGRESS CHECK

Solve by Cramer's rule.

$$2x + 3y = -4$$
$$3x + 4y = -7$$

Answer
$x = -5$, $y = 2$

The steps outlined in Example 6 can be applied to solve any system of linear equations in which the number of equations is the same as the number of variables and in which $\|D\| \neq 0$. Here is an example with three equations and three unknowns.

EXAMPLE 7

Solve by Cramer's rule.

$$3x \qquad + 2z = -2$$
$$2x - y \qquad = 0$$
$$2y + 6z = -1$$

Solution

We form the determinant of coefficients.

$$\|D\| = \begin{vmatrix} 3 & 0 & 2 \\ 2 & -1 & 0 \\ 0 & 2 & 6 \end{vmatrix}$$

Then

$$x = \frac{|D_1|}{|D|} \qquad y = \frac{|D_2|}{|D|} \qquad z = \frac{|D_3|}{|D|}$$

where $|D_1|$ is obtained from $|D|$ by replacing its first column by the column of right-hand sides, $|D_2|$ is obtained from $|D|$ by replacing its second column by the column of right-hand sides, and $|D_3|$ is obtained from $|D|$ by replacing its third column by the column of right-hand sides. Thus

$$x = \frac{\begin{vmatrix} -2 & 0 & 2 \\ 0 & -1 & 0 \\ -1 & 2 & 6 \end{vmatrix}}{|D|} \qquad y = \frac{\begin{vmatrix} 3 & -2 & 2 \\ 2 & 0 & 0 \\ 0 & -1 & 6 \end{vmatrix}}{|D|} \qquad z = \frac{\begin{vmatrix} 3 & 0 & -2 \\ 2 & -1 & 0 \\ 0 & 2 & -1 \end{vmatrix}}{|D|}$$

Expanding by cofactors we calculate $|D| = -10$, $|D_1| = 10$, $|D_2| = 20$, and $|D_3| = -5$, obtaining

$$x = \frac{10}{-10} = -1 \qquad y = \frac{20}{-10} = -2 \qquad z = \frac{-5}{-10} = \frac{1}{2}$$

PROGRESS CHECK
Solve by Cramer's rule.

$$
\begin{aligned}
3x \qquad - z &= 1 \\
-6x + 2y \qquad &= -5 \\
-4y + 3z &= 5
\end{aligned}
$$

Answer

$x = \dfrac{2}{3}, \, y = -\dfrac{1}{2}, \, z = 1$

WARNING
(a) Each equation of the linear system must be written in the form $Ax + By + Cz = k$ before using Cramer's rule.
(b) If $|D| = 0$, Cramer's rule cannot be used.

EXERCISE SET 7.4
In Exercises 1–6 evaluate the given determinant.

1. $\begin{vmatrix} 2 & -3 \\ 4 & 5 \end{vmatrix}$ 2. $\begin{vmatrix} 3 & 4 \\ -1 & 2 \end{vmatrix}$ 3. $\begin{vmatrix} -4 & 1 \\ 0 & 2 \end{vmatrix}$

4. $\begin{vmatrix} 2 & 2 \\ 3 & 3 \end{vmatrix}$ 5. $\begin{vmatrix} 0 & 0 \\ 1 & 3 \end{vmatrix}$ 6. $\begin{vmatrix} -4 & -1 \\ -2 & 3 \end{vmatrix}$

In Exercises 7–10 let

$$A = \begin{bmatrix} 3 & -1 & 2 \\ 4 & 1 & -3 \\ 5 & -2 & 0 \end{bmatrix}$$

7. Compute the minor of each of the following elements.
   (a) a_{11} (b) a_{23} (c) a_{31} (d) a_{33}
8. Compute the minor of each of the following elements.
   (a) a_{12} (b) a_{22} (c) a_{23} (d) a_{32}
9. Compute the cofactor of each of the following elements.
   (a) a_{11} (b) a_{23} (c) a_{31} (d) a_{33}
10. Compute the cofactor of each of the following elements.
    (a) a_{12} (b) a_{22} (c) a_{23} (d) a_{32}

In Exercises 11–20 evaluate the given determinant.

11. $\begin{vmatrix} 4 & -2 & 5 \\ 5 & 2 & 0 \\ 2 & 0 & 4 \end{vmatrix}$
12. $\begin{vmatrix} 4 & 1 & 2 \\ 0 & 2 & 3 \\ 0 & 0 & -4 \end{vmatrix}$
13. $\begin{vmatrix} -1 & 2 & 0 \\ 3 & 4 & 1 \\ 6 & 5 & 2 \end{vmatrix}$

14. $\begin{vmatrix} -1 & 3 & 2 \\ 0 & 7 & 7 \\ 2 & 1 & 3 \end{vmatrix}$
15. $\begin{vmatrix} 2 & 2 & 4 \\ 3 & 8 & 1 \\ 1 & 1 & 2 \end{vmatrix}$
16. $\begin{vmatrix} 0 & 1 & 3 \\ 2 & 5 & -1 \\ 4 & 2 & -2 \end{vmatrix}$

17. $\begin{vmatrix} 3 & 2 & 1 & 0 \\ -1 & -3 & -1 & 0 \\ 0 & 0 & 2 & 2 \\ 4 & 1 & 3 & 3 \end{vmatrix}$
18. $\begin{vmatrix} -1 & 2 & 4 & 0 \\ 3 & -2 & -3 & 0 \\ 0 & 4 & 2 & 5 \\ 0 & -3 & 1 & 4 \end{vmatrix}$

19. $\begin{vmatrix} 2 & -3 & 2 & -4 \\ 0 & 4 & -1 & 9 \\ 0 & 1 & 2 & 0 \\ 0 & 1 & 3 & -1 \end{vmatrix}$
20. $\begin{vmatrix} 1 & 1 & 0 & 1 \\ 0 & -1 & 4 & -1 \\ -2 & 3 & 1 & -4 \\ 0 & 2 & 0 & 2 \end{vmatrix}$

In Exercises 21–28 solve the given linear system by use of Cramer's rule.

21. $\begin{aligned} 2x + y + z &= -1 \\ 2x - y + 2z &= 2 \\ x + 2y + z &= -4 \end{aligned}$
22. $\begin{aligned} x - y + z &= -5 \\ 3x + y + 2z &= -5 \\ 2x - y - z &= -2 \end{aligned}$

23. $\begin{aligned} 2x + y - z &= 9 \\ x - 2y + 2z &= -3 \\ 3x + 3y + 4z &= 11 \end{aligned}$
24. $\begin{aligned} 2x + y - z &= -2 \\ -2x - 2y + 3z &= 2 \\ 3x + y - z &= -4 \end{aligned}$

25. $\begin{aligned} -x - y + 2z &= 7 \\ x + 2y - 2z &= -7 \\ 2x - y + z &= -4 \end{aligned}$
26. $\begin{aligned} 4x + y - z &= -1 \\ x - y + 2z &= 3 \\ -x + 2y - z &= 0 \end{aligned}$

27. $\begin{aligned} x + y - z + 2w &= 0 \\ 2x + y - w &= -2 \\ 3x + 2z &= -3 \\ -x + 2y + 3w &= 1 \end{aligned}$
28. $\begin{aligned} 2x + y - 3w &= -7 \\ 3x + 2z + w &= -1 \\ -x + 2y + 3w &= 0 \\ -2x - 3y + 2z - w &= 8 \end{aligned}$

29. Show that
$$\begin{vmatrix} a_1 + b_1 & a_2 + b_2 \\ c & d \end{vmatrix} = \begin{vmatrix} a_1 & a_2 \\ c & d \end{vmatrix} + \begin{vmatrix} b_1 & b_2 \\ c & d \end{vmatrix}$$

30.   Show that

$$\begin{vmatrix} ka_{11} & ka_{12} \\ a_{21} & a_{22} \end{vmatrix} = \begin{vmatrix} a_{11} & a_{12} \\ ka_{21} & ka_{22} \end{vmatrix} = k \begin{vmatrix} a_{11} & a_{12} \\ a_{21} & a_{22} \end{vmatrix}$$

31.   Prove that if a row or column of a square matrix consists entirely of zeros, then the determinant of the matrix is zero. (*Hint:* Expand by cofactors.)

32.   Prove that if matrix B is obtained by multiplying each element of a row of a square matrix A by a constant k, then $|B| = k|A|$.

33.   Prove that if A is an $n \times n$ matrix and $B = kA$, where k is a constant, then $|B| = k^n|A|$.

34.   Prove that if matrix B is obtained from a square matrix A by interchanging the rows and columns of A, then $|B| = |A|$.

TERMS AND SYMBOLS

matrix (p. 260)	**elementary row operations** (p. 262)	**identity matrix,** I_n (p. 276)
entries, elements (p. 260)	**pivot element** (p. 263)	**additive inverse** (p. 276)
dimension (p. 260)	**pivot row** (p. 263)	**invertible or nonsingular matrix** (p. 277)
square matrix (p. 260)	**Gauss–Jordan elimination** (p. 265)	**inverse** (p. 277)
order (p. 260)	**equality of matrices** (p. 268)	A^{-1} (p. 277)
row matrix (p. 260)	**scalar** (p. 270)	**determinant** (p. 286)
column matrix (p. 260)	**scalar multiplication** (p. 270)	**minor** (p. 287)
$[a_{ij}]$ (p. 261)	**zero matrix** (p. 276)	**cofactor** (p. 288)
coefficient matrix (p. 261)		**expansion by cofactors** (p. 289)
augmented matrix (p. 261)		**Cramer's rule** (p. 291)

KEY IDEAS FOR REVIEW

■ A matrix is simply a rectangular array of numbers.

■ Systems of linear equations can be conveniently handled in matrix notation. By dropping the names of the variables, matrix notation focuses on the coefficients and the right-hand side of the system. The elementary row operations are then seen to be an abstraction of those operations that produce equivalent systems of equations.

■ Gaussian elimination and Gauss–Jordan elimination both involve the use of elementary row operations on the augmented matrix corresponding to a linear system. In the case of a system of three equations with three unknowns, the final matrices will be of this form:

$$\begin{bmatrix} * & * & * & \vdots & * \\ 0 & * & * & \vdots & * \\ 0 & 0 & * & \vdots & * \end{bmatrix} \qquad \begin{bmatrix} 1 & 0 & 0 & \vdots & c_1 \\ 0 & 1 & 0 & \vdots & c_2 \\ 0 & 0 & 1 & \vdots & c_3 \end{bmatrix}$$

Gaussian elimination   Gauss–Jordan elimination

If Gauss–Jordan elimination is used, the solution can be read from the final matrix; if Gaussian elimination is used, back-substitution is then performed with the final matrix.

■ A linear system can be written in the form $AX = B$ where A is the coefficient matrix, X is a column matrix of the unknowns, and B is the column matrix of the right-hand sides.

■ The sum and difference of two matrices A and B can be formed only if A and B are of the same dimension. The product AB can be formed only if the number of columns of A is the same as the number of rows of B.

■ The $n \times n$ matrix B is said to be the inverse of the $n \times n$ matrix A if $AB = I_n$ and $BA = I_n$. We denote the inverse of A by A^{-1}. The inverse can be computed by using elementary row operations to transform the matrix $[A \mid I_n]$ to the form $[I_n \mid B]$. Then $B = A^{-1}$.

■ If the linear system $AX = B$ has a unique solution, then $X = A^{-1}B$.

■ Associated with every square matrix is a number called a determinant. The rule for evaluating a determinant of order 2 is

$$\begin{vmatrix} a & b \\ c & d \end{vmatrix} = ad - bc$$

■ For determinants of order greater than 2, the method of expansion by cofactors may be used to reduce the problem to that of evaluating determinants of order 2.

■ When expanding by cofactors, choose the row or column which contains the most zeros. This will ease the arithmetic burden.

■ Cramer's rule provides a means for solving a linear system by expressing the value of each unknown as a quotient of determinants.

REVIEW EXERCISES

Exercises 1–4 refer to the matrix

$$A = \begin{bmatrix} -1 & 4 & 2 & 0 & 8 \\ 2 & 0 & -3 & -1 & 5 \\ 4 & -6 & 9 & 1 & -2 \end{bmatrix}$$

1. Determine the dimension of the matrix A.

2. Find a_{24} 3. Find a_{31} 4. Find a_{15}

Exercises 5–6 refer to the linear system

$$3x - 7y = 14$$
$$x + 4y = 6$$

5. Write the coefficient matrix of the linear system.
6. Write the augmented matrix of the linear system.

In Exercises 7–8 write a linear system corresponding to the augmented matrix.

7. $\begin{bmatrix} 4 & -1 & \mid & 3 \\ 2 & 5 & \mid & 0 \end{bmatrix}$

8. $\begin{bmatrix} -2 & 4 & 5 & \mid & 0 \\ 6 & -9 & 4 & \mid & 0 \\ 3 & 2 & -1 & \mid & 0 \end{bmatrix}$

In Exercises 9–12 use back-substitution to solve the linear systems corresponding to the given augmented matrix.

9. $\begin{bmatrix} 1 & -2 & \mid & 7 \\ 0 & 1 & \mid & -4 \end{bmatrix}$

10. $\begin{bmatrix} 1 & 2 & \mid & \frac{21}{2} \\ 0 & 1 & \mid & 5 \end{bmatrix}$

11. $\begin{bmatrix} 1 & -4 & 2 & | & -18 \\ 0 & 1 & -2 & | & 5 \\ 0 & 0 & 1 & | & -1 \end{bmatrix}$ 12. $\begin{bmatrix} 1 & -2 & 2 & | & -9 \\ 0 & 1 & 3 & | & -8 \\ 0 & 0 & 1 & | & -3 \end{bmatrix}$

In Exercises 13–16 use matrix methods to solve the given linear system.

13. $\begin{aligned} x + y &= 2 \\ 2x - 4y &= -5 \end{aligned}$ 14. $\begin{aligned} 3x - y &= -17 \\ 2x + 3y &= -4 \end{aligned}$

15. $\begin{aligned} x + 3y + 2z &= 0 \\ -2x + 3z &= -12 \\ 2x - 6y - z &= 6 \end{aligned}$ 16. $\begin{aligned} 2x - y - 2z &= 3 \\ -2x + 3y + z &= 3 \\ 2y - z &= 6 \end{aligned}$

In Exercises 17–18 solve for x.

17. $\begin{bmatrix} 5 & -1 \\ 3 & 2x \end{bmatrix} = \begin{bmatrix} 5 & -1 \\ 3 & -6 \end{bmatrix}$ 18. $\begin{bmatrix} 6 & x^2 \\ 4x & -2 \end{bmatrix} = \begin{bmatrix} 6 & 9 \\ -12 & -2 \end{bmatrix}$

Exercises 19–28 refer to the following matrices.

$$A = \begin{bmatrix} 2 & -1 \\ 3 & 2 \end{bmatrix} \quad B = \begin{bmatrix} -1 & 5 \\ 4 & -3 \end{bmatrix} \quad C = \begin{bmatrix} -1 & 0 \\ 0 & 4 \\ 2 & -2 \end{bmatrix} \quad D = \begin{bmatrix} 1 & 3 & 4 \\ -1 & 0 & -6 \end{bmatrix}$$

If possible, find the following.

19. $A + B$ 20. $B - A$ 21. $A + C$
22. $5D$ 23. CD 24. DC
25. BC 26. CB 27. $A + 2B$
28. $-AB$

In Exercises 29–30 find the inverse of the given matrix.

29. $\begin{bmatrix} -2 & 3 \\ 1 & 4 \end{bmatrix}$ 30. $\begin{bmatrix} 1 & 1 & -4 \\ -5 & -2 & 0 \\ 4 & 2 & -1 \end{bmatrix}$

In Exercises 31–32 solve the given system by finding the inverse of the coefficient matrix.

31. $\begin{aligned} 2x - y &= 1 \\ x + y &= 5 \end{aligned}$ 32. $\begin{aligned} x + 2y - 2z &= -4 \\ 3x - y &= -2 \\ y + 4z &= 1 \end{aligned}$

In Exercises 33–38 evaluate the given determinant.

33. $\begin{vmatrix} 3 & 1 \\ -4 & 2 \end{vmatrix}$ 34. $\begin{vmatrix} -1 & 2 \\ 0 & 6 \end{vmatrix}$ 35. $\begin{vmatrix} 2 & -1 \\ 6 & -3 \end{vmatrix}$

36. $\begin{vmatrix} 1 & 0 & -1 \\ 2 & 3 & -5 \\ 0 & 4 & 0 \end{vmatrix}$ 37. $\begin{vmatrix} 1 & -1 & 2 \\ 0 & 5 & 4 \\ 2 & 3 & 8 \end{vmatrix}$ 38. $\begin{vmatrix} 1 & 2 & -1 \\ 0 & 3 & 4 \\ 0 & 0 & -1 \end{vmatrix}$

In Exercises 39–44 use Cramer's rule to solve the given linear system.

39. $\begin{aligned} 2x - y &= -3 \\ -2x + 3y &= 11 \end{aligned}$ 40. $\begin{aligned} 3x - y &= 7 \\ 2x + 5y &= -18 \end{aligned}$

41. $\begin{aligned} x + 2y &= 2 \\ 2x - 7y &= 48 \end{aligned}$

42. $\begin{aligned} 2x + 3y - z &= -3 \\ -3x \quad\quad + 4z &= 16 \\ 2y + 5z &= 9 \end{aligned}$

43. $\begin{aligned} 3x \quad\quad + z &= 0 \\ x + y + z &= 0 \\ -3y + 2z &= -4 \end{aligned}$

44. $\begin{aligned} 2x + 3y + z &= -5 \\ 2y + 2z &= -3 \\ 4x + y - 2z &= -2 \end{aligned}$

PROGRESS TEST 7A

Problems 1–2 refer to the matrix

$$A = \begin{bmatrix} -1 & 2 \\ -2 & 4 \\ 0 & 7 \end{bmatrix}$$

1. Find the dimension of the matrix A.
2. Find a_{31}.
3. Write the augmented matrix of the linear system

$$\begin{aligned} -7x \quad\quad + 6z &= 3 \\ 2y - z &= 10 \\ x - y + z &= 5 \end{aligned}$$

4. Write a linear system corresponding to the augmented matrix.

$$\begin{bmatrix} -5 & 2 & | & 4 \\ 3 & -4 & | & 4 \end{bmatrix}$$

5. Use back-substitution to solve the linear system corresponding to the augmented matrix

$$\begin{bmatrix} 1 & 1 & | & 0 \\ 0 & 1 & | & \frac{1}{2} \end{bmatrix}$$

6. Solve the linear system

$$\begin{aligned} -x + 2y &= 10 \\ \frac{1}{2}x + 2y &= -7 \end{aligned}$$

by applying Gaussian elimination to the augmented matrix.

7. Solve the linear system

$$\begin{aligned} 2x - y + 3z &= 2 \\ x + 2y - z &= 1 \\ -x + y + 4z &= 2 \end{aligned}$$

by applying Gauss–Jordan elimination to the augmented matrix.

8. Solve for x.

$$\begin{bmatrix} 2x - 1 & 0 \\ 1 & -3 \end{bmatrix} = \begin{bmatrix} 5 & 0 \\ 1 & -3 \end{bmatrix}$$

Problems 9–12 refer to the matrices

$$A = \begin{bmatrix} -4 & 0 & 3 \\ 6 & 2 & -3 \end{bmatrix} \qquad B = \begin{bmatrix} -1 \\ -3 \end{bmatrix} \qquad C = \begin{bmatrix} 4 & 2 \\ -2 & 0 \\ 3 & -1 \end{bmatrix} \qquad D = \begin{bmatrix} 1 & -6 \\ 0 & 2 \\ 4 & -1 \end{bmatrix}$$

If possible, find the following.

9. $C - 2D$ 10. AC 11. CB 12. BA

13. Find the inverse of the matrix.

$$\begin{bmatrix} -1 & 0 & 4 \\ 2 & 1 & -1 \\ 1 & -3 & 2 \end{bmatrix}$$

14. Solve the given linear system by finding the inverse of the coefficient matrix.

$$3x - 2y = -8$$
$$2x + 3y = -1$$

In Problems 15–16 evaluate the given determinant.

15. $\begin{vmatrix} -6 & -2 \\ 2 & 1 \end{vmatrix}$ 16. $\begin{vmatrix} 0 & -1 & 2 \\ 2 & -2 & 3 \\ 1 & 4 & 5 \end{vmatrix}$

17. Use Cramer's rule to solve the linear system.

$$x + 2y = -2$$
$$-2x - 3y = 1$$

PROGRESS TEST 7B

Problems 1–2 refer to the matrix

$$B = \begin{bmatrix} -1 & 5 & 0 & 6 \\ 4 & -2 & 1 & -2 \end{bmatrix}$$

1. Determine the dimension of the matrix A.
2. Find b_{23}.
3. Write the coefficient matrix of the linear system.

$$2x - 6y = 5$$
$$x + 3y = -2$$

4. Write a linear system corresponding to the augmented matrix

$$\begin{bmatrix} 16 & 0 & 6 & | & 10 \\ -4 & -2 & 5 & | & 8 \\ 2 & 3 & -1 & | & -6 \end{bmatrix}$$

5. Use back-substitution to solve the linear system corresponding to the augmented matrix.

$$\begin{bmatrix} 1 & 3 & -1 & | & 0 \\ 0 & 1 & -2 & | & 5 \\ 0 & 0 & 1 & | & -3 \end{bmatrix}$$

6. Solve the linear system

$$2x + 3y = -11$$
$$3x - 2y = 3$$

by applying Gaussian elimination to the augmented matrix.

7. Solve the linear system

$$2x + y - 2z = 7$$
$$-3x - 5y + 4z = -3$$
$$5x + 4y = 17$$

by applying Gauss–Jordan elimination to the augmented matrix.

8. Solve for y.

$$\begin{bmatrix} 2 & -5 & 1 \\ -3 & 1-y & 2 \end{bmatrix} = \begin{bmatrix} 2 & -5 & 1 \\ -3 & 6 & 2 \end{bmatrix}$$

Problems 9–12 refer to the matrices

$$A = \begin{bmatrix} 2 & -3 \\ -1 & 0 \\ 2 & 1 \end{bmatrix} \qquad B = \begin{bmatrix} 1 & -2 & 3 \end{bmatrix} \qquad C = \begin{bmatrix} 0 & -4 \\ 3 & 1 \\ -2 & 5 \end{bmatrix}$$

If possible, find the following.

9. BA 10. $2C + 3A$ 11. CB 12. $BC - A$

13. Find the inverse of the matrix.

$$\begin{bmatrix} 2 & -5 & -2 \\ -1 & 3 & 0 \\ -2 & 0 & 4 \end{bmatrix}$$

14. Solve the given linear system by finding the inverse of the coefficient matrix.

$$5x - 2y = -6$$
$$-2x + 2y = 3$$

In Problems 15–16 evaluate the given determinant.

15. $\begin{vmatrix} -2 & 4 \\ 3 & 5 \end{vmatrix}$ 16. $\begin{vmatrix} -2 & 0 & 1 \\ 1 & 2 & 0 \\ 2 & -1 & -1 \end{vmatrix}$

17. Use Cramer's rule to solve the linear system.

$$x + y = -1$$
$$2x - 4y = -5$$

CHAPTER EIGHT
ROOTS
OF
POLYNOMIALS

In Section 3.3 we observed that the polynomial function

$$f(x) = ax + b \tag{1}$$

is called a linear function, and the polynomial function

$$g(x) = ax^2 + bx + c, \, a \neq 0 \tag{2}$$

is called a quadratic function. To facilitate the study of polynomial functions in general, we now introduce the notation

$$P(x) = a_n x^n + a_{n-1} x^{n-1} + \cdots + a_1 x + a_0, \quad a_n \neq 0 \tag{3}$$

to represent a **polynomial function of degree n**. Note that the coefficients a_k may be real or complex numbers and that the subscript k of the coefficient a_k is the same as the exponent of x in x^k.

If $a \neq 0$ in Equation (1), we set the polynomial function equal to zero and obtain the linear equation

$$ax + b = 0$$

which has precisely one solution, $-b/a$. If we set the polynomial function in Equation (2) equal to zero, we have the quadratic equation

$$ax^2 + bx + c = 0$$

that has the two solutions given by the quadratic formula. If we set the polynomial function in Equation (3) equal to zero we have the **polynomial equation of degree n**

$$a_n x^n + a_{n-1} x^{n-1} + \cdots + a_1 x + a_0 = 0 \qquad (4)$$

Our attention in this chapter will turn to finding the roots or solutions of Equation (4). These solutions are also known as the **roots** or **zeros of the polynomial.** We will attempt to answer the following questions for a polynomial equation of degree n.

- How many roots does a polynomial have in the field of complex numbers?
- How many of the roots of a polynomial are real numbers?
- If the coefficients of a polynomial are integers, how many of the roots are rational numbers?
- Is there a relationship between the roots and factors of a polynomial?

These questions have attracted the attention of mathematicians since the sixteenth century. A method for finding the roots of polynomial equations of degree 3 was published around 1535 and is known as Cardan's formula despite the fact that Girolamo Cardano stole the result from his friend, Nicolo Tartaglia. Shortly afterward a method that is attributed to Ferrari was published for solving polynomial equations of degree 4.

The search for formulas giving the roots of a polynomial of degree 5 or higher, in terms of its coefficients, continued into the nineteenth century. At that time the Norwegian mathematician N. H. Abel and the French mathematician Evariste Galois proved that no such formulas are possible. Galois' work on this problem was completed a year before his death in a duel at age 21. His proof made use of the new concepts of group theory. The work was so advanced that his teachers wrote it off as being unintelligible gibberish.

8.1
POLYNOMIAL DIVISION AND SYNTHETIC DIVISION

To find the roots of a polynomial, it will be necessary to divide the polynomial by a second polynomial. There is a procedure for polynomial division that parallels the long division process of arithmetic. In arithmetic, if we divide a real number p by the real number $d \neq 0$, we obtain a quotient q and a remainder r so that we can write

$$\frac{p}{d} = q + \frac{r}{d} \qquad (1)$$

where

$$r < d \qquad (2)$$

This result can also be written in the form

$$p = qd + r, \qquad r < d$$

For example,

$$\frac{7284}{13} = 560 + \frac{4}{13}$$

or

$$7284 = (560)(13) + 4$$

In the long division process for polynomials, we divide the dividend $P(x)$ by the divisor $D(x) \neq 0$ to obtain a quotient $Q(x)$ and a remainder $R(x)$. We then have

$$\frac{P(x)}{D(x)} = Q(x) + \frac{R(x)}{D(x)} \tag{3}$$

where

$$\text{degree of } R(x) < \text{degree of } D(x) \tag{4}$$

Note that Equations (1) and (3) have the same form. Equation (2) requires that the remainder be less than the divisor, and the parallel requirement for polynomials in Equation (4) is that the *degree* of the remainder be less than that of the divisor.

We illustrate the long division process for polynomials by an example.

EXAMPLE 1
Divide $3x^3 - 7x^2 + 1$ by $x - 2$.

Polynomial Division	
Step 1. Arrange the terms of both polynomials by descending powers of x. If a power is missing, write the term with a zero coefficient.	$x - 2 \overline{)\, 3x^3 - 7x^2 + 0x + 1}$
Step 2. Divide the first term of the dividend by the first term of the divisor. The answer is written above the first term of the dividend.	$\begin{array}{r} 3x^2 \qquad\qquad\qquad \\ x - 2 \overline{)\, 3x^3 - 7x^2 + 0x + 1} \end{array}$
Step 3. Multiply the divisor by the quotient obtained in Step 2 and then subtract the product.	$\begin{array}{r} 3x^2 \qquad\qquad\qquad \\ x - 2 \overline{)\, 3x^3 - 7x^2 + 0x + 1} \\ 3x^3 - 6x^2 \qquad\qquad \\ \hline -x^2 + 0x + 1 \end{array}$
Step 4. Repeat Steps 2 and 3 until the degree of the remainder is less than the degree of the divisor.	$\begin{array}{r} 3x^2 - x - 2 \\ x - 2 \overline{)\, 3x^3 - 7x^2 + 0x + 1} \\ 3x^3 - 6x^2 \qquad\qquad \\ \hline -x^2 + 0x + 1 \\ -x^2 + 2x \qquad \\ \hline -2x + 1 \\ -2x + 4 \\ \hline -3 \end{array}$

Step 5. Write the answer in the form of Equation (3).	$\dfrac{3x^3 - 7x^2 + 0x + 1}{x - 2}$
	$= 3x^2 - x - 2 - \dfrac{3}{x - 2}$

PROGRESS CHECK

Divide $4x^2 - 3x + 6$ by $x + 2$

Answer

$4x - 11 + \dfrac{28}{x + 2}$

Our work in this chapter will frequently require division of a polynomial by a first-degree polynomial $x - r$ where r is a constant. Fortunately, there is a shortcut called **synthetic division** that simplifies this task. To demonstrate synthetic division we will do Example 1 again, writing only the coefficients.

$$
\begin{array}{r}
\mathbf{3} \quad \mathbf{-1} \quad \mathbf{-2} \\
-2\ )\overline{\mathbf{3} \quad -7 \quad\ \ 0 \quad\ \ 1} \\
\underline{\mathbf{3} \quad -6} \\
-\mathbf{1} \quad\ \ 0 \quad\ \ 1 \\
\underline{-\mathbf{1} \quad\ \ 2} \\
-\mathbf{2} \quad\ \ 1 \\
\underline{-2 \quad\ \ 4} \\
-3
\end{array}
$$

Note that the boldface numerals are duplicated. We can use this to our advantage and simplify the process as follows.

$$
\begin{array}{r}
-2\underline{|}\quad 3 \quad -7 \quad\ \ 0 \quad\ \ 1 \\
\underline{-6 \quad\ \ 2 \quad\ \ 4} \\
3 \quad -1 \quad -2 \quad -3
\end{array}
$$

$$\underbrace{\qquad\qquad\qquad}_{\substack{\text{coefficients} \\ \text{of the} \\ \text{quotient}}} \quad \underset{\text{remainder}}{|}$$

We copied the leading coefficient (3) of the dividend in the third row, multiplied it by the divisor (-2), and wrote the result (-6) in the second row under the next coefficient. The numbers in the second column were subtracted to obtain $-7-(-6) = -1$. The procedure is repeated until the third row is of the same length as the first row.

Since subtraction is more apt to produce errors than is addition, we can modify this process slightly. If the divisor is $x - r$, we will write r instead of $-r$ in the box and use addition in each step instead of subtraction. Repeating our example, we have

$$
\begin{array}{r}
2\underline{|}\quad 3 \quad -7 \quad\ \ 0 \quad\ \ 1 \\
\underline{6 \quad -2 \quad -4} \\
3 \quad -1 \quad -2 \quad -3
\end{array}
$$

EXAMPLE 2
Divide $4x^3 - 2x + 5$ by $x + 2$ using synthetic division.

Synthetic Division	
Step 1. If the divisor is $x - r$, write r in the box. Arrange the coefficients of the dividend by descending power of x, supplying a zero coefficient for every missing power.	*Step 1.* $\underline{-2}\rfloor$ 4 0 −2 5
Step 2. Copy the leading coefficient in the third row.	*Step 2.* $\underline{-2}\rfloor$ 4 0 −2 5 $\overline{\;4}$
Step 3. Multiply the last entry in the third row by the number in the box and write the result in the second row under the next coefficient. Add the numbers in that column.	*Step 3.* $\underline{-2}\rfloor$ 4 0 −2 5 −8 $\overline{\;4\;\;-8}$
Step 4. Repeat Step 3 until there is an entry in the third row for each entry in the first row. The last number in the third row is the remainder; the other numbers are the coefficients of the quotient.	*Step 4.* $\underline{-2}\rfloor$ 4 0 −2 5 −8 16 −28 $\overline{\;4\;\;-8\;\;\;14\;\;-23}$ $\dfrac{4x^3 - 2x + 5}{x + 2}$ $= 4x^2 - 8x + 14 - \dfrac{23}{x + 2}$

PROGRESS CHECK
Use synthetic division to obtain the quotient $Q(x)$ and the constant remainder R when $2x^4 - 10x^2 - 23x + 6$ is divided by $x - 3$.

Answer
$Q(x) = 2x^3 + 6x^2 + 8x + 1; R = 9$

WARNING ———————————————————————————
In synthetic division, when dividing by $x - r$ we place r in the box. Thus, when the divisor is $x + 3$, we place -3 in the box since $x + 3 = x - (-3)$. Similarly, when the divisor is $x - 3$, we place $+3$ in the box since $x - 3 = x - (+3)$.

———————————————————————————————————————

EXERCISE SET 8.1
In Exercises 1–10 use polynomial division to find the quotient $Q(x)$ and the remainder $R(x)$ when the first polynomial is divided by the second polynomial.

1. $x^2 - 7x + 12$, $x - 5$
2. $x^2 + 3x + 3$, $x + 2$
3. $2x^3 - 2x$, $x^2 + 2x - 1$
4. $3x^3 - 2x^2 + 4$, $x^2 - 2$
5. $3x^4 - 2x^2 + 1$, $x + 3$
6. $x^5 - 1$, $x^2 - 1$

7. $2x^3 - 3x^2,\quad x^2 + 2$

8. $3x^3 - 2x - 1,\quad x^2 - x$

9. $x^4 - x^3 + 2x^2 - x + 1,\quad x^2 + 1$

10. $2x^4 - 3x^3 - x^2 - x - 2,\quad x - 2$

In Exercises 11–20 use synthetic division to find the quotient $Q(x)$ and the constant remainder R when the first polynomial is divided by the second polynomial.

11. $x^3 - x^2 - 6x + 5,\quad x + 2$

12. $2x^3 - 3x^2 - 4,\quad x - 2$

13. $x^4 - 81,\quad x - 3$

14. $x^4 - 81,\quad x + 3$

15. $3x^3 - x^2 + 8,\quad x + 1$

16. $2x^4 - 3x^3 - 4x - 2,\quad x - 1$

17. $x^5 + 32,\quad x + 2$

18. $x^5 + 32,\quad x - 2$

19. $6x^4 - x^2 + 4,\quad x - 3$

20. $8x^3 + 4x^2 - x - 5,\quad x + 3$

8.2
THE REMAINDER AND FACTOR THEOREMS

From our work with the division process in the previous section, we may surmise that division of a polynomial $P(x)$ by $x - r$ results in a quotient $Q(x)$ and a constant remainder R such that

$$P(x) = (x - r) \cdot Q(x) + R$$

Since this identity holds for all real values of x, it must hold when $x = r$. Consequently,

$$P(r) = (r - r) \cdot Q(r) + R$$
$$P(r) = 0 \cdot Q(r) + R$$

or

$$P(r) = R$$

We have proved the Remainder Theorem.

Remainder Theorem

If a polynomial $P(x)$ is divided by $x - r$, then the remainder is $P(r)$.

EXAMPLE 1
Determine the remainder when $P(x) = 2x^3 - 3x^2 - 2x + 1$ is divided by $x - 3$.

Solution
By the Remainder Theorem, the remainder is $R = P(3)$. We then have

$$R = P(3) = 2(3)^3 - 3(3)^2 - 2(3) + 1 = 22$$

We may verify this result by using synthetic division.

$$
\begin{array}{r|rrrr}
3 & 2 & -3 & -2 & 1 \\
  &   & 6 & 9 & 21 \\
\hline
  & 2 & 3 & 7 & \mathbf{22}
\end{array}
$$

The numeral in boldface is the remainder, so we have verified that $R = 22$.

PROGRESS CHECK.

Determine the remainder when $3x^2 - 2x - 6$ is divided by $x + 2$.

Answer
10

The Remainder Theorem can be used to tabulate values from which we can sketch the graph of a function. The most efficient scheme for performing the calculations is a streamlined form of synthetic division in which the addition is performed without writing the middle row. Repeating Example 1 in this condensed form we have

$$\begin{array}{r|rrrr} & 2 & -3 & -2 & 1 \\ \underline{3|} & 2 & 3 & 7 & 22 \end{array}$$

Then the point (3,22) lies on the graph of $y = 2x^3 - 3x^2 - 2x + 1$. In general, we may choose a value a of the independent variable and use synthetic division to find the remainder $P(a)$. Then $(a, P(a))$ is a point on the graph of $P(x)$.

EXAMPLE 2

Sketch the graph of $P(x) = 2x^3 - 3x^2 - 2x + 1$ on the interval $[-3,3]$.

Solution

To sketch the graph of $y = P(x)$, we will allow x to assume integer values from -3 to $+3$. The remainder is found by using the condensed form of synthetic division and is the y-coordinate corresponding to the chosen value of x.

	2	-3	-2	1	(x,y)
-3	2	-9	25	-74	$(-3,-74)$
-2	2	-7	12	-23	$(-2,-23)$
-1	2	-5	3	-2	$(-1,-2)$
0	2	-3	-2	1	$(0,1)$
1	2	-1	-3	-2	$(1,-2)$
2	2	1	0	1	$(2,1)$
3	2	3	7	22	$(3,22)$

The ordered pairs shown at the right of each row are the coordinates of points on the graph shown in Figure 1.

Let's assume that a polynomial $P(x)$ can be written as a product of polynomials, that is,

$$P(x) = D_1(x) D_2(x) \ldots D_n(x)$$

where $D_i(x)$ is a polynomial of degree greater than zero. Then $D_i(x)$ is called a **factor** of $P(x)$. If we focus on $D_1(x)$ and let

$$Q(x) = D_2(x) D_3(x) \ldots D_n(x)$$

then

$$P(x) = D_1(x) Q(x)$$

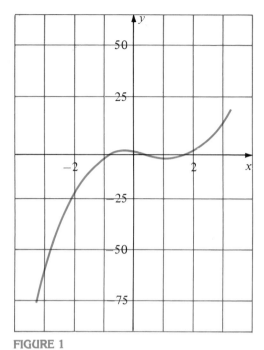

FIGURE 1

which demonstrates the following rule.

The polynomial $D(x)$ is a factor of a polynomial $P(x)$ if and only if the division of $P(x)$ by $D(x)$ results in a remainder of zero.

We can now combine this rule and the Remainder Theorem to prove the Factor Theorem.

Factor Theorem

A polynomial $P(x)$ has a factor $x - r$ if and only if

$$P(r) = 0.$$

If $x - r$ is a factor of $P(x)$ then division of $P(x)$ by $x - r$ must result in a remainder of 0. By the Remainder Theorem, the remainder is $P(r)$, and hence $P(r) = 0$. Conversely, if $P(r) = 0$, then the remainder is 0 and $P(x) = (x - r) Q(x)$ for some polynomial $Q(x)$ of degree one less than that of $P(x)$. By definition, $x - r$ is then a factor of $P(x)$.

EXAMPLE 3
Show that $x + 2$ is a factor of

$$P(x) = x^3 - x^2 - 2x + 8$$

Solution

By the Factor Theorem, $x + 2$ is a factor if $P(-2) = 0$. Using synthetic division to evaluate $P(-2)$,

$$
\begin{array}{r|rrrr}
-2 & 1 & -1 & -2 & 8 \\
   &   & -2 & 6 & -8 \\
\hline
   & 1 & -3 & 4 & 0
\end{array}
$$

we see that $P(-2) = 0$. Thus, $x + 2$ is a factor of $P(x)$.

PROGRESS CHECK

Show that $x - 1$ is a factor of $P(x) = 3x^6 - 3x^5 - 4x^4 + 6x^3 - 2x^2 - x + 1$.

EXERCISE SET 8.2

In Exercises 1–6 use the Remainder Theorem and synthetic division to find $P(r)$.

1. $P(x) = x^3 - 4x^2 + 1$, $r = 2$ 2. $P(x) = x^4 - 3x^2 - 5x$, $r = -1$
3. $P(x) = x^5 - 2$, $r = -2$ 4. $P(x) = 2x^4 - 3x^3 + 6$, $r = 2$
5. $P(x) = x^6 - 3x^4 + 2x^3 + 4$, $r = -1$ 6. $P(x) = x^6 - 2$, $r = 1$

In Exercises 7–12 use the Remainder Theorem to determine the remainder when $P(x)$ is divided by $x - r$.

7. $P(x) = x^3 - 2x^2 + x - 3$, $x - 2$ 8. $P(x) = 2x^3 + x^2 - 5$, $x + 2$
9. $P(x) = -4x^3 + 6x - 2$, $x - 1$ 10. $P(x) = 6x^5 - 3x^4 + 2x^2 + 7$, $x + 1$
11. $P(x) = x^5 - 30$, $x + 2$ 12. $P(x) = x^4 - 16$, $x - 2$

In Exercises 13–18 use the Remainder Theorem and synthetic division to sketch the graph of the given polynomial for $-3 \le x \le 3$.

13. $P(x) = x^3 + x^2 + x + 1$ 14. $P(x) = 3x^4 + 5x^3 + x^2 + 5x - 2$
15. $P(x) = 2x^3 + 3x^2 - 5x - 6$ 16. $P(x) = x^3 + 3x^2 - 4x - 12$
17. $P(x) = x^4 - 10x^3 + 1$ 18. $P(x) = 4x^4 + 4x^3 - 9x^2 - x + 2$

In Exercises 19–26 use the Factor Theorem to decide whether or not the first polynomial is a factor of the second polynomial.

19. $x - 2$, $x^3 - x^2 - 5x + 6$ 20. $x - 1$, $x^3 + 4x^2 - 3x + 1$
21. $x + 2$, $x^4 - 3x - 5$ 22. $x + 1$, $2x^3 - 3x^2 + x + 6$
23. $x + 3$, $x^3 + 27$ 24. $x + 2$, $x^4 + 16$
25. $x + 2$, $x^4 - 16$ 26. $x - 3$, $x^3 + 27$

In Exercises 27–30, use synthetic division to determine the value of k or r as requested.

27. Determine the values of r for which division of $x^2 - 2x - 1$ by $x - r$ has a remainder of 2.
28. Determine the values of r so that

$$\frac{x^2 - 6x - 1}{x - r}$$

has a remainder of -9.

29. Determine the values of k for which $x - 2$ is a factor of $x^3 - 3x^2 + kx - 1$.
30. Determine the values of k for which $2k^2x^3 + 3kx^2 - 2$ is divisible by $x - 1$.
31. Use the Factor Theorem to show that $x - 2$ is a factor of $P(x) = x^8 - 256$.

32. Use the Factor Theorem to show that $P(x) = 2x^4 + 3x^2 + 2$ has no factor of the form $x - r$, where r is a real number.
33. Use the Factor Theorem to show that $x - y$ is a factor of $x^n - y^n$, where n is a natural number.

8.3
FACTORS AND ROOTS

In Chapter Two we saw that the roots of quadratic equations may be complex numbers. Since complex numbers play a key role in providing solutions of polynomial equations, we will now explore further properties of this number system.

The complex number $a - bi$ is called the **complex conjugate** (or simply the **conjugate**) of the complex number $a + bi$. For example, $3 - 2i$ is the conjugate of $3 + 2i$, $4i$ is the conjugate of $-4i$, and 2 is the conjugate of 2. Forming the product $(a + bi)(a - bi)$ we have

$$(a + bi)(a - bi) = a^2 - abi + abi - b^2i^2$$

$$= a^2 + b^2 \qquad \text{Since } i^2 = -1$$

Because a and b are real numbers, $a^2 + b^2$ is also a real number. We can summarize this result as follows.

> The product of a complex number and its conjugate is a real number.
> $$(a + bi)(a - bi) = a^2 + b^2$$

We can now demonstrate that the quotient of two complex numbers is also a complex number. The quotient

$$\frac{q + ri}{s + ti}$$

can be written in the form $a + bi$ by multiplying both numerator and denominator by $s - ti$, the conjugate of the denominator. For example,

$$\frac{-2 + 3i}{3 - 2i} = \frac{-2 + 3i}{3 - 2i} \cdot \frac{3 + 2i}{3 + 2i}$$

$$= \frac{-6 - 4i + 9i + 6i^2}{3^2 + 2^2} = \frac{-6 + 5i + 6(-1)}{9 + 4}$$

$$= \frac{-12 + 5i}{13} = -\frac{12}{13} + \frac{5}{13}i$$

Of course, the reciprocal of the complex number $s + ti$ is the quotient $1/(s + ti)$, which can also be written as a complex number by using the same technique. We have therefore demonstrated that every nonzero complex number has a multiplicative inverse in the set of complex numbers.

EXAMPLE 1
Write the reciprocal of $2 - 5i$ in the form $a + bi$.

Solution
The reciprocal is $1/(2 - 5i)$. Multiplying both numerator and denominator by the conjugate $2 + 5i$, we have

$$\frac{1}{2 - 5i} \cdot \frac{2 + 5i}{2 + 5i} = \frac{2 + 5i}{2^2 + 5^2} = \frac{2 + 5i}{29} = \frac{2}{29} + \frac{5}{29} i$$

Verify that $(2 - 5i)\left(\dfrac{2}{29} + \dfrac{5}{29} i\right) = 1$.

PROGRESS CHECK
Write the following in the form $a + bi$.

(a) $\dfrac{4 - 2i}{5 + 2i}$ 　　　(b) $\dfrac{1}{2 - 3i}$ 　　　(c) $\dfrac{-3i}{3 + 5i}$

Answers

(a) $\dfrac{16}{29} - \dfrac{18}{29} i$ 　　(b) $\dfrac{2}{13} + \dfrac{3}{13} i$ 　　(c) $-\dfrac{15}{34} - \dfrac{9}{34} i$

　　If we let $z = a + bi$, it is customary to write the conjugate $a - bi$ as $\bar{z}$. We will have need to use the following properties of complex numbers and their conjugates.

If z and w are complex numbers, then

(1) $\bar{z} = \bar{w}$ if and only if $z = w$.
(2) $\bar{z} = z$ if and only if z is a real number.
(3) $\overline{z + w} = \bar{z} + \bar{w}$
(4) $\overline{z \cdot w} = \bar{z} \cdot \bar{w}$
(5) $\overline{z^n} = \bar{z}^n$, 　 n a positive integer

To prove properties (1)–(5), let $z = a + bi$ and $w = c + di$. Properties (1) and (2) follow directly from the definition of equality of complex numbers. To prove property (3), we note that $z + w = (a + c) + (b + d)i$. Then, by the definition of a complex conjugate,

$$\overline{z + w} = (a + c) - (b + d)i$$
$$= (a - bi) + (c - di)$$
$$= \bar{z} + \bar{w}$$

Properties (4) and (5) can be proved in a similar manner, although a rigorous proof of property (5) requires the use of mathematical induction, a method we will discuss in a later chapter.

EXAMPLE 2

If $z = 1 + 2i$ and $w = 3 - i$, verify that

(a) $\overline{z + w} = \overline{z} + \overline{w}$ (b) $\overline{z \cdot w} = \overline{z} \cdot \overline{w}$ (c) $\overline{z^2} = \overline{z}^2$

Solution

(a) Adding, we get $z + w = 4 + i$. Therefore $\overline{z + w} = 4 - i$.
     Also,

$$\overline{z} + \overline{w} = (1 - 2i) + (3 + i) = 4 - i$$

Thus, $\overline{z + w} = \overline{z} + \overline{w}$.

(b) Multiplying, we get $z \cdot w = (1 + 2i)(3 - i) = 5 + 5i$.
     Therefore $\overline{z \cdot w} = 5 - 5i$.
     Also,

$$\overline{z} \cdot \overline{w} = (1 - 2i)(3 + i) = 5 - 5i$$

Thus, $\overline{z \cdot w} = \overline{z} \cdot \overline{w}$.

(c) Squaring, we get

$$z^2 = (1 + 2i)(1 + 2i) = -3 + 4i$$

Therefore $\overline{z^2} = -3 - 4i$.

Also,

$$\overline{z}^2 = (1 - 2i)(1 - 2i) = -3 - 4i$$

Thus, $\overline{z^2} = \overline{z}^2$.

PROGRESS CHECK

If $z = 2 + 3i$ and $w = \frac{1}{2} - 2i$, verify that

(a) $\overline{z + w} = \overline{z} + \overline{w}$ (b) $\overline{z \cdot w} = \overline{z} \cdot \overline{w}$ (c) $\overline{z^2} = \overline{z}^2$ (d) $\overline{w^3} = \overline{w}^3$

We are now in a position to answer some of the questions posed in the introduction to this chapter. By using the Factor Theorem, we can show that there is a close relationship between the factors and the roots of the polynomial $P(x)$. By definition, r is a root of $P(x)$ if and only if $P(r) = 0$. But the Factor Theorem tells us that $P(r) = 0$ if and only if $x - r$ is a factor of $P(x)$. This leads to the following alternate statement of the Factor Theorem.

Factor Theorem

A polynomial $P(x)$ has a root r if and only if $x - r$ is a factor of $P(x)$.

EXAMPLE 3

Find a polynomial $P(x)$ of degree 3 whose roots are -1, 1, and -2.

Solution

By the Factor Theorem, $x + 1, x - 1$, and $x + 2$ are factors of $P(x)$. The product

$$P(x) = (x + 1)(x - 1)(x + 2) = x^3 + 2x^2 - x - 2$$

is a polynomial of degree 3 with the desired roots. Note that multiplying $P(x)$ by any nonzero real number results in another polynomial that has the same roots. For example, the polynomial

$$R(x) = 5x^3 + 10x^2 - 5x - 10$$

also has -1, 1, and -2 as its roots. Thus, the answer is not unique.

PROGRESS CHECK

Find a polynomial $P(x)$ of degree 3 whose roots are 2, 4, and -3.

Answer
$x^3 - 3x^2 - 10x + 24$

Does a polynomial always have a root? The answer was supplied by Carl Friedrich Gauss in 1799. The proof of his theorem, however, is beyond the scope of this book.

The Fundamental Theorem of Algebra

Every polynomial $P(x)$ of degree $n \geq 1$ has at least one complex root.

Gauss, who is considered by many to have been the greatest mathematician of all time, supplied the proof at age 22. The importance of the theorem is reflected in its title. We now see why it was necessary to create the complex numbers and that we need not create any other number system beyond the complex numbers in order to solve polynomial equations.

How many roots does a polynomial of degree n have? The next theorem will bring us closer to an answer.

Linear Factor Theorem

A polynomial $P(x)$ of degree $n \geq 1$ can be written as the product of n linear factors.

$$P(x) = a(x - r_1)(x - r_2) \ldots (x - r_n)$$

Note that a is the leading coefficient of $P(x)$ and that $r_1, r_2, \ldots, r_n$ are, in general, complex numbers.

To prove this theorem, we first note that the Fundamental Theorem of Algebra guarantees us the existence of a root r_1. By the Factor Theorem, $x - r_1$ is a factor and consequently

$$P(x) = (x - r_1)Q_1(x) \tag{1}$$

where $Q_1(x)$ is a polynomial of degree $n - 1$. If $n - 1 \geq 1$, then $Q_1(x)$ must have a root r_2. Thus

$$Q_1(x) = (x - r_2)Q_2(x) \tag{2}$$

where $Q_2(x)$ is of degree $n - 2$. Substituting in Equation (1) for $Q_1(x)$ we have

$$P(x) = (x - r_1)(x - r_2)Q_2(x) \tag{3}$$

This process is repeated n times until $Q_n(x) = a$ is of degree 0. Hence,

$$P(x) = a(x - r_1)(x - r_2) \ldots (x - r_n) \tag{4}$$

Since a is the leading coefficient of the polynomial on the right side of Equation (4), it must also be the leading coefficient of $P(x)$.

It is now easy to establish the following which may be thought of as an alternate form of the Fundamental Theorem of Algebra.

Number of Roots Theorem

If $P(x)$ is a polynomial of degree $n \geq 1$, then $P(x)$ has precisely n roots among the complex numbers.

We may prove this theorem as follows. If we write $P(x)$ in the form of Equation (4), we see that $r_1, r_2, \ldots, r_n$ are roots of the equation $P(x) = 0$ and hence there exist n roots. If there is an additional root r that is distinct from the roots $r_1, r_2, \ldots, r_n$, then $r - r_1, r - r_2, \ldots, r - r_n$ are all different from 0. Substituting r for x in Equation (4) yields

$$P(r) = a(r - r_1)(r - r_2) \ldots (r - r_n) \tag{5}$$

which cannot equal 0 since the product of nonzero numbers cannot equal 0. Thus, $r_1, r_2, \ldots, r_n$ are roots of $P(x)$ and there are no other roots. Hence, $P(x)$ has precisely n roots.

It is important to recognize that the roots of a polynomial need not be distinct from each other. The polynomial

$$P(x) = x^2 - 2x + 1$$

can be written in the factored form

$$P(x) = (x - 1)(x - 1)$$

which shows that the roots of $P(x)$ are 1 and 1. Since a root is associated with a factor and a factor may be repeated, we may have repeated roots. If the factor $x - r$ appears k times, we say that r is a **root of multiplicity k**.

EXAMPLE 4

Find all roots of the polynomial

$$P(x) = \left(x - \frac{1}{2}\right)^3 (x + i)(x - 5)^4$$

Solution

The distinct roots are $\frac{1}{2}$, $-i$, and 5. Further, $\frac{1}{2}$ is a root of multiplicity 3; $-i$ is a root of multiplicity 1; 5 is a root of multiplicity 4.

If we know that r is a root of $P(x)$, then we may write

$$P(x) = (x - r) Q(x)$$

and note that the roots of $Q(x)$ are also roots of $P(x)$. We call $Q(x) = 0$ the **depressed equation** since $Q(x)$ is of lower degree than $P(x)$. In the next example we illustrate the use of the depressed equation in finding the roots of a polynomial.

EXAMPLE 5

If 4 is a root of the polynomial $P(x) = x^3 - 8x^2 + 21x - 20$, find the other roots.

Solution

Since 4 is a root of $P(x)$, $x - 4$ is a factor of $P(x)$. Therefore,

$$P(x) = (x - 4) Q(x)$$

To find the depressed equation, we compute $Q(x) = P(x)/(x - 4)$ by synthetic division.

$$
\begin{array}{r|rrrr}
4 & 1 & -8 & 21 & -20 \\
  &   & 4 & -16 & 20 \\
\hline
  & 1 & -4 & 5 & 0
\end{array}
$$

$$\underbrace{}_{\substack{\text{Coefficients} \\ \text{of } Q(x)}} \quad \overset{|}{\text{Remainder}}$$

The depressed equation is

$$x^2 - 4x + 5 = 0$$

Using the quadratic formula, the roots of the depressed equation are found to be $2 + i$ and $2 - i$. The roots of $P(x)$ are then seen to be 4, $2 + i$, and $2 - i$.

PROGRESS CHECK

If -2 is a root of the polynomial $P(x) = x^3 - 7x - 6$, find the remaining roots.

Answer
$-1, 3$

EXAMPLE 6

If -1 is a root of multiplicity 2 of $P(x) = x^4 + 4x^3 + 2x^2 - 4x - 3$, find the remaining roots and write $P(x)$ as a product of linear factors.

Solution

Since -1 is a double root of $P(x)$, then $(x + 1)^2$ is a factor of $P(x)$. Therefore,

$$P(x) = (x + 1)^2 Q(x)$$

or

$$P(x) = (x^2 + 2x + 1) Q(x)$$

Using polynomial division, we can divide both sides of the last equation by $x^2 + 2x + 1$ to obtain

$$
\begin{aligned}
Q(x) &= \frac{x^4 + 4x^3 + 2x^2 - 4x - 3}{x^2 + 2x + 1} \\
&= x^2 + 2x - 3 \\
&= (x - 1)(x + 3)
\end{aligned}
$$

The roots of the depressed equation $Q(x) = 0$ are 1 and -3 and these are the remaining roots of $P(x)$. By the Linear Factor Theorem,

$$P(x) = (x + 1)^2 (x - 1)(x + 3)$$

PROGRESS CHECK

If -2 is a root of multiplicity 2 of $P(x) = x^4 + 4x^3 + 5x^2 + 4x + 4$, write $P(x)$ as a product of linear factors.

Answer

$P(x) = (x + 2)(x + 2)(x + i)(x - i)$

We know from the quadratic formula that if a quadratic equation with real coefficients has a complex root $a + bi$, then the conjugate $a - bi$ is the other root. The following theorem extends this result to a polynomial of degree n with real coefficients.

> ### Conjugate Roots Theorem
>
> If $P(x)$ is a polynomial of degree $n \geq 1$ with real coefficients, and if $a + bi, b \neq 0$, is a root of $P(x)$, then the complex conjugate $a - bi$ is also a root of $P(x)$.

PROOF OF CONJUGATE ROOTS THEOREM (Optional)

To prove the Conjugate Roots Theorem, we let $z = a + bi$ and make use of the properties of complex conjugates developed earlier in this section. We may write

$$P(x) = a_n x^n + a_{n-1} x^{n-1} + \ldots + a_1 x + a_0 \tag{6}$$

and, since z is a root of $P(x)$,

$$a_n z^n + a_{n-1} z^{n-1} + \ldots + a_n z + a_0 = 0 \tag{7}$$

But if $z = w$, then $\bar{z} = \bar{w}$. Applying this property of complex numbers to both sides of Equation (7), we have

$$\overline{a_n z^n + a_{n-1} z^{n-1} + \ldots + a_1 z + a_0} = \bar{0} = 0 \tag{8}$$

We also know that $\overline{z + w} = \bar{z} + \bar{w}$. Applying this property to the left side of Equation (8) we see that

$$\overline{a_n z^n} + \overline{a_{n-1} z^{n-1}} + \ldots + \overline{a_1 z} + \overline{a_0} = 0 \tag{9}$$

Further, $\overline{z \cdot w} = \bar{z} \cdot \bar{w}$ so that we may rewrite Equation (9) as

$$\overline{a_n} \overline{z^n} + \overline{a_{n-1}} \overline{z^{n-1}} + \ldots + \overline{a_1} \bar{z} + \overline{a_0} = 0 \tag{10}$$

Since a_i are all real numbers, we know that $\overline{a_i} = a_i$. Finally, we use the property $\overline{z^n} = \bar{z}^n$ to rewrite Equation (10) as

$$a_n \bar{z}^n + a_{n-1} \bar{z}^{n-1} + \ldots a_1 \bar{z} + a_0 = 0$$

which establishes that $\bar{z}$ is a root of $P(x)$.

EXAMPLE 7

Find a polynomial $P(x)$ with real coefficients that is of degree 3 and whose roots include -2 and $1 - i$.

Solution

Since $1 - i$ is a root, it follows from the Conjugate Roots Theorem that $1 + i$ is also a root of $P(x)$. By the Factor Theorem, $(x + 2)$, $[x - (1 - i)]$ and $[x - (1 + i)]$ are factors of $P(x)$. Therefore,

$$P(x) = (x + 2) [x - (1 - i)][x - (1 + i)]$$
$$= (x + 2) (x^2 - 2x + 2)$$
$$= x^3 - 2x + 4$$

PROGRESS CHECK

Find a polynomial $P(x)$ with real coefficients that is of degree 4 and whose roots include i and $-3 + i$.

Answer

$P(x) = x^4 + 6x^3 + 11x^2 + 6x + 10$

The following is a corollary of the Conjugate Roots Theorem.

> A polynomial $P(x)$ of degree $n \geq 1$ with real coefficients can be written as a product of linear and quadratic factors with real coefficients so that the quadratic factors have no real roots.

By the Linear Factor Theorem, we may write

$$P(x) = a(x - r_1) (x - r_2) \ldots (x - r_n)$$

where $r_1, r_2, \ldots, r_n$ are the n roots of $P(x)$. Of course, some of these roots may be complex numbers. A complex root $a + bi$, $b \neq 0$, may be paired with its conjugate $a - bi$ to provide the quadratic factor

$$[x - (a + bi)][x - (a - bi)] = x^2 - 2ax + a^2 + b^2$$

that has real coefficients. Thus, a quadratic factor with real coefficients results from each pair of complex conjugate roots; a linear factor with real coefficients results from each real root. Further, the discriminant of the quadratic factor $x^2 - 2ax + a^2 + b^2$ is $-4b^2$ and is therefore always negative, which shows that the quadratic factor has no real roots.

EXERCISE SET 8.3

In Exercises 1–6 multiply by the conjugate and simplify.

1. $2 - i$ 2. $3 + i$ 3. $3 + 4i$

4. $2 - 3i$ 5. $-4 - 2i$ 6. $5 + 2i$

In Exercises 7–15 perform the indicated operations and write the answer in the form $a + bi$.

7. $\dfrac{2 + 5i}{1 - 3i}$ 8. $\dfrac{1 + 3i}{2 - 5i}$ 9. $\dfrac{3 - 4i}{3 + 4i}$

10. $\dfrac{4 - 3i}{4 + 3i}$ 11. $\dfrac{3 - 2i}{2 - i}$ 12. $\dfrac{2 - 3i}{3 - i}$

13. $\dfrac{2 + 5i}{3i}$ 14. $\dfrac{5 - 2i}{-3i}$ 15. $\dfrac{4i}{2 + i}$

In Exercises 16–21 find the reciprocal and write the answer in the form $a + bi$.

16. $3 + 2i$ 17. $4 + 3i$ 18. $\frac{1}{2} - i$

19. $1 - \frac{1}{3}i$ 20. $-7i$ 21. $-5i$

22. Prove that the multiplicative inverse of the complex number $a + bi$ (a and b not both 0) is

$$\frac{a}{a^2 + b^2} - \frac{b}{a^2 + b^2}\, i$$

23. If z and w are complex numbers, prove that $\overline{z \cdot w} = \bar{z} \cdot \bar{w}$.

24. If z is a complex number, verify that $\overline{z^2} = \bar{z}^2$ and $\overline{z^3} = \bar{z}^3$.

In Exercises 25–30 find a polynomial $P(x)$ of lowest degree that has the indicated roots.

25. $2, -4, 4$ 26. $5, -5, 1, -1$ 27. $-1, -2, -3$

28. $-3, \sqrt{2}, -\sqrt{2}$ 29. $4, 1 \pm \sqrt{3}$ 30. $1, 2, 2 \pm \sqrt{2}$

In Exercises 31–34 find the polynomial $P(x)$ of lowest degree that has the indicated roots and satisfies the given condition. (*Hint:* Write $P(x)$ in the form

$$P(x) = a(x - r_1)(x - r_2) \ldots (x - r_n)$$

where $r_1, r_2, \ldots, r_n$, are the indicated roots and a is a real number to be determined.)

31. $\frac{1}{2}, \frac{1}{2}, -2; P(2) = 3$ 32. $3, 3, -2, 2; P(4) = 12$

33. $\sqrt{2}, -\sqrt{2}, 4; P(-1) = 5$ 34. $\frac{1}{2}, -2, 5; P(0) = 5$

In Exercises 35–42 find the roots of the given equation.

35. $(x - 3)(x + 1)(x - 2) = 0$ 36. $(x - 3)(x^2 - 3x - 4) = 0$

37. $(x + 2)(x^2 - 16) = 0$ 38. $(x^2 - x)(x^2 - 2x + 5) = 0$

39. $(x^2 + 3x + 2)(2x^2 + x) = 0$ 40. $(x^2 + x + 4)(x - 3)^2 = 0$

41. $(x - 5)^3 (x + 5)^2 = 0$ 42. $(x + 1)^2 (x + 3)^4 (x - 2) = 0$

In Exercises 43–46 find a polynomial that has the indicated roots and no others.

43. -2 of multiplicity 3

44. 1 of multiplicity 2, -4 of multiplicity 1

45. $\frac{1}{2}$ of multiplicity 2, -1 of multiplicity 2

46. -1 of multiplicity 2, 0 and 2 each of multiplicity 1

In Exercises 47–52 use the given root(s) to help in finding the remaining roots of the equation.

47. $x^3 - 3x - 2 = 0;\quad -1$ 48. $x^3 - 7x^2 + 4x + 24 = 0;\quad 3$

49. $x^3 - 8x^2 + 18x - 15 = 0;\quad 5$ 50. $x^3 - 2x^2 - 7x - 4 = 0;\quad -1$

51. $x^4 + x^3 - 12x^2 - 28x - 16 = 0;\quad -2$

52. $x^4 - 2x^2 + 1;\quad$ 1 is a double root

In Exercises 53–58 find a polynomial that has the indicated roots and no others.

53. $1 + 3i, -2$ 54. $1, -1, 2 - i$

55. $1 + i, 2 - i$ 56. $-2, 3, 1 + 2i$

57. -2 is a root of multiplicity 2, $3 - 2i$

58. 3 is a triple root, $-i$

In Exercises 59–64 use the given root(s) to help in writing the given equation as a product of linear and quadratic factors with real coefficients.

59. $x^3 - 7x^2 + 16x - 10 = 0$; $3 - i$
60. $x^3 + x^2 - 7x + 65 = 0$; $2 + 3i$
61. $x^4 + 4x^3 + 13x^2 + 18x + 20 = 0$; $-1 - 2i$
62. $x^4 + 3x^3 - 5x^2 - 29x - 30 = 0$; $-2 + i$
63. $x^5 + 3x^4 - 12x^3 - 42x^2 + 32x + 120 = 0$; $-3 - i, -2$
64. $x^5 - 8x^4 + 29x^3 - 54x^2 + 48x - 16 = 0$; $2 + 2i, 2$
65. Write a polynomial $P(x)$ with complex coefficients that has the root $a + bi$, $b \neq 0$, and does not have $a - bi$ as a root.
66. Prove that a polynomial equation of degree 4 with real coefficients has 4 real roots, 2 real roots, or no real roots.
67. Prove that a polynomial equation of odd degree with real coefficients has at least one real root.

8.4
REAL AND RATIONAL ROOTS

In this section we will restrict our investigation to polynomials with real coefficients. Our first objective is to obtain some information concerning the number of positive real roots and the number of negative real roots of such polynomials.

If the terms of a polynomial with real coefficients are written in descending order, then a **variation in sign** occurs whenever two successive terms have opposite signs. In determining the number of variations of sign, we ignore terms with zero coefficients. The polynomial

$$4x^5 - 3x^4 - 2x^2 + 1$$

has two variations in sign. The French mathematician René Descartes (1596–1650), who provided us with the foundations of analytic geometry, also gave us a theorem that relates the nature of the real roots of polynomials to the variations in sign. The proof of Descartes' theorem is outlined in Exercises 39–44.

Descartes' Rule of Signs

If $P(x)$ is a polynomial with real coefficients, then
(I)   the number of positive roots is either equal to the number of variations in sign of $P(x)$, or is less than the number of variations in sign by an even number, and
(II) the number of negative roots is either equal to the number of variations in sign of $P(-x)$ or is less than the number of variations in sign by an even number.

If it is determined that a polynomial of degree n has r real roots, then the remaining $n - r$ roots must be complex numbers.

To apply Descartes' Rule of Signs to the polynomial

$$P(x) = 3x^5 + 2x^4 - x^3 + 2x - 3$$

we first note that there are 3 variations in sign as indicated. Thus, there are either 3 positive roots or there is 1 positive root. Next, we form $P(-x)$,

$$P(-x) = 3(-x)^5 + 2(-x)^4 - (-x)^3 + 2(-x) - 3$$
$$= -3x^5 + 2x^4 + x^3 - 2x - 3$$

which can be obtained by negating the coefficients of the odd power terms. We see that $P(-x)$ has two variations in sign and conclude that $P(x)$ has either 2 negative roots or no negative roots.

EXAMPLE 1
Use Descartes' Rule of Signs to analyze the roots of the equation

$$2x^5 + 7x^4 + 3x^2 - 2 = 0$$

Solution
Since

$$P(x) = 2x^5 + 7x^4 + 3x^2 - 2$$

has 1 variation in sign, there is precisely 1 positive root. The polynomial $P(-x)$ is formed

$$P(-x) = -2x^5 + 7x^4 + 3x^2 - 2$$

and is seen to have 2 variations in sign, so that $P(-x)$ has either 2 negative roots or no negative roots. Since $P(x)$ has 5 roots, the possibilities are

1 positive root, 2 negative roots, 2 complex roots

1 positive root, 0 negative roots, 4 complex roots

PROGRESS CHECK
Use Descartes' Rule of Signs to analyze the nature of the roots of the equation

$$x^6 + 5x^4 - 4x^2 - 3 = 0$$

Answer
1 positive root, 1 negative root, 4 complex roots

The following theorem provides the basis for a systematic search for the rational roots of polynomials with integer coefficients.

Rational Root Theorem

If the coefficients of the polynomial

$$P(x) = a_n x^n + a_{n-1} x^{n-1} + \ldots a_1 x + a_0 \quad (a_n \neq 0)$$

are all integers and p/q is a rational root, in lowest terms, then
(I) p is a factor of the constant term a_0, and
(II) q is a factor of the leading coefficient a_n.

PROOF OF RATIONAL ROOT
THEOREM (Optional)

Since p/q is a root of $P(x)$, then $P(p/q) = 0$. Thus,

$$a_n\left(\frac{p}{q}\right)^n + a_{n-1}\left(\frac{p}{q}\right)^{n-1} + \cdots + a_1\left(\frac{p}{q}\right) + a_0 = 0 \qquad (1)$$

Multiplying Equation (1) by q^n, we have

$$a_n p^n + a_{n-1}p^{n-1}q + \ldots + a_1 pq^{n-1} + a_0 q^n = 0 \qquad (2)$$

or

$$a_n p^n + a_{n-1}p^{n-1}q + \ldots + a_1 pq^{n-1} = -a_0 q^n \qquad (3)$$

Taking the common factor p out of the left-hand side of Equation (3) yields

$$p(a_n p^{n-1} + a_{n-1}p^{n-2}q + \ldots + a_1 q^{n-1}) = -a_0 q^n \qquad (4)$$

Since $a_1, a_2, \ldots, a_n, p$, and q are all integers, the quantity in parentheses in the left-hand side of Equation (4) is an integer. Division of the left-hand side by p results in an integer and we conclude that p must also be a factor of the right-hand side, $-a_0 q^n$. But p and q have no common factors since, by hypothesis, p/q is in lowest terms. Hence, p must be a factor of a_0 which proves part (I) of the Rational Root Theorem.

We may also rewrite Equation (2) in the form

$$q(a_{n-1}p^{n-1} + a_{n-2}p^{n-2}q + \ldots + a_1 pq^{n-2} + a_0 q^{n-1}) = -a_n p^n \qquad (5)$$

An argument similar to the preceding one now establishes part (II) of the theorem.

EXAMPLE 2
Find the rational roots of the equation

$$8x^4 - 2x^3 + 7x^2 - 2x - 1 = 0$$

Solution
If p/q is a rational root in lowest terms, then p is a factor of 1 and q is a factor of 8. We can now list the possibilities:

possible numerators: ± 1 (the factors of 1)

possible denominators: $\pm 1, \pm 2, \pm 4, \pm 8$ (the factors of 8)

possible rational roots: $\pm 1, \pm\dfrac{1}{2}, \pm\dfrac{1}{4}, \pm\dfrac{1}{8}$

Synthetic division can be used to test if these numbers are roots. Trying $x = 1$ and $x = -1$, we find that they are not roots. Trying $\frac{1}{2}$ we have

$$
\begin{array}{r|rrrrr}
\tfrac{1}{2} & 8 & -2 & 7 & -2 & -1 \\
 &   & 4 & 1 & 4 & 1 \\
\hline
 & 8 & 2 & 8 & 2 & 0
\end{array}
$$

which demonstrates that $\frac{1}{2}$ is a root. Similarly,

$$
\begin{array}{r|rrrrr}
-\frac{1}{4} & 8 & -2 & 7 & -2 & -1 \\
& & -2 & 1 & -2 & 1 \\
\hline
& 8 & -4 & 8 & -4 & 0
\end{array}
$$

which shows that $-\frac{1}{4}$ is also a root. The student may verify that these roots are not repeated and that none of the other possible rational roots will result in a zero remainder when synthetic division is employed. We can conclude that the other two roots are a pair of complex conjugates.

PROGRESS CHECK
Find the rational roots of the equation

$$9x^4 - 12x^3 + 13x^2 - 12x + 4 = 0$$

Answer

$\dfrac{2}{3}, \dfrac{2}{3}$

EXAMPLE 3
Find all roots of the equation

$$8x^5 + 12x^4 + 14x^3 + 13x^2 + 6x + 1 = 0$$

Solution
We first list the possible rational roots.

> possible numerators: ± 1 (factors of 1)

> possible denominators: $\pm 1, \pm 2, \pm 4, \pm 8$ (factors of 8)

> possible rational roots: $\pm 1, \pm \dfrac{1}{2}, \pm \dfrac{1}{4}, \pm \dfrac{1}{8}$

We next employ Descartes' Rule of Signs. Since $P(x)$ has no variations in sign, there are no positive roots. $P(-x)$ has 5 variations in sign, indicating that there are either 5 negative roots, 3 negative roots, or 1 negative root. Using synthetic division to test the possible negative rational roots, we find (verify) that $-\frac{1}{2}$ is a root.

$$
\begin{array}{r|rrrrrr}
-\frac{1}{2} & 8 & 12 & 14 & 13 & 6 & 1 \\
& & -4 & -4 & -5 & -4 & -1 \\
\hline
& 8 & 8 & 10 & 8 & 2 & 0
\end{array}
$$

coefficients of depressed equation

We can now use the depressed equation and continue testing, with the same list of possible negative roots. Once again, $-\frac{1}{2}$ is seen to be a root.

$$
\begin{array}{r|rrrrr}
-\frac{1}{2} & 8 & 8 & 10 & 8 & 2 \\
& & -4 & -2 & -4 & -2 \\
\hline
& 8 & 4 & 8 & 4 & 0
\end{array}
$$

coefficients of depressed equation

This illustrates an important point: A rational root may be a multiple root! Applying the same technique to the resulting depressed equation, we see that $-\frac{1}{2}$ is once again a root.

$$\begin{array}{r|rrrr}
-\frac{1}{2} & 8 & 4 & 8 & 4 \\
 & & -4 & 0 & -4 \\
\hline
 & 8 & 0 & 8 & 0
\end{array}$$

$$\underbrace{}$$

coefficients of depressed equation

The final depressed equation

$$8x^2 + 8 = 0 \quad \text{or} \quad x^2 + 1 = 0$$

has the roots $\pm i$. Thus, the original equation has the roots

$$-\frac{1}{2}, -\frac{1}{2}, -\frac{1}{2}, i, \text{ and } -i$$

PROGRESS CHECK

Find all roots of the polynomial

$$P(x) = 9x^4 - 3x^3 + 16x^2 - 6x - 4$$

Answer

$$\frac{2}{3}, -\frac{1}{3}, \pm \sqrt{2}i$$

EXAMPLE 4
Prove that $\sqrt{3}$ is not a rational number.

Solution
If we let $x = \sqrt{3}$, then $x^2 = 3$ or $x^2 - 3 = 0$. By the Rational Root Theorem, the only possible rational roots are ± 1, ± 3. Synthetic division can be used to show that none of these are roots. However, $\sqrt{3}$ is a root of $x^2 - 3 = 0$. Hence, $\sqrt{3}$ is not a rational number.

A number that is a root of some polynomial equation with integer coefficients is said to be **algebraic**. The requirement that the coefficients be integers is critical since any real number a will satisfy the equation $x - a = 0$. We see that $\frac{2}{3}$ is algebraic since it is a root of the equation $3x - 2 = 0$; $\sqrt{2}$ is also algebraic since it satisfies the equation $x^2 - 2 = 0$.

Are all real numbers algebraic? To show that a real number a is *not* algebraic we must show that there is *no* polynomial equation with integer coefficients that has a as one of its roots. Although this appears to be an impossible task, Georg Cantor (1845–1918), in his brilliant work on infinite sets, provided an answer. There are indeed numbers which are not algebraic. We call such numbers **transcendental**. You are already familiar with two transcendental numbers: π and e. Thus, the numbers π and e are not roots of any polynomial equations with integer coefficients.

EXERCISE SET 8.4

In Exercises 1–12 use Descartes' Rule of Signs to analyze the nature of the roots of the given equation. List all possibilities.

1. $3x^4 - 2x^3 + 6x^2 + 5x - 2 = 0$
2. $2x^6 + 5x^5 + x^3 - 6 = 0$

3. $x^6 + 2x^4 + 4x^2 + 1 = 0$

4. $3x^3 - 2x + 2 = 0$

5. $x^5 - 4x^3 + 7x - 4 = 0$

6. $2x^3 - 5x^2 + 8x - 2 = 0$

7. $5x^3 + 2x^2 + 7x - 1 = 0$

8. $x^5 + 6x^4 - x^3 - 2x - 3 = 0$

9. $x^4 - 2x^3 + 5x^2 + 2 = 0$

10. $3x^4 - 2x^3 - 1 = 0$

11. $x^8 + 7x^3 + 3x - 5 = 0$

12. $x^7 + 3x^5 - x^3 - x + 2 = 0$

In Exercises 13–22 find all rational roots of the given equation.

13. $x^3 - 2x^2 - 5x + 6 = 0$

14. $3x^3 - x^2 - 3x + 1 = 0$

15. $6x^4 - 7x^3 - 13x^2 + 4x + 4 = 0$

16. $36x^4 - 15x^3 - 26x^2 + 3x + 2 = 0$

17. $5x^6 - x^5 - 5x^4 + 6x^3 - x^2 - 5x + 1 = 0$

18. $16x^4 - 16x^3 - 29x^2 + 32x - 6 = 0$

19. $4x^4 - x^3 + 5x^2 - 2x - 6 = 0$

20. $6x^4 + 2x^3 + 7x^2 + x + 2 = 0$

21. $2x^5 - 13x^4 + 26x^3 - 22x^2 + 24x - 9 = 0$

22. $8x^5 - 4x^4 + 6x^3 - 3x^2 - 2x + 1 = 0$

In Exercises 23–30 find all roots of the given equation.

23. $4x^4 + x^3 + x^2 + x - 3 = 0$

24. $x^4 + x^3 + x^2 + 3x - 6 = 0$

25. $5x^5 - 3x^4 - 10x^3 + 6x^2 - 40x + 24 = 0$

26. $12x^4 - 52x^3 + 75x^2 - 16x - 5 = 0$

27. $6x^4 - x^3 - 5x^2 + 2x = 0$

28. $2x^4 - \dfrac{3}{2}x^3 + \dfrac{11}{2}x^2 + \dfrac{23}{2}x + \dfrac{5}{2} = 0$

29. $2x^4 - x^3 - 28x^2 + 30x - 8 = 0$

30. $12x^4 + 4x^3 - 17x^2 + 6x = 0$

In Exercises 31–34 find the integer value(s) of k for which the given equation has rational roots, and find the roots. (*Hint:* Use synthetic division.)

31. $x^3 + kx^2 + kx + 2 = 0$

32. $x^4 - 4x^3 - kx^2 + 6kx + 9 = 0$

33. $x^4 - 3x^3 + kx^2 - 4x - 1 = 0$

34. $x^3 - 3kx^2 + k^2x + 4 = 0$

35.   If $P(x)$ is a polynomial with real coefficients and has one variation in sign, prove that $P(x)$ has exactly one positive root.

36.   If $P(x)$ is a polynomial with integer coefficients and the leading coefficient is $+1$ or -1, prove that the rational roots of $P(x)$ are all integers and are factors of the constant term.

37. Prove that $\sqrt{5}$ is not a rational number.
38. If p is a prime, prove that $\sqrt{p}$ is not a rational number.
39. Prove that if $P(x)$ is a polynomial with real coefficients and r is a positive root of $P(x)$, then the reduced equation

$$Q(x) = \frac{P(x)}{(x - r)}$$

has at least one fewer variation in sign than $P(x)$. (*Hint:* Assume the leading coefficient of $P(x)$ to be positive and use synthetic division to obtain $Q(x)$. Note that the coefficients of $Q(x)$ remain positive at least until there is a variation in sign in $P(x)$.)

40. Prove that if $P(x)$ is a polynomial with real coefficients then the number of positive roots is not greater than the number of variations in sign in $P(x)$. (*Hint:* Let $r_1, r_2, \ldots, r_k$ be the positive roots of $P(x)$, and let

$$P(x) = (x - r_1)(x - r_2) \ldots (x - r_k)\, Q(x)$$

Use the result of Exercise 39 to show that $Q(x)$ has at least k fewer variations in sign than does $P(x)$.)

41. Prove that if $r_1, r_2, \ldots, r_k$ are positive numbers, then

$$P(x) = (x - r_1)(x - r_2) \ldots (x - r_k)$$

has alternating signs. (*Hint:* Use the result of Exercise 40.)

42. Prove that the number of variations in sign of a polynomial with real coefficients is even if the first and last coefficients have the same sign and is odd if they are of opposite sign.

43. Prove that if the number of positive roots of the polynomial $P(x)$ with real coefficients is less than the number of variations in sign, then it is less by an even number. (*Hint:* Write $P(x)$ as a product of linear factors corresponding to the positive and negative roots, and of quadratic factors corresponding to complex roots. Apply the results of Exercises 41 and 42.)

44. Prove that the positive roots of $P(-x)$ correspond to the negative roots of $P(x)$, that is, if $a > 0$ is a root of $P(-x)$, then $-a$ is a root of $P(x)$.

8.5
RATIONAL FUNCTIONS AND THEIR GRAPHS

A function f of the form

$$f(x) = \frac{P(x)}{Q(x)}$$

where $P(x)$ and $Q(x)$ are polynomials and $Q(x) \neq 0$ is called a **rational function.** We will study the behavior of rational functions with the objective of sketching their graphs.

We first note that the polynomials $P(x)$ and $Q(x)$ are defined for all real values of x. Since we must avoid division by zero, the domain of the function f will consist of all real numbers except those for which $Q(x) = 0$.

EXAMPLE 1

Determine the domain of each function.

(a) $f(x) = \dfrac{x + 1}{x - 1}$ (b) $g(x) = \dfrac{x^2 + 9}{x^2 - 4}$ (c) $h(x) = \dfrac{x^2}{x^2 + 1}$

Solution

(a) We must exclude all values for which the denominator $x - 1 = 0$. Thus, the domain of f consists of all real numbers except $x = 1$.

(b) Since $x^2 - 4 = 0$ when $x = \pm 2$, the domain of g consists of all real numbers except $x = \pm 2$.

(c) Since $x^2 + 1 = 0$ has no real solutions, the domain of h is the set of all real numbers.

PROGRESS CHECK

Determine the domain of each function.

(a) $S(x) = \dfrac{x}{2x^2 - 3x - 2}$ (b) $T(x) = \dfrac{-1}{x^4 + x^2 + 2}$

Answers

(a) $x \neq -\dfrac{1}{2}, 2$ (b) *All real numbers*

Let's first consider rational functions for which the numerator is a constant, for example,

$$f(x) = \frac{1}{x} \quad \text{and} \quad g(x) = \frac{1}{x^2}$$

The domain of both f and g is the set of all nonzero real numbers. Furthermore, the graph of f is symmetric with respect to the origin since the equation $y = 1/x$ remains unchanged when x and y are replaced by $-x$ and $-y$, respectively. Similarly, the graph of g is symmetric with respect to the y-axis since the equation $y = 1/x^2$ is unchanged when x is replaced by $-x$. We therefore need only plot those points corresponding to positive values of x and utilize symmetry to obtain the graphs of Figure 2.

x	$\dfrac{1}{x}$	$\dfrac{1}{x^2}$
0.001	1000	1,000,000
0.01	100	10,000
0.1	10	100
1	1	1
2	0.5	0.25
4	0.25	0.06

When a graph gets closer and closer to a line we say that the line is an **asymptote** of the graph. Note the behavior of the graphs of f and g (Figure 2) as x gets closer and closer to 0. We say that the line $x = 0$ is a **vertical asymptote** for each of these graphs. Similarly, we note that the line $y = 0$ is

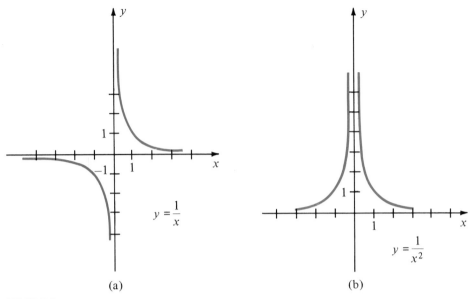

FIGURE 2

a **horizontal asymptote** in both cases. We will later show that the x-axis is a horizontal asymptote for any rational function for which the numerator is a constant and the denominator is a polynomial of degree 1 or higher.

The determination of asymptotes is extremely helpful in the graphing of rational functions. The following theorem provides the means for finding all vertical asymptotes.

Vertical Asymptote Theorem

The graph of the rational function

$$f(x) = \frac{P(x)}{Q(x)}$$

has a vertical asymptote at $x = r$ if and only if r is a real root of $Q(x)$ but not of $P(x)$.

EXAMPLE 2
Determine the vertical asymptotes of the graph of the function

$$T(x) = \frac{2}{x^3 - 2x^2 - 3x}$$

Solution
Factoring the denominator, we have

$$T(x) = \frac{2}{x(x + 1)(x - 3)}$$

and we conclude that $x = 0$, $x = -1$, and $x = 3$ are vertical asymptotes of the graph of T.

Let's examine the behavior of the function $T(x)$ of Example 2 when x is in the neighborhood of $+3$. When x is slightly more than $+3$, $x - 3$ is positive as are also x and $x + 1$; therefore, $T(x)$ is positive and growing larger and larger as x gets closer and closer to $+3$. When x is slightly less than $+3$, $x - 3$ is negative but both x and $x + 1$ are positive; therefore, $T(x)$ is negative and growing smaller and smaller as x gets closer and closer to $+3$. This leads to the portion of the graph of $T(x)$ shown in Figure 3a. Similarly, when x is slightly more than 0, $T(x)$ is negative and when x is slightly less than 0, $T(x)$ is positive (Figure 3b). The behavior of $T(x)$ when x is close to -1 is shown in Figure 3c. Since the numerator of $T(x)$ is

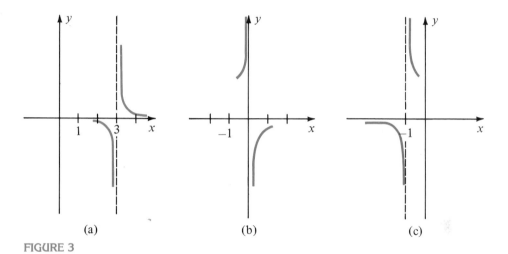

(a)   (b)   (c)

FIGURE 3

constant, $T(x) \neq 0$ for any value of x and the graph of $T(x)$ does not cross the x-axis. Moreover, since $T(x)$ is of the form $k/Q(x)$ where k is a constant, the x-axis is a horizontal asymptote. Combining these observations with the portions of the graph of $T(x)$ sketched in Figure 3 leads to the graph of $T(x)$ sketched in Figure 4.

EXAMPLE 3
Sketch the graphs of the rational functions

(a) $F(x) = \dfrac{1}{x - 1}$ (b) $G(x) = \dfrac{1}{(x + 2)^2}$

Solution
The graphs are shown in Figure 5. Note that the graphs are identical to those of Figure 2 with all points moved right one unit in the case of F and moved two units left in the case of G. In both cases we say that the y-axis has been **translated.**

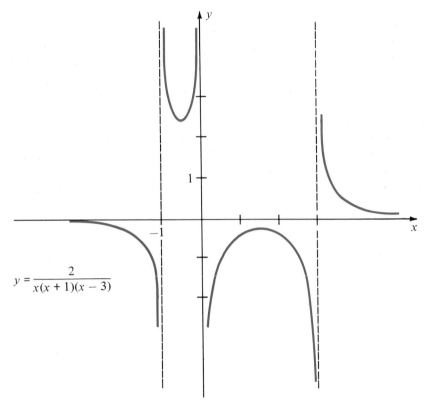

$$y = \frac{2}{x(x + 1)(x - 3)}$$

FIGURE 4

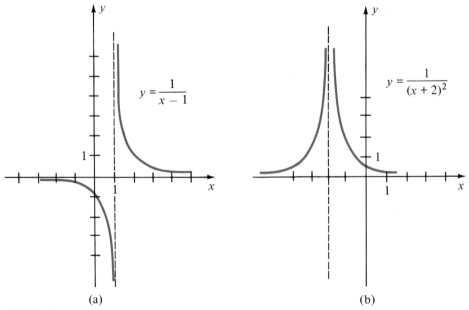

$$y = \frac{1}{x - 1}$$

$$y = \frac{1}{(x + 2)^2}$$

(a)   (b)

FIGURE 5

The x-axis is a horizontal asymptote for each of the rational functions sketched in Figures 2, 3, and 4. In general, we can determine the existence of a horizontal asymptote by studying the behavior of a rational function as x approaches $+\infty$ and $-\infty$, that is, as $|x|$ becomes very large. We first note that the expression

$$\frac{k}{x^n}, \qquad n \text{ a positive integer and } k \text{ a constant}$$

will become very small as $|x|$ becomes very large, that is, $1/x^n$ approaches 0 as $|x|$ approaches ∞. We now describe the procedure for determining horizontal asymptotes.

EXAMPLE 4
Determine the horizontal asymptotes of the function

$$f(x) = \frac{2x^2 - 5}{3x^2 + 2x - 4}$$

Solution
We illustrate the steps of the procedure.

Horizontal Asymptotes	
Step 1. Factor out the highest power of x found in the numerator; factor out the highest power of x found in the denominator.	*Step 1.* $$f(x) = \frac{x^2\left(2 - \dfrac{5}{x^2}\right)}{x^2\left(3 + \dfrac{2}{x} - \dfrac{4}{x^2}\right)}$$
Step 2. Since we are interested in large values of $|x|$, we may cancel common factors in the numerator and denominator.	*Step 2.* $$f(x) = \frac{2 - \dfrac{5}{x^2}}{3 + \dfrac{2}{x} - \dfrac{4}{x^2}}, x \neq 0$$
Step 3. Let $|x|$ increase. Then all terms of the form k/x^n approach 0 and may be discarded.	*Step 3.* The terms $-\dfrac{5}{x^2}$, $\dfrac{2}{x}$, and $-\dfrac{4}{x^2}$ approach 0 as $|x|$ approaches ∞.
Step 4. If what remains is a real number c, then $y = c$ is the horizontal asymptote. Otherwise there is no horizontal asymptote.	*Step 4.* Discarding these terms, we have $y = \dfrac{2}{3}$ as the horizontal asymptote.

EXAMPLE 5
Determine the horizontal asymptote of the function

$$f(x) = \frac{2x^3 + 3x - 2}{x^2 + 5}$$

if there is one.

Solution

Factoring, we have

$$f(x) = \frac{x^3\left(2 + \dfrac{3}{x^2} - \dfrac{2}{x^3}\right)}{x^2\left(1 + \dfrac{5}{x^2}\right)}$$

$$= \frac{x\left(2 + \dfrac{3}{x^2} - \dfrac{2}{x^3}\right)}{1 + \dfrac{5}{x^2}}, \qquad x \neq 0$$

As $|x|$ increases, the terms $3/x^2$, $-2/x^3$, and $5/x^2$ approach zero and can be discarded. What remains is $2x$, which becomes larger and larger as $|x|$ increases. Thus, there is no horizontal asymptote and $|y|$ becomes larger and larger as $|x|$ approaches infinity.

The following theorem can be proved by utilizing the procedure of Example 4.

Horizontal Asymptote Theorem

The graph of the rational function

$$f(x) = \frac{P(x)}{Q(x)}$$

has a horizontal asymptote if the degree of $P(x)$ is less than or equal to the degree of $Q(x)$.

Note that the graph of a rational function may have many vertical asymptotes but at most one horizontal asymptote.

PROGRESS CHECK

Determine the horizontal asymptote of the graph of each function.

(a) $f(x) = \dfrac{x - 1}{2x^2 + 1}$

(b) $g(x) = \dfrac{4x^2 - 3x + 1}{-3x^2 + 1}$

(c) $h(x) = \dfrac{3x^3 - x + 1}{2x^2 - 1}$

Answers

(a) $y = 0$ (b) $y = -\dfrac{4}{3}$ (c) *No horizontal asymptote*

We now summarize the information that can be gathered in preparation for sketching the graph of a rational function.

- symmetry with respect to the axes and the origin
- x-intercepts
- vertical asymptotes

- horizontal asymptotes
- brief table of values including points near the vertical asymptotes

EXAMPLE 6
Sketch the graph of

$$f(x) = \frac{x^2}{x^2 - 1}$$

Solution
Symmetry.   Since $f(-x) = f(x)$ the graph of f is symmetric with respect to the y-axis.

Intercepts.   Setting the numerator equal to 0, we see that the graph of f crosses the x-axis at the point $(0, 0)$.

Vertical asymptotes.   Setting the denominator equal to zero, we find that $x = 1$ and $x = -1$ are vertical asymptotes of the graph of f.

Horizontal asymptotes.   We note that

$$f(x) = \frac{x^2}{x^2\left(1 - \frac{1}{x^2}\right)} = \frac{1}{1 - \frac{1}{x^2}}, \quad x \neq 0$$

As $|x|$ approaches infinity, $1/x^2$ approaches 0 and the values of $f(x)$ approach 1. Thus, $y = 1$ is the horizontal asymptote. Plotting a few points, we sketch the graph of Figure 6.

x	y
1/2	−0.33
3/4	−1.29
5/4	2.78
3/2	1.80
2	1.33

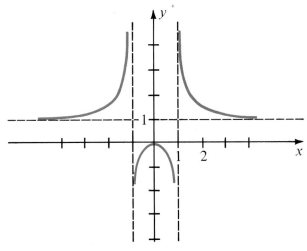

FIGURE 6

PROGRESS CHECK

Sketch the graph of

$$f(x) = \frac{x^2 - x - 6}{x^2 - 2x}$$

Answer
See Figure 7.

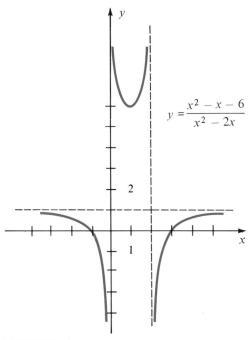

$$y = \frac{x^2 - x - 6}{x^2 - 2x}$$

FIGURE 7

We conclude this section with an example of a rational function that is not in reduced form, that is, the numerator and denominator have a common factor.

EXAMPLE 7

Sketch the graph of the function

$$f(x) = \frac{x^2 - 1}{x - 1}$$

Solution
We observe that

$$f(x) = \frac{x^2 - 1}{x - 1} = \frac{(x + 1)(x - 1)}{x - 1} = x + 1, \qquad x \neq 1$$

Thus, the graph of the function f coincides with the straight line $y = x + 1$ with the exception that f is undefined when $x = 1$ (Figure 8).

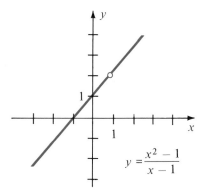

FIGURE 8

PROGRESS CHECK
Sketch the graph of the function

$$f(x) = \frac{8 - 2x^2}{x + 2}$$

Answer
See Figure 9.

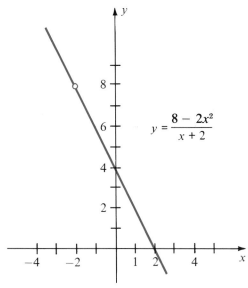

FIGURE 9

EXERCISE SET 8.5

In Exercises 1–6 determine the domain of the given function.

1. $f(x) = \dfrac{x^2}{x - 1}$

2. $f(x) = \dfrac{x - 1}{x^2 + x - 2}$

3. $g(x) = \dfrac{x^2 + 1}{x^2 - 2x}$

4. $g(x) = \dfrac{x^2 + 2}{x^2 - 2}$

5. $F(x) = \dfrac{x^2 - 3}{x^2 + 3}$

6. $T(x) = \dfrac{3x + 2}{2x^3 - x^2 - x}$

In Exercises 7–21 determine the vertical and horizontal asymptotes of the graph of the given function. Sketch the graph.

7. $f(x) = \dfrac{1}{x - 4}$ 8. $f(x) = \dfrac{-2}{x - 3}$ 9. $f(x) = \dfrac{3}{x + 2}$

10. $f(x) = \dfrac{-1}{(x - 1)^2}$ 11. $f(x) = \dfrac{1}{(x + 1)^2}$ 12. $f(x) = \dfrac{-1}{x^2 + 1}$

13. $f(x) = \dfrac{x + 2}{x - 2}$ 14. $f(x) = \dfrac{x}{x + 2}$ 15. $f(x) = \dfrac{2x^2 + 1}{x^2 - 4}$

16. $f(x) = \dfrac{x^2 - 1}{x^2 + 2x - 3}$ 17. $f(x) = \dfrac{x^2 + 2}{2x^2 - x - 6}$

18. $f(x) = \dfrac{x^2 - 1}{x + 2}$ 19. $f(x) = \dfrac{x^2}{4x - 4}$ 20. $f(x) = \dfrac{x - 1}{2x^3 - 2x}$

21. $f(x) = \dfrac{x^3 + 4x^2 + 3x}{x^2 - 25}$

In Exercises 22–27 determine the domain and sketch the graph of the reducible function.

22. $f(x) = \dfrac{x^2 - 25}{2x - 10}$ 23. $f(x) = \dfrac{2x^2 - 8}{x + 2}$ 24. $f(x) = \dfrac{2x^2 + 2x - 12}{3x - 6}$

25. $f(x) = \dfrac{x^2 + 2x - 8}{2x^2 - 8x + 8}$ 26. $f(x) = \dfrac{x + 2}{x^2 - x - 6}$ 27. $f(x) = \dfrac{2x}{x^2 + x}$

TERMS AND SYMBOLS

polynomial function of degree n (p. 302)	$\bar{z}$ (p. 312)	**transcendental numbers** (p. 324)
polynomial equation of degree n (p. 303)	**root of multiplicity k** (p. 315)	**rational function** (p. 326)
roots or zeros of a polynomial (p. 303)	**depressed equation** (p. 315)	**vertical asymptote** (p. 327)
synthetic division (p. 305)	**variation in sign** (p. 320)	**horizontal asymptote** (p. 328)
complex conjugate (p. 311)	**algebraic numbers** (p. 324)	

KEY IDEAS FOR REVIEW

■ Polynomial division results in a quotient and a remainder, both of which are polynomials. The degree of the remainder must be less than the degree of the divisor.

■ Synthetic division is a quick way to divide a polynomial by a first-degree polynomial $x - r$, where r is a real constant.

■ The following are the primary theorems concerning polynomials and their roots.

Remainder Theorem
If a polynomial $P(x)$ is divided by $x - r$, then the remainder is $P(r)$.

Factor Theorem
A polynomial $P(x)$ has a root r if and only if $x - r$ is a factor of $P(x)$.

Linear Factor Theorem
A polynomial $P(x)$ of degree $n \geq 1$ can be written as the product of n linear factors.

$$P(x) = a(x - r_1)(x - r_2) \ldots (x - r_n)$$

where $r_1, r_2, \ldots, r_n$ are the complex roots of $P(x)$ and a is the leading coefficient of $P(x)$.

Number of Roots Theorem

If $P(x)$ is a polynomial of degree $n \geq 1$, then $P(x)$ has precisely n roots among the complex numbers.

Conjugate Roots Theorem

If $a + bi$, $b \neq 0$, is a root of the polynomial $P(x)$ with real coefficients, then $a - bi$ is also a root of $P(x)$.

Rational Root Theorem

If p/q is a rational root (in lowest terms) of the polynomial $P(x)$ with integer coefficients, then p is a factor of the constant term a_0 of $P(x)$ and q is a factor of the leading coefficient a_n of $P(x)$.

■ If r is a real root of the polynomial $P(x)$, then the roots of the depressed equation are the other roots of $P(x)$. The depressed equation can be found by using synthetic division.

■ Descartes' Rule of Signs tells us the maximum number of positive roots and the maximum number of negative roots of a polynomial $P(x)$ with real coefficients.

■ If $P(x)$ has integer coefficients, then the Rational Root Theorem enables us to list all possible rational roots of $P(x)$. Synthetic division can then be used to test these potential rational roots, since r is a root if and only if the remainder is zero, that is, if and only if $P(r) = 0$.

■ Always determine the vertical and horizontal asymptotes of a rational function before attempting to sketch its graph.

REVIEW EXERCISES

In Exercises 1–2 use synthetic division to find the quotient $Q(x)$ and the constant remainder R when the first polynomial is divided by the second polynomial.

1. $2x^3 + 6x - 4$, $x - 1$ 2. $x^4 - 3x^3 + 2x - 5$, $x + 2$

In Exercises 3–4 use synthetic division to find $P(2)$ and $P(-1)$.

3. $7x^3 - 3x^2 + 2$ 4. $x^5 - 4x^3 + 2x$

In Exercises 5–6 use the Factor Theorem to show that the second polynomial is a factor of the first polynomial.

5. $2x^4 + 4x^3 + 3x^2 + 5x - 2$, $x + 2$

6. $2x^3 - 5x^2 + 6x - 2$, $x - \dfrac{1}{2}$

In Exercises 7–9 write the given quotient in the form $a + bi$.

7. $\dfrac{3 - 2i}{4 + 3i}$ 8. $\dfrac{2 + i}{-5i}$ 9. $\dfrac{-5}{1 + i}$

In Exercises 10–12 write the reciprocal of the given complex number in the form $a + bi$.

10. $1 + 3i$ 11. $-4i$ 12. $2 - 5i$

In Exercises 13–15 find a polynomial of lowest degree that has the indicated roots.

13. $-3, -2, -1$ 14. $3, \pm\sqrt{-3}$ 15. $-2, \pm\sqrt{3}, 1$

In Exercises 16–18 find a polynomial that has the indicated roots and no others.

16. $\frac{1}{2}$ of multiplicity 2, -1 of multiplicity 2

17. $i, -i$, each of multiplicity 2

18. -1 of multiplicity 3, 3 of multiplicity 1

In Exercises 19–21 use the given root(s) to assist in finding the remaining roots of the equation.

19. $2x^3 - x^2 - 13x - 6 = 0$; -2

20. $x^3 - 2x^2 - 9x + 4 = 0$;   4

21. $2x^4 - 15x^3 + 34x^2 - 19x - 20 = 0$; $2 + i$

In Exercises 22–25 use Descartes' Rule of Signs to determine the maximum number of positive and negative real roots of the given equation.

22. $x^4 - 2x - 1 = 0$ 23. $x^5 - x^4 + 3x^3 - 4x^2 + x - 5 = 0$

24. $x^3 - 5 = 0$ 25. $3x^4 - 2x^2 + 1 = 0$

In Exercises 26–28 find all the rational roots of the given equation.

26. $6x^3 - 5x^2 - 33x - 18 = 0$

27. $6x^4 - 7x^3 - 19x^2 + 32x - 12 = 0$

28. $x^4 + 3x^3 + 2x^2 + x - 1 = 0$

In Exercises 29–30 find all roots of the given equation.

29. $6x^3 + 15x^2 - x - 10 = 0$

30. $2x^4 - 3x^3 - 10x^2 + 19x - 6 = 0$

In Exercises 31–32 sketch the graph of the given function.

31. $f(x) = \dfrac{x}{x + 1}$ 32. $f(x) = \dfrac{x^2}{x + 1}$

PROGRESS TEST 8A

1.   Find the quotient and remainder when $2x^4 - x^2 + 1$ is divided by $x^2 + 2$.
2.   Use synthetic division to find the quotient and remainder when $3x^4 - x^3 - 2$ is divided by $x + 2$.
3.   If $P(x) = x^3 - 2x^2 + 7x + 5$, use synthetic division to find $P(-2)$.
4.   Determine the remainder when $4x^5 - 2x^4 - 5$ is divided by $x + 2$.
5.   Use the Factor Theorem to show that $x - 3$ is a factor of

$$2x^4 - 9x^3 + 9x^2 + x - 3$$

In Problems 6–7 find a polynomial of lowest degree that has the indicated roots.

6. $-2, 1, 3$ 7. $-1, 1, 3 \pm \sqrt{2}$

In Problems 8–9 find the roots of the given equation.

8. $(x^2 + 1)(x - 2) = 0$ 9. $(x + 1)^2(x^2 - 3x - 2) = 0$

In Problems 10–12 find a polynomial that has the indicated roots and no others.

10. -3 of multiplicity 2; 1 of multiplicity 3

11. $-\frac{1}{4}$ of multiplicity 2; $i, -i$, and 1

12. $i, 1 + i$

In Problems 13–14 use the given root(s) to help in finding the remaining roots of the equation.

13. $4x^3 - 3x + 1 = 0$; -1 14. $x^4 - x^2 - 2x + 2 = 0$;   1

15. If $2 + i$ is a root of $x^3 - 6x^2 + 13x - 10 = 0$, write the equation as a product of linear and quadratic factors with real coefficients.

In Problems 16–17 determine the maximum number of roots of the given equation of the type indicated.

16. $2x^5 - 3x^4 + 1 = 0$; positive real roots

17. $3x^4 + 2x^3 - 2x^2 - 1 = 0$; negative real roots

In Problems 18–19 find all rational roots of the given equation.

18. $6x^3 - 17x^2 + 14x + 3 = 0$

19. $2x^5 - x^4 - 4x^3 + 2x^2 + 2x - 1 = 0$

20. Find all roots of the equation $3x^4 + 7x^3 - 3x^2 + 7x - 6 = 0$

21. Sketch the graph of the function $f(x) = \dfrac{x^2 + 2}{x^2 - 1}$.

PROGRESS TEST 8B

1. Find the quotient and remainder when $3x^5 + 2x^3 - x^2 - 2$ is divided by $2x^2 - x - 1$.

2. Use synthetic division to find the quotient and remainder when $-2x^3 + 3x^2 - 1$ is divided by $x - 1$.

3. If $P(x) = 2x^4 - 2x^3 + x - 4$, use synthetic division to find $P(-1)$.

4. Determine the remainder when $3x^4 - 5x^3 + 3x^2 + 4$ is divided by $x - 2$.

5. Use the Factor Theorem to show that $x + 2$ is a factor of $x^3 - 4x^2 - 9x + 6$.

In Problems 6–7 find a polynomial of lowest degree that has the indicated roots.

6. $-\frac{1}{2}, 1, 1, -1$　　　　　　　　　7. $2, 1 \pm \sqrt{3}$

In Problems 8–9 find the roots of the given equation.

8. $(x^2 - 3x + 2)(x - 2)^2$　　　　　　　9. $(x^2 + 3x - 1)(x - 2)(x + 3)^2$

In Problems 10–12 find a polynomial that has the indicated roots and no others.

10. $\frac{1}{2}$ of multiplicity 3; -2 of multiplicity 1

11. -3 of multiplicity 2; $1 + i, 1 - i$

12. $3 \pm \sqrt{-1}, -1$ of multiplicity 2

In Problems 13–14 use the given root(s) to help in finding the remaining roots of the equation.

13. $x^3 - x^2 - 8x - 4 = 0$; -2　　　　14. $x^4 - 3x^3 - 22x^2 + 68x - 40 = 0$; 2, 5

15. If $1 - i$ is a root of $2x^4 - x^3 - 4x^2 + 10x - 4 = 0$, write the equation as a product of linear and quadratic factors with real coefficients.

In Problems 16–17 determine the maximum number of roots, of the type indicated, of the given equation.

16. $3x^4 + 3x - 1 = 0$; positive real roots

17. $2x^4 + x^3 - 3x^2 + 2x + 1 = 0$; negative real roots

In Problems 18–19 find all rational roots of the given equation.

18. $3x^3 + 7x^2 - 4 = 0$　　　　　　　19. $4x^4 - 4x^3 + x^2 - 4x - 3 = 0$

20. Find all roots of the equation $2x^4 - x^3 - 4x^2 + 2x = 0$.

21. Sketch the graph of the function $f(x) = \dfrac{2x}{x^2 - 1}$.

CHAPTER NINE
TOPICS IN ALGEBRA

This chapter presents several topics in algebra that are somewhat independent of the flow of ideas in the earlier chapters of this book. Some of these topics, such as sequences, deal with functions whose domain is the set of natural numbers. An important reason for studying sequences and series is that the underlying concepts can be used as an introduction to calculus.

Another of these topics, mathematical induction, provides a means of proving certain theorems involving the natural numbers that appear to resist other means of proof. As an example, we will use mathematical induction to prove that the sum of the first n consecutive integers is $n(n + 1)/2$.

Another topic is the binomial theorem, which gives us a way to expand the expression $(a + b)^n$. Those students who proceed to a study of calculus will find this theorem used when they begin to study the derivative.

Probability theory, a very useful topic in algebra, enables us to state the likelihood of occurrence of a given event and has obvious applications to games of chance. The theory of permutations and combinations enables us to count the ways in which we can arrange or select a subset of a set of objects, and is necessary background to a study of probability theory.

9.1
ARITHMETIC PROGRESSIONS

Can you see a pattern or relationship that describes this string of numbers?

$$1, 4, 9, 16, 25, \ldots$$

If we rewrite this string as

$$1^2, 2^2, 3^2, 4^2, 5^2, \ldots$$

it is clear that these are the squares of successive natural numbers. Each number in the string is called a **term.** We could write the nth term of the list as a function a defined by

$$a(n) = n^2$$

where n is a natural number. Such a string of numbers is called an **infinite sequence** since the list is infinitely long.

> An **infinite sequence** is a function whose domain is the set of all natural numbers.

The range of the function a is

$$a(1), a(2), a(3) \ldots$$

which we write as

$$a_1, a_2, a_3, \ldots$$

That is, we indicate a sequence by using subscript notation rather than function notation. We say that a_1 is the first term of the sequence, a_2 is the second term, and so on, and we write the nth term as a_n.

A string of numbers whose domain is a finite subset of the natural numbers is called a **finite sequence.**

EXAMPLE 1
Write the first four terms of a sequence whose nth term is

$$a_n = \frac{n}{n + 1}$$

Solution
To find a_1, we substitute $n = 1$ in the formula for a_n.

$$a_1 = \frac{1}{1 + 1} = \frac{1}{2}$$

Similarly, we have

$$a_2 = \frac{2}{2 + 1} = \frac{2}{3}, \quad a_3 = \frac{3}{3 + 1} = \frac{3}{4}, \quad a_4 = \frac{4}{4 + 1} = \frac{4}{5}$$

PROGRESS CHECK
Write the first four terms of a sequence whose nth term is

$$a_n = \frac{n - 1}{n^2}$$

Answer

$0, \dfrac{1}{4}, \dfrac{2}{9}, \dfrac{3}{16}$

Now let's try to find a pattern or relationship for the sequence

$$2, 5, 8, 11, \ldots$$

You may notice that the nth term can be written as $a_n = 3n - 1$. But there is a simpler way to describe this sequence. Each term after the first can be obtained by adding 3 to the preceding term.

$$a_1 = 2, \quad a_2 = a_1 + 3, \quad a_3 = a_2 + 3, \ldots$$

A sequence in which each successive term is obtained by adding a fixed number to the previous term is called an **arithmetic progression** or **arithmetic sequence.**

In an arithmetic progression, there is a number d such that

$$a_n = a_{n-1} + d$$

for all $n > 1$. The number d is called the **common difference.**

Returning to the sequence

$$2, 5, 8, 11, \ldots$$

the nth term can be written as

$$a_n = a_{n-1} + 3 \quad \text{with} \quad a_1 = 2$$

This is an arithmetic progression with the first term equal to 2 and a common difference of 3. The formula $a_n = a_{n-1} + 3$ is said to be a **recursive formula** since it defines the nth term by reference to preceding terms. Beginning with $a_1 = 2$, the formula is used "recursively" (over and over) to obtain a_2, then a_3, then a_4, and so on.

EXAMPLE 2

Which of the following are arithmetic progressions?

(a) $5, 7, 9, 11, \ldots$. Since each term can be obtained from the preceding term by adding 2, this is an arithmetic progression with the first term equal to 5 and a common difference of 2.

(b) $4, 8, 11, 13, \ldots$. This is not an arithmetic progression since there is not a common difference between terms. The difference between the first and second terms is 4, while that between the second and third terms is 3.

(c) $3, -1, -5, -9, \ldots$. The difference between terms is -4, that is,

$$a_n = a_{n-1} - 4$$

This is an arithmetic progression with the first term equal to 3 and a common difference of -4.

(d) $1, \frac{3}{2}, 2, \frac{5}{2}, \ldots$. This is an arithmetic progression with a common difference of $\frac{1}{2}$.

PROGRESS CHECK
Which of the following are arithmetic progressions?

(a)   16, 17, 16, 17, . . .

(b) $-2, -1, 0, 1, . . .$

(c) $6, \frac{9}{2}, 3, \frac{3}{2}, . . .$

(d)   2, 4, 8, 12, . . .

Answer
(b) and (c)

EXAMPLE 3
Write the first four terms of an arithmetic progression whose first term is -4 and whose common difference is -3.

Solution
Beginning with -4, we add the common difference -3 to obtain

$$-4 + (-3) = -7; \quad -7 + (-3) = -10; \quad -10 + (-3) = -13$$

Alternatively, we note that the sequence is defined by

$$a_n = a_{n-1} - 3, \quad a_1 = -4$$

which leads to the terms

$$a_1 = -4, \quad a_2 = -7, \quad a_3 = -10, \quad a_4 = -13$$

PROGRESS CHECK
Write the first four terms of an arithmetic progression whose first term is 4 and whose common difference is $-\frac{1}{3}$.

Answer
$4, \dfrac{11}{3}, \dfrac{10}{3}, 3, . . .$

For a given arithmetic progression, it's easy to find a formula for the nth term, a_n, in terms of n and the first term a_1. Since

$$a_2 = a_1 + d$$

and

$$a_3 = a_2 + d$$

we see that

$$a_3 = (a_1 + d) + d = a_1 + 2d$$

Similarly, we can show that

$$a_4 = a_3 + d = (a_1 + 2d) + d = a_1 + 3d$$
$$a_5 = a_4 + d = (a_1 + 3d) + d = a_1 + 4d$$

In general,

> The nth term a_n of an arithmetic progression is given by
> $$a_n = a_1 + (n - 1)d$$

A rigorous proof of this formula requires mathematical induction (see Exercise Set of Section 3 of this chapter).

EXAMPLE 4

Find the seventh term of the arithmetic progression whose first term is 2 and whose common difference is 4.

Solution

We substitute $n = 7$, $a_1 = 2$, $d = 4$ in the formula

$$a_n = a_1 + (n - 1)d$$

obtaining

$$a_7 = 2 + (7 - 1)4 = 2 + 24 = 26$$

PROGRESS CHECK

Find the 16th term of the arithmetic progression whose first term is -5 and whose common difference is $\frac{1}{2}$.

Answer

$$\frac{5}{2}$$

EXAMPLE 5

Find the 25th term of the arithmetic progression whose first and 20th terms are -7 and 31, respectively.

Solution

We can apply the given information to find d.

$$a_n = a_1 + (n - 1)d$$
$$a_{20} = a_1 + (20 - 1)d$$
$$31 = -7 + 19d$$
$$d = 2$$

Now we use the formula for a_n to find a_{25}.

$$a_n = a_1 + (n - 1)d$$
$$a_{25} = -7 + (25 - 1)2$$
$$a_{25} = 41$$

PROGRESS CHECK

Find the 60th term of the arithmetic progression whose first and 10th terms are 3 and $-\frac{3}{2}$, respectively.

Answer

$$-\frac{53}{2}$$

In many applications of sequences, we wish to *add* the terms. A sum of terms of a sequence is called a **series.** When we are dealing with an arithmetic progression, the associated series is called an **arithmetic series.**

We denote the sum of the first n terms of the arithmetic progression a_1, $a_2, a_3, \ldots$ by S_n.

$$S_n = a_1 + a_2 + a_3 + \cdots + a_{n-2} + a_{n-1} + a_n$$

Since an arithmetic progression has a common difference d, we may write

$$S_n = a_1 + (a_1 + d) + (a_1 + 2d) + \cdots + (a_n - 2d) + (a_n - d) + a_n \quad (1)$$

where we write $a_2, a_3, \ldots$, in terms of a_1 and we write $a_{n-1}, a_{n-2}, \ldots$, in terms of a_n. Rewriting the right-hand side of Equation (1) in reverse order, we have

$$S_n = a_n + (a_n - d) + (a_n - 2d) + \cdots + (a_1 + 2d) + (a_1 + d) + a_1 \quad (2)$$

Summing the corresponding sides of Equations (1) and (2),

$$2S_n = (a_1 + a_n) + (a_1 + a_n) + (a_1 + a_n) + \cdots \qquad \text{Repeated } n \text{ times}$$

$$= n(a_1 + a_n)$$

Thus,

$$S_n = \frac{n}{2}(a_1 + a_n)$$

Since $a_n = a_1 + (n-1)d$, we see that

$$S_n = \frac{n}{2}[a_1 + a_1 + (n-1)d] \qquad \text{Substituting for } a_n$$

$$= \frac{n}{2}[2a_1 + (n-1)d]$$

We now have two useful formulas.

Arithmetic Series

(1) $S_n = \frac{n}{2}(a_1 + a_n)$

(2) $S_n = \frac{n}{2}[2a_1 + (n-1)d]$

The choice as to which formula to use depends upon the available information. The following examples illustrate the use of the formulas.

EXAMPLE 6

Find the sum of the first 30 terms of an arithmetic sequence whose first term is -20 and whose common difference is 3.

Solution

We know that $n = 30$, $a_1 = -20$, and $d = 3$. Substituting in

$$S_n = \frac{n}{2}[2a_1 + (n-1)d]$$

we obtain

$$S_{30} = \frac{30}{2}[2(-20) + (30 - 1)3]$$

$$= 15(-40 + 87)$$

$$= 705$$

PROGRESS CHECK

Find the sum of the first 10 terms of the arithmetic sequence whose first term is 2 and whose common difference is $-\frac{1}{2}$.

Answer

$$-\frac{5}{2}$$

EXAMPLE 7

The first term of an arithmetic series is 2, the last term is 58, and the sum is 450. Find the number of terms and the common difference.

Solution

We have $a_1 = 2$, $a_n = 58$, and $S_n = 450$. Substituting in

$$S_n = \frac{n}{2}(a_1 + a_n)$$

we have

$$450 = \frac{n}{2}(2 + 58)$$

$$900 = 60n$$

$$n = 15$$

Now we substitute in

$$a_n = a_1 + (n - 1)d$$

$$58 = 2 + (14)d$$

$$56 = 14d$$

$$d = 4$$

PROGRESS CHECK

The first term of an arithmetic series is 6, the last term is 1, and the sum is 77/2. Find the number of terms and the common difference.

Answer

$$n = 11, \quad d = -\frac{1}{2}$$

EXERCISE SET 9.1

In each of the following write the first four terms of the sequence whose nth term is given as a_n.

1. $a_n = 2n$ 2. $a_n = 2n + 1$ 3. $a_n = 2n + 3$

4. $a_n = \dfrac{n^2 - 1}{n^2 + 1}$ 5. $a_n = \dfrac{n^2}{2n + 1}$ 6. $a_n = \dfrac{2n + 1}{n^2}$

Which of the following are arithmetic progressions?

7. $3, 6, 9, 12, \ldots$ 8. $4, \dfrac{11}{2}, 7, \dfrac{17}{2}, \ldots$

9. $1, -2, -6, -10, \ldots$ 10. $-\dfrac{1}{2}, -1, -\dfrac{3}{2}, -2, \ldots$

11. $0, \dfrac{1}{4}, \dfrac{1}{2}, \dfrac{3}{4}, \ldots$ 12. $2, 0, 1, 4, \ldots$

13. $-1, 2, 5, 8, \ldots$ 14. $12, 8, 4, 1, \ldots$

Write the first four terms of the arithmetic progression whose first term is a_1 and whose common difference is d.

15. $a_1 = 2, d = 4$ 16. $a_1 = -2, d = -5$

17. $a_1 = 3, d = -\dfrac{1}{2}$ 18. $a_1 = \dfrac{1}{2}, d = 2$

19. $a_1 = \dfrac{1}{3}, d = -\dfrac{1}{3}$ 20. $a_1 = 6, d = \dfrac{5}{2}$

Find the specified term of the arithmetic progression whose first term is a_1 and whose common difference is d.

21. $a_1 = 4, d = 3$; 8th term 22. $a_1 = -3, d = \dfrac{1}{4}$; 14th term

23. $a_1 = 14, d = -2$; 12th term 24. $a_1 = 6, d = -\dfrac{1}{3}$; 9th term

Given two terms of an arithmetic progression, find the specified term.

25. $a_1 = -2, a_{20} = -2$; 24th term 26. $a_1 = \dfrac{1}{2}, a_{12} = 6$; 30th term

27. $a_1 = 0, a_{61} = 20$; 20th term 28. $a_1 = 23, a_{15} = -19$; 6th term

29. $a_1 = -\dfrac{1}{4}, a_{41} = 10$; 22nd term 30. $a_1 = -3, a_{18} = 65$; 30th term

Find the sum of the specified number of terms of the arithmetic progression whose first term is a_1 and whose common difference is d.

31. $a_1 = 3, d = 2$; 20 terms 32. $a_1 = -4, d = \dfrac{1}{2}$; 24 terms

33. $a_1 = \dfrac{1}{2}, d = -2$; 12 terms 34. $a_1 = -3, d = -\dfrac{1}{3}$; 18 terms

35. $a_1 = 82, d = -2$, 40 terms 36. $a_1 = 6, d = 4$; 16 terms

37. How many terms of the arithmetic progression $2, 4, 6, 8, \ldots$ add up to 930?
38. How many terms of the arithmetic progression $44, 41, 38, 35, \ldots$ add up to 340?
39. The first term of an arithmetic series is 3, the last term is 90, and the sum is 1395. Find the number of terms and the common difference.
40. The first term of an arithmetic series is -3, the last term is $\frac{5}{2}$, and the sum is -3. Find the number of terms and the common difference.
41. The first term of an arithmetic series is $\frac{1}{2}$, the last term is $\frac{7}{4}$, and the sum is $\frac{27}{4}$. Find the number of terms and the common difference.

42. The first term of an arithmetic series is 20, the last term is -14, and the sum is 54. Find the number of terms and the common difference.
43. Find the sum of the first 16 terms of an arithmetic progression whose 4th and 10th terms are $-\frac{5}{4}$ and $\frac{1}{4}$, respectively.
44. Find the sum of the first 12 terms of an arithmetic progression whose 3rd and 6th terms are 9 and 18, respectively.

9.2
GEOMETRIC PROGRESSIONS

The sequence

$$3, 6, 12, 24, 48$$

has a distinct pattern. Each term after the first is obtained by multiplying the preceding one by 2. Thus, we could rewrite the sequence as

$$3, \quad 3 \cdot 2, \quad (3 \cdot 2) \cdot 2, \quad (3 \cdot 2 \cdot 2) \cdot 2, \ldots$$

Such a sequence is called a **geometric progression** or **geometric sequence.** Each successive term is found by multiplying the previous term by a fixed number.

> In a geometric progression, there is a number r such that
> $$a_n = ra_{n-1} \quad \text{for all } n > 1$$

The constant r is called the **common ratio** and can be found by dividing any term a_k by the preceding term, a_{k-1}.

$$r = \frac{a_k}{a_{k-1}}$$

EXAMPLE 1
If the sequence is a geometric progression, find the common ratio.

(a) $2, -4, 8, -16, \ldots$. Since each term can be obtained by multiplying the preceding one by -2, this is a geometric progression with common ratio of -2.
(b) $1, 2, 6, 24, \ldots$. The ratio between successive terms is not constant. This is not a geometric progression.
(c) $\frac{1}{4}, \frac{1}{8}, \frac{1}{16}, \frac{1}{32}, \ldots$. This is a geometric progression with common ratio of $\frac{1}{2}$.

PROGRESS CHECK
If the sequence is a geometric progression, find the common ratio.

(a) $3, -9, 27, -81, \ldots$

(b) $4, 1, -2, -5, \ldots$

(c) $6, 2, \dfrac{2}{3}, \dfrac{2}{9}, \ldots$

(d) $4, 16, 48, 96, \ldots$

Answers
Sequence (a) is a geometric progression with $r = -3$.
Sequence (c) is a geometric progression with $r = \frac{1}{3}$.

Let's look at successive terms of a geometric progression whose first term is a_1 and whose common ratio is r. We have

$$a_2 = ra_1$$
$$a_3 = ra_2 = r(ra_1) = r^2a_1$$
$$a_4 = ra_3 = r(r^2a_1) = r^3a_1$$

The pattern suggests that the exponent of r is one less than the subscript of a in the left-hand side.

The nth term of a geometric progression is given by

$$a_n = a_1 r^{n-1}$$

Once again, mathematical induction is required to prove that the formula holds for all natural numbers (see Exercise Set of Section 3 of this chapter).

EXAMPLE 2
Find the seventh term of the geometric progression $-4, -2, -1, \ldots$.

Solution
Since

$$r = \frac{a_k}{a_{k-1}}$$

we see that

$$r = \frac{a_3}{a_2} = \frac{-1}{-2} = \frac{1}{2}$$

Substituting $a_1 = -4$, $r = \frac{1}{2}$, and $n = 7$, we have

$$a_n = a_1 r^{n-1}$$
$$a_7 = (-4)\left(\frac{1}{2}\right)^{7-1} = (-4)\left(\frac{1}{2}\right)^6$$
$$= (-4)\left(\frac{1}{64}\right) = -\frac{1}{16}$$

PROGRESS CHECK
Find the sixth term of the geometric progression $2, -6, 18, \ldots$.

Answer
-486

In a geometric progression, the terms between the first and last terms are called **geometric means.** We will illustrate the method of calculating such means.

EXAMPLE 3
Insert three geometric means between 3 and 48.

Solution

The geometric progression must look like this.

$$3, a_2, a_3, a_4, 48, \ldots$$

Thus, $a_1 = 3$, $a_5 = 48$, and $n = 5$. Substituting in

$$a_n = a_1 r^{n-1}$$

$$48 = 3r^4$$

$$r^4 = 16$$

$$r = \pm 2$$

Thus there are two geometric progressions with three geometric means between 3 and 48.

$$3, 6, 12, 24, 48, \ldots \qquad r = 2$$

$$3, -6, 12, -24, 48, \ldots \qquad r = -2$$

PROGRESS CHECK

Insert two geometric means between 5 and $\frac{8}{25}$.

Answer

$$2, \frac{4}{5}$$

If $a_1, a_2, \ldots$ is a geometric progression, then the corresponding sum of the first n terms

$$S_n = a_1 + a_2 + \cdots + a_n \qquad (1)$$

is called a **geometric series.** Since each term of the series can be rewritten as $a_k = a_1 r^{k-1}$, we can rewrite Equation (1) as

$$S_n = a_1 + a_1 r + a_1 r^2 + \cdots + a_1 r^{n-2} + a_1 r^{n-1} \qquad (2)$$

Multiplying each term in Equation (2) by r we have

$$r S_n = a_1 r + a_1 r^2 + a_1 r^3 + \cdots + a_1 r^{n-1} + a_1 r^n \qquad (3)$$

Subtracting Equation (2) from Equation (3) produces

$$r S_n - S_n = a_1 r^n - a_1$$

$$(r - 1)S_n = a_1(r^n - 1) \qquad \text{Factoring}$$

$$S_n = \frac{a_1(r^n - 1)}{r - 1} \qquad \begin{array}{l}\text{Dividing by } r - 1 \\ (\text{if } r \neq 1)\end{array}$$

Changing the signs in both the numerator and the denominator gives us the following equation for the sum of n terms.

Geometric Series

$$S_n = \frac{a_1(1 - r^n)}{1 - r}$$

EXAMPLE 4

Find the sum of the first six terms of the geometric progression whose first three terms are 12, 6, 3.

Solution

The common ratio can be found by dividing any term by the preceding term.

$$r = \frac{a_k}{a_{k-1}} = \frac{a_2}{a_1} = \frac{6}{12} = \frac{1}{2}$$

Substituting $a_1 = 12$, $r = \frac{1}{2}$, $n = 6$ in the formula for S_n, we have

$$S_n = \frac{a_1(1 - r^n)}{1 - r}$$

$$= \frac{12\left[1 - \left(\frac{1}{2}\right)^6\right]}{1 - \frac{1}{2}} = \frac{12\left(1 - \frac{1}{64}\right)}{\frac{1}{2}}$$

$$= 24\left(\frac{63}{64}\right) = \frac{189}{8}$$

PROGRESS CHECK

Find the sum of the first five terms of the geometric progression whose first three terms are 2, $-\frac{4}{3}$, $\frac{8}{9}$.

Answer

$$\frac{110}{81}$$

EXAMPLE 5

A father promises to give each child 2 cents on the first day, 4 cents on the second day, and to continue doubling the amount each day for a total of eight days. How much will each child receive on the last day? How much will each child have received in total after eight days?

Solution

The daily payout to each child forms a geometric progression 2, 4, 8, . . . with $a_1 = 2$ and $r = 2$. The term a_8 is given by substituting in

$$a_n = a_1 r^{n-1}$$

$$a_8 = a_1 r^{8-1} = 2 \cdot 2^7 = 256$$

Thus, each child received $2.56 on the last evening. The total received by each child is given by

$$S_n = \frac{a_1(1 - r^n)}{1 - r}$$

$$S_8 = \frac{a_1(1 - r^8)}{1 - r} = \frac{2(1 - 2^8)}{1 - 2}$$

$$= \frac{2(1 - 256)}{-1} = 510$$

Each child receives a total of $5.10 after eight evenings.

PROGRESS CHECK

A ball is dropped from a height of 64 feet. On each bounce, it rebounds half the height it fell (Figure 1). How high is the ball at the top of the fifth bounce? What is the total distance the ball has traveled?

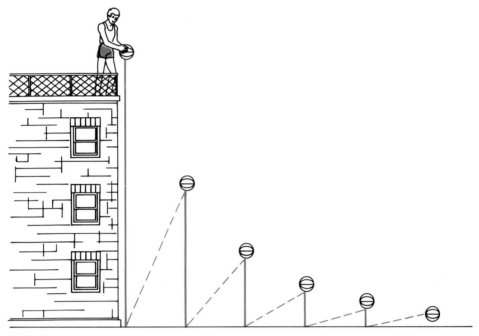

FIGURE 1

Answer
2 feet;   186 feet

We now want to focus on a geometric series for which $|r| < 1$, say

$$\frac{1}{2} + \frac{1}{4} + \frac{1}{8} + \cdots + \frac{1}{2^n} + \cdots$$

To see how the sum increases as n increases, let's form a table of values of S_n.

n	1	2	3	4	5	6	7	8	9
S_n	0.500	0.750	0.875	0.938	0.969	0.984	0.992	0.996	0.998

We begin to suspect that S_n gets closer and closer to 1 as n increases. To see that this is really so, let's look at the formula

$$S_n = \frac{a_1(1 - r^n)}{1 - r}$$

when $|r| < 1$. When a number r that is less than 1 in absolute value is raised to higher and higher positive integer powers, the value of r^n gets smaller and smaller. Thus, the term r^n can be made as small as we like by choosing

n sufficiently large. Since we are dealing with an infinite series, we say that "r^n approaches zero as n approaches infinity." We then replace r^n with 0 in the formula and denote the sum by S.

Sum of an Infinite Geometric Series

$$S = \frac{a_1}{1 - r}, \quad \text{when } |r| < 1$$

Applying this formula to the preceding series, we see that

$$S = \frac{\dfrac{1}{2}}{1 - \dfrac{1}{2}} = 1$$

which justifies the conjecture resulting from the examination of the above table. It is appropriate to remark that the ideas used in deriving the formula for an infinite geometric series have led us to the very border of the beginning concepts of calculus.

EXAMPLE 6

Evaluate the sum $\dfrac{3}{2} + 1 + \dfrac{2}{3} + \dfrac{4}{9} + \cdots$

Solution

The common ratio $r = \frac{2}{3}$. The sum of the infinite geometric series, with $|r| < 1$, is given by

$$S = \frac{a_1}{1 - r} = \frac{\dfrac{3}{2}}{1 - \dfrac{2}{3}} = \frac{9}{2}$$

PROGRESS CHECK

Evaluate the sum $4 - 1 + \dfrac{1}{4} - \dfrac{1}{16} + \cdots$

Answer
$\dfrac{16}{5}$

The notation

$$0.6525\overline{52}$$

indicates a repeating decimal with a pattern in which 52 is repeated indefinitely. Every repeating decimal can be written as a rational number. We will apply the formula for the sum of an infinite geometric series to find the rational number equal to a repeating decimal.

EXAMPLE 7

Find the rational number that is equal to $0.6525\overline{52}$.

Solution
Note that

$$0.65252\overline{52} = 0.6 + 0.052 + 0.00052 + 0.0000052 + \cdots$$

We treat the sum

$$0.052 + 0.00052 + 0.0000052 + \cdots$$

as an infinite geometric series with $a = 0.052$ and $r = 0.01$. Then

$$S = \frac{a}{1-r} = \frac{0.052}{1-0.01} = \frac{0.052}{0.99} = \frac{52}{990}$$

and the repeating decimal is equal to

$$0.6 + \frac{52}{990} = \frac{6}{10} + \frac{52}{990} = \frac{646}{990} = \frac{323}{495}$$

PROGRESS CHECK
Write the repeating decimal $2.5454\overline{54}$ as a rational number.

Answer
$\dfrac{252}{99}$

EXERCISE SET 9.2
In each of the following determine if the given sequence is a geometric progression. If it is, find the common ratio.

1. $3, 6, 12, 24, \ldots$
2. $-4, 12, -36, 108, \ldots$
3. $-4, 3, -\dfrac{9}{4}, \dfrac{27}{16}, \ldots$
4. $3, -1, \dfrac{1}{2}, -\dfrac{1}{4}, \ldots$
5. $1.2, 0.24, 0.048, 0.0096, \ldots$
6. $\dfrac{1}{4}, \dfrac{1}{2}, 2, 8, \ldots$

Write the first four terms of the geometric progression whose first term is a_1 and whose common ratio is r.

7. $a_1 = 3, \quad r = 3$
8. $a_1 = -4, \quad r = 2$
9. $a_1 = 4, \quad r = \dfrac{1}{2}$
10. $a_1 = 16, \quad r = -\dfrac{3}{2}$
11. $a_1 = -3, \quad r = 2$
12. $a_1 = 3, \quad r = -\dfrac{2}{3}$
13. If $a_1 = 3$ and $r = -2$, find a_8.
14. If $a_1 = 18$ and $r = -\dfrac{1}{2}$, find a_6.
15. Given the sequence $16, 8, 4, \ldots$, find a_7.
16. Given the sequence $15, -10, \dfrac{20}{3}, \ldots$, find a_6.
17. If $a_1 = 3$ and $a_5 = \dfrac{1}{27}$, find a_7.
18. If $a_1 = 2$ and $a_6 = \dfrac{1}{16}$, find a_3.
19. If $a_1 = \dfrac{16}{81}$ and $a_6 = \dfrac{3}{2}$, find a_8.

20. If $a_4 = \frac{1}{4}$ and $a_7 = 1$, find r.

21. If $a_2 = 4$ and $a_8 = 256$, find r.

22. If $a_3 = 3$ and $a_6 = -81$, find a_8.

23. If $a_1 = \frac{1}{2}$, $r = 2$, and $a_n = 32$, find n.

24. If $a_1 = -2$, $r = 3$, and $a_n = -162$, find n.

25. Insert two geometric means between $\frac{1}{3}$ and 9.

26. Insert two geometric means between -3 and 192.

27. Insert two geometric means between 1 and $\frac{1}{64}$.

28. Insert three geometric means between $\frac{2}{3}$ and $\frac{32}{243}$.

29. Find the sum of the first seven terms of the geometric progression whose first three terms are 3, 1, $\frac{1}{3}$.

30. Find the sum of the first six terms of the geometric progression whose first three terms are $\frac{1}{3}$, 1, 3.

31. Find the sum of the first five terms of the geometric progression whose first three terms are $-3, \frac{6}{5}, -\frac{12}{25}$.

32. Find the sum of the first six terms of the geometric progression whose first three terms are $2, \frac{4}{3}, \frac{8}{9}$.

33. If $a_1 = 4$ and $r = 2$, find S_8.

34. If $a_1 = -\frac{1}{2}$ and $r = -3$, find S_{10}.

35. If $a_1 = 2$ and $a_4 = -\frac{54}{8}$, find S_5.

36. If $a_1 = 64$ and $a_7 = 1$, find S_6.

37. A Christmas Club calls for savings of $5 in January, and twice as much on each successive month as in the previous month. How much money will have been saved by the end of November?

38. A city had 20,000 people in 1980. If the population increases 5% per year, how many people will the city have in 1990?

39. A city had 30,000 people in 1980. If the population increases 25% every ten years, how many people will the city have in the year 2010?

40. For good behavior a child is offered a reward consisting of 1 cent on the first day, 2 cents on the second day, 4 cents on the third day, and so on. If the child behaves properly for two weeks, what is the total amount that the child will receive?

Evaluate the sum of each geometric series.

41. $1 + \frac{1}{2} + \frac{1}{4} + \frac{1}{8} + \cdots$

42. $\frac{4}{5} + \frac{1}{5} + \frac{1}{20} + \frac{1}{80} + \cdots$

43. $1 - \frac{1}{3} + \frac{1}{9} - \frac{1}{27} + \cdots$

44. $\frac{1}{2} - \frac{1}{4} + \frac{1}{8} - \frac{1}{16} + \cdots$

45. $2 + \dfrac{1}{2} + \dfrac{1}{8} + \dfrac{1}{32} + \cdots$ 46. $1 + 0.1 + 0.01 + 0.001 + \cdots$

47. $0.5 + (0.5)^2 + (0.5)^3 + (0.5)^4 + \cdots$

48. $\dfrac{2}{5} + \dfrac{4}{25} + \dfrac{8}{125} + \dfrac{16}{625} + \cdots$ 49. $\dfrac{1}{3} - \dfrac{2}{9} + \dfrac{4}{27} - \dfrac{8}{81} + \cdots$

50. Find the rational number equal to $3.666\overline{6}$.

51. Find the rational number equal to $0.3676\overline{767}$.

52. Find the rational number equal to $4.1414\overline{14}$.

53. Find the rational number equal to $0.325\overline{325}$.

9.3
MATHEMATICAL INDUCTION

Mathematical induction is a method of proof that serves as one of the most powerful tools available to the mathematician. Viewed another way, mathematical induction is a property of the natural numbers that enables us to prove theorems that would otherwise appear unmanageable.

We begin by considering the sums of consecutive odd integers

$$1 = 1$$
$$1 + 3 = 4$$
$$1 + 3 + 5 = 9$$
$$1 + 3 + 5 + 7 = 16$$
$$1 + 3 + 5 + 7 + 9 = 25$$

We instantly recognize that the sequence 1, 4, 9, 16, 25 consists of the squares of the integers 1, 2, 3, 4, and 5. Is this coincidental or do we have a general rule? Is the sum of the first n consecutive odd integers always equal to n^2? Curiosity leads us to try yet one more case.

$$1 + 3 + 5 + 7 + 9 + 11 = 36 = 6^2$$

Indeed, the sum of the first six odd integers is 6^2. This strengthens our *suspicion* that the result may hold in general but we cannot possibly verify a theorem for *all* positive integers by testing one integer at a time, since the set of positive integers is an infinite set. At this point we need to turn to the

Principle of Mathematical Induction

If a statement involving a natural number n
  (I) is true when $n = 1$ and
  (II) whenever the statement is true for some value k of n, it is also true
       for $n = k + 1$,
then the statement is true for all positive integer values of n.

Let's examine the logic of the principle of mathematical induction. Part (I) says that we must verify the statement for $n = 1$. Then, by Part (II), the statement is also true for $n = 1 + 1 = 2$. But Part (II) then implies that the statement must also be true for $n = 2 + 1 = 3$, and so on. The effect is similar to an endless string of dominoes whereby each domino causes the next to fall. Thus, it is plausible that the principle has established the validity of the statement for *all* positive integer values of n.

We outline the steps involved in applying the principles of mathematical induction in the following example.

EXAMPLE 1
Prove that the sum of the first n consecutive integers is given by $n(n + 1)/2$.

Solution

Mathematical Induction	Example 1
Step 1. Verify that the statement is true for $n = 1$.	*Step 1*. The "sum" of the first integer is 1. Evaluating the formula for $n = 1$ yields $$\frac{1(1 + 1)}{2} = \frac{2}{2} = 1$$ which verifies the formula for $n = 1$.
Step 2. Assume the statement is true for $n = k$. Show it is true for $n = k + 1$.	*Step 2*. For $n = k$ we have $$1 + 2 + 3 + \cdots + k = \frac{k(k + 1)}{2}$$ Adding the next consecutive integer, $k + 1$, to both sides, we obtain $$1 + 2 + \cdots + k + (k + 1) = \frac{k(k + 1)}{2} + (k + 1)$$ $$= (k + 1)\left(\frac{k}{2} + 1\right)$$ $$= (k + 1)\left(\frac{k + 2}{2}\right)$$ $$= \frac{1}{2}(k + 1)(k + 2)$$ Thus, the formula holds for $n = k + 1$. By the principle of mathematical induction, it is then true for all positive integer values of n.

EXAMPLE 2
Prove that the sum of the first n consecutive odd integers is given by n^2.

Solution
To verify the formula for $n = 1$ we need only observe that $1 = 1^2$.

The following table shows the correspondence between the natural numbers and the odd integers. We see that when $n = k$, the value of the nth consecutive

n	1	2	3	4	$\cdots$	k
nth odd integer	1	3	5	7	$\cdots$	$2k - 1$

odd integer is $2k - 1$. Since the formula is assumed to be true for $n = k$, we have

$$1 + 3 + 5 + \cdots + (2k - 1) = k^2$$

Adding the next consecutive odd integer, $2k + 1$, to both sides, we obtain

$$1 + 3 + \cdots + (2k - 1) + (2k + 1) = k^2 + (2k + 1)$$

or

$$1 + 3 + \cdots + (2k + 1) = (k + 1)^2$$

Thus, the sum of the first $k + 1$ consecutive odd integers is $(k + 1)^2$. By the principle of mathematical induction, the formula is true for all positive integer values of n.

The student should be aware that many of the theorems that were used in this book can be proved formally by using mathematical induction. Here is an example of a basic property of positive integer exponents that yields to this type of proof.

EXAMPLE 3
Prove that $(xy)^n = x^n y^n$ for all positive integer values of n.

Solution
For $n = 1$, we have

$$(xy)^1 = xy = x^1 y^1$$

which verifies the validity of the statement for $n = 1$. Assuming the statement holds for $n = k$, we have

$$(xy)^k = x^k y^k$$

To show that the statement holds for $n = k + 1$, we write

$$
\begin{aligned}
(xy)^{k+1} &= (xy)^k(xy) & \text{Definition of exponents} \\
&= (x^k y^k)(xy) & \text{Statement holds for } n = k \\
&= (x^k x)(y^k y) & \text{Associative and commutative laws} \\
&= x^{k+1} y^{k+1} & \text{Definition of exponents}
\end{aligned}
$$

Thus, the statement holds for $n = k + 1$, and by the principle of mathematical induction the statement holds for all integer values of n.

EXERCISE SET 9.3

In Exercises 1–10 prove that the statement is true for all positive integer values of n by using the principle of mathematical induction.

1. $2 + 4 + 6 + \cdots + 2n = n(n + 1)$

2. $1^2 + 3^2 + 5^2 + \cdots + (2n - 1)^2 = \dfrac{n(2n + 1)(2n - 1)}{3}$

3. $2 + 5 + 8 + \cdots + (3n - 1) = \dfrac{n(3n + 1)}{2}$

4. $4 + 8 + 12 + \cdots + 4n = 2n(n + 1)$

5. $5 + 10 + 15 + \cdots + 5n = \dfrac{5n(n + 1)}{2}$

6. $1^2 + 2^2 + 3^2 + \cdots + n^2 = \dfrac{n(n + 1)(2n + 1)}{6}$

7. $1 \cdot 2 + 2 \cdot 3 + 3 \cdot 4 + \cdots + n(n + 1) = \dfrac{n(n + 1)(n + 2)}{3}$

8. $1^3 + 2^3 + 3^3 + \cdots + n^3 = \dfrac{n^2(n + 1)^2}{4}$

9. $1 + 5 + 9 + \cdots + (4n - 3) = n(2n - 1)$

10. $\left(\dfrac{x}{y}\right)^n = \dfrac{x^n}{y^n}$

11. Prove that the nth term a_n of an arithmetic progression whose first term is a_1 and common difference is d is given by $a_n = a_1 + (n - 1)d$.

12. Prove that the nth term a_n of a geometric progression whose first term is a_1 and common ratio is r is given by $a_n = a_1 r^{n-1}$.

13. Prove that $2 + 2^2 + 2^3 + \cdots + 2^n = 2^{n+1} - 2$.

14. Prove that $a + ar + ar^2 + \cdots + ar^{n-1} = \dfrac{a(1 - r^n)}{1 - r}$.

15. Prove that $x^n - 1$ is divisible by $x - 1$, $x \neq 1$. [*Hint:* Recall that divisibility requires the existence of a polynomial $Q(x)$ such that $x^n - 1 = (x - 1)Q(x)$.]

16. Prove that $x^n - y^n$ is divisible by $x - y$, $x \neq y$. [*Hint:* Note that $x^{n+1} - y^{n+1} = (x^{n+1} - xy^n) + (xy^n - y^{n+1})$.]

9.4
THE BINOMIAL THEOREM

By sequential multiplication by $(a + b)$ you may verify that

$$(a + b)^1 = a + b$$
$$(a + b)^2 = a^2 + 2ab + b^2$$
$$(a + b)^3 = a^3 + 3a^2b + 3ab^2 + b^3$$
$$(a + b)^4 = a^4 + 4a^3b + 6a^2b^2 + 4ab^3 + b^4$$
$$(a + b)^5 = a^5 + 5a^4b + 10a^3b^2 + 10a^2b^3 + 5ab^4 + b^5$$

The expression on the right-hand side of the equation is called the **expansion** of the left-hand side. If we were to predict the form of the expansion of $(a + b)^n$, where n is a natural number, the preceding example would lead us to conclude that it has the following properties.

(a) The expansion has $n + 1$ terms.

(b) The first term is a^n and the last term is b^n.

(c) The sum of the exponents of a and b in each term is n.

(d) In each successive term after the first, the exponent of a decreases by 1 and the exponent of b increases by 1.

(e) The coefficients may be obtained from the following array, which is known as **Pascal's triangle.** Each number, with the exception of those at the ends of the rows, is the sum of the two nearest numbers in the row above. The numbers at the ends of the row are always 1.

$$
\begin{array}{ccccccccccc}
 & & & & 1 & & 1 & & & & \\
 & & & 1 & & 2 & & 1 & & & \\
 & & 1 & & 3 & & 3 & & 1 & & \\
 & 1 & & 4 & & 6 & & 4 & & 1 & \\
1 & & 5 & & 10 & & 10 & & 5 & & 1
\end{array}
$$

Pascal's triangle is not a convenient means for determining the coefficients of the expansion when n is large. Here is an alternate method.

(e′) The coefficient of any term (after the first) can be found by the following rule. In the preceding term, multiply the coefficient by the exponent of a and then divide by one more than the exponent of b.

EXAMPLE 1
Write the expansion of $(a + b)^6$.

Solution
From Property (b) we know that the first term is a^6. Thus,

$$(a + b)^6 = a^6 + \cdots$$

From Property (e′) the next coefficient is

$$\frac{1 \cdot 6}{1} = 6$$

(since the coefficient of b is 0). By Property (d) the exponents of a and b in this term are 5 and 1, respectively, so we have

$$(a + b)^6 = a^6 + 6a^5b + \cdots$$

Applying Property (e′) again, the next coefficient is

$$\frac{6 \cdot 5}{2} = 15$$

and by Property (d) the exponents of a and b in this term are 4 and 2, respectively. Thus,

$$(a + b)^6 = a^6 + 6a^5b + 15a^4b^2 + \cdots$$

Continuing in this manner we see that

$$(a + b)^6 = a^6 + 6a^5b + 15a^4b^2 + 20a^3b^3 + 15a^2b^4 + 6ab^5 + b^6$$

PROGRESS CHECK
Write the first five terms in the expansion of $(a + b)^{10}$.

Answer
$a^{10} + 10a^9b + 45a^8b^2 + 120a^7b^3 + 210a^6b^4$

The expansion of $(a + b)^n$ that we have described is called the **binomial**

theorem or **binomial formula** and can be written

$$(a + b)^n = a^n + \frac{n}{1} a^{n-1}b + \frac{n(n-1)}{1 \cdot 2} a^{n-2}b^2 + \frac{n(n-1)(n-2)}{1 \cdot 2 \cdot 3} a^{n-3}b^3$$

$$+ \cdots + \frac{n(n-1)(n-2) \cdots (n-r+1)}{1 \cdot 2 \cdot 3 \cdots r} a^{n-r}b^r + \cdots + b^n$$

The binomial formula can be proved by the method of mathematical induction discussed in the preceding section, but we shall not present the proof in this book.

EXAMPLE 2
Find the expansion of $(2x - 1)^4$.

Solution
Let $a = 2x$, $b = -1$, and apply the binomial formula.

$$(2x - 1)^4 = (2x)^4 + \frac{4}{1}(2x)^3(-1) + \frac{4 \cdot 3}{1 \cdot 2}(2x)^2(-1)^2 + \frac{4 \cdot 3 \cdot 2}{1 \cdot 2 \cdot 3}(2x)(-1)^3 + (-1)^4$$

$$= 16x^4 - 32x^3 + 24x^2 - 8x + 1$$

PROGRESS CHECK
Find the expansion of $(x^2 - 2)^4$.

Answer
$x^8 - 8x^6 + 24x^4 - 32x^2 + 16$

Note that the denominator of the coefficient in the binomial formula is always the product of the first n natural numbers. We use the symbol $n!$, which is read as n factorial, to indicate this type of product. For example,

$$4! = 4 \cdot 3 \cdot 2 \cdot 1 = 24$$

$$6! = 6 \cdot 5 \cdot 4 \cdot 3 \cdot 2 \cdot 1 = 720$$

and

$$n! = n(n-1)(n-2) \cdots 4 \cdot 3 \cdot 2 \cdot 1$$

Since $(n-1)! = (n-1)(n-2)(n-3) \cdots 4 \cdot 3 \cdot 2 \cdot 1$, we see that

$$n! = n(n-1)!$$

Thus, $15! = 15 \cdot 14!$, $8! = 8 \cdot 7!$, and $1! = 1 \cdot 0!$. Since $1! = 1$, we have $1! = 1 \cdot 0! = 1$, which can be true only if $0! = 1$. For consistency, we must therefore *define* $0!$ as

$$0! = 1$$

EXAMPLE 3
Evaluate each of the following.

(a) $\dfrac{5!}{3!}$

Since $5! = 5 \cdot 4 \cdot 3!$ we may write

$$\frac{5!}{3!} = \frac{5 \cdot 4 \cdot 3!}{3!} = 5 \cdot 4 = 20$$

(b) $\dfrac{9!}{8!} = \dfrac{9 \cdot 8!}{8!} = 9$

(c) $\dfrac{10!4!}{12!} = \dfrac{10!4!}{12 \cdot 11 \cdot 10!} = \dfrac{4!}{12 \cdot 11} = \dfrac{4 \cdot 3 \cdot 2 \cdot 1}{12 \cdot 11} = \dfrac{2}{11}$

(d) $\dfrac{n!}{(n-2)!} = \dfrac{n(n-1)(n-2)!}{(n-2)!} = n(n-1) = n^2 - n$

(e) $\dfrac{(2-2)!}{3!} = \dfrac{0!}{3 \cdot 2} = \dfrac{1}{6}$

PROGRESS CHECK
Evaluate each of the following.

(a) $\dfrac{12!}{10!}$ (b) $\dfrac{6!}{4!2!}$ (c) $\dfrac{10!8!}{9!7!}$

(d) $\dfrac{n!(n-1)!}{(n+1)!(n-2)!}$ (e) $\dfrac{8!}{6!(3-3)!}$

Answers

(a) 132   (b) 15   (c) 80   (d) $\dfrac{n-1}{n+1}$ (e) 56

Here is what the binomial formula looks like in factorial notation.

$$(a + b)^n = a^n + \frac{n!}{1!(n-1)} a^{n-1}b + \frac{n!}{2!(n-2)!} a^{n-2}b^2$$

$$+ \frac{n!}{3!(n-3)!} a^{n-3}b^3 + \cdots + \frac{n!}{r!(n-r)!} a^{n-r}b^r$$

$$+ \cdots + b^n$$

Sometimes we merely want to find a certain term in the expansion of $(a + b)^n$. We shall use the following observation to answer this question. In the binomial formula for the expansion of $(a + b)^n$, b occurs in the second term, b^2 occurs in the third term, b^3 occurs in the fourth term and, in general, b^k occurs in the $(k + 1)$th term. The exponents of a and b must add up to n in each term. Since the exponent of b in the $(k + 1)$th term is k, we conclude that the exponent of a must be $n - k$.

EXAMPLE 4
Find the fourth term in the expansion of $(x - 1)^5$.

Solution

The exponent of b in the fourth term is 3, and the exponent of a is then $5 - 3 = 2$. From the binomial formula we see that the coefficient of the term a^2b^3 is

$$\frac{5!}{3!2!}$$

Since $a = x$ and $b = -1$, the fourth term is

$$\frac{5!}{3!2!} x^2(-1)^3 = -10x^2$$

PROGRESS CHECK

Find the third term in the expansion of

$$\left(\frac{x}{2} - 1\right)^8$$

Answer

$\dfrac{7}{16} x^6$

EXAMPLE 5

Find the term in the expansion of $(x^2 - y^2)^6$ that involves y^8.

Solution

Since $y^8 = (-y^2)^4$, we seek that term which involves b^4 in the expansion of $(a + b)^6$. Thus, $b^4 = (-y^2)^4 = y^8$ occurs in the fifth term. In this term the exponent of a is $6 - 4 = 2$. By the binomial formula the corresponding coefficient is

$$\frac{6!}{4!2!} = 15$$

Since $a = x^2$ and $b = -y^2$, the desired term is

$$15(x^2)^2(-y^2)^4 = 15x^4y^8$$

PROGRESS CHECK

Find the term in the expansion of $(x^3 - \sqrt{2})^5$ that involves x^6.

Answer
$-20\sqrt{2}x^6$

EXERCISE SET 9.4

Expand and simplify.

1. $(3x + 2y)^5$
2. $(2a - 3b)^6$
3. $(4x - y)^4$
4. $\left(3 + \frac{1}{2}x\right)^4$
5. $(2 - xy)^5$
6. $(3a^2 + b)^4$
7. $(a^2b + 3)^4$
8. $(x - y)^7$
9. $(a - 2b)^8$
10. $\left(\frac{x}{y} + y\right)^6$
11. $\left(\frac{1}{3}x + 2\right)^3$
12. $\left(\frac{x}{y} + \frac{y}{x}\right)^5$

Find the first four terms in the given expansion and simplify.

13. $(2 + x)^{10}$
14. $(x - 3)^{12}$
15. $(3 - 2a)^9$

16. $(a^2 + b^2)^{11}$ 17. $(2x - 3y)^{14}$ 18. $\left(a - \dfrac{1}{a^2}\right)^8$

19. $(2x - yz)^{13}$ 20. $\left(x - \dfrac{1}{y}\right)^{15}$

Evaluate.

21. $5!$ 22. $7!$ 23. $\dfrac{12!}{11!}$

24. $\dfrac{13!}{12!}$ 25. $\dfrac{11!}{8!}$ 26. $\dfrac{7!}{9!}$

27. $\dfrac{10!}{6!}$ 28. $\dfrac{9!}{6!}$ 29. $\dfrac{6!}{3!}$

30. $\dfrac{8!}{5!3!}$ 31. $\dfrac{10!}{6!4!}$ 32. $\dfrac{(n + 1)!}{(n - 1)!}$

In each expansion find only the term specified.

33.   The fourth term in $(2x - 4)^7$.

34.   The third term in $(4a + 3b)^{11}$.

35.   The fifth term in $\left(\dfrac{1}{2}x - y\right)^{12}$.

36.   The sixth term in $(3x - 2y)^{10}$.

37.   The fifth term in $\left(\dfrac{1}{x} - 2\right)^9$.

38.   The next to last term in $(a + 4b)^5$.

39.   The middle term in $(x - 3y)^6$.

40.   The middle term in $\left(2a + \dfrac{1}{2}b\right)^6$.

41.   The term involving x^4 in $(3x + 4y)^7$.

42.   The term involving x^6 in $(2x^2 - 1)^9$.

43.   The term involving x^6 in $(2x^3 - 1)^9$.

44.   The term involving x^8 in $\left(x^2 + \dfrac{1}{y}\right)^8$

45.   The term involving x^{12} in $\left(x^3 + \dfrac{1}{2}\right)^7$.

46.   The term involving x^{-4} in $\left(y + \dfrac{1}{x^2}\right)^8$.

47.   Evaluate $(1.3)^6$ to four decimal places by writing it as $(1 + 0.3)^6$ and using the binomial formula.

48.   Using the method of Example 47, evaluate
      (a) $(3.4)^4$ (b) $(48)^5$ (*Hint:* $48 = 50 - 2$)

9.5
COUNTING: PERMUTATIONS AND COMBINATIONS

How many arrangements can be made using the letters a, b, c, and d three at a time? One way to solve this problem is to enumerate all the possible arrangements. The tree diagram shown in Figure 2 is a graphic device that yields precisely what we need. The letters a, b, c, and d are listed at the top and represent the candidates for the first letter. The three branches

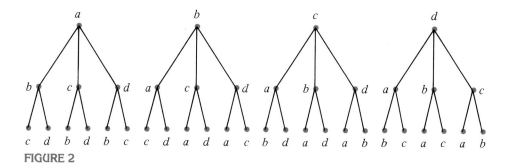

FIGURE 2

emanating from these lead to the possible choices for the second letter, and so on. For example, the portion of the tree shown in Figure 3 illustrates the arrangements *bda* and *bdc*. In this way we determine that there are a total of 24 arrangements.

FIGURE 3

There is a more efficient way to solve this problem. Each arrangement consists of a choice of candidates to fill three positions in Figure 4. Any one of the four candidates *a*, *b*, *c*, or *d* can be assigned to the first position;

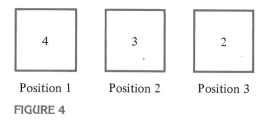

FIGURE 4

once a candidate is assigned to the first position, any one of the three remaining candidates can be assigned to the second position; finally, any one of the remaining two candidates can be assigned to the third position. Since each candidate in each position can be associated with every other candidate in the other positions, the *product*

$$4 \cdot 3 \cdot 2 = 24$$

yields the total number of arrangements. This simple example illustrates a very important principle.

Counting Principle

If one event can occur in *m* different ways, and, after it has happened in one of these ways, a second event can occur in *n* different ways, then both events can occur in *mn* different ways.

Note that the order or sequence of events is significant since each arrangement is counted as one of the "*mn* different ways."

EXAMPLE 1

In how many ways can five students be seated in a row of five seats?

Solution

We have five positions to be filled. Any one of the five students may occupy the first position, after which any one of the remaining four students can occupy the next position. Reapplying the counting principle to the other positions, we see that the number of arrangements is

$$5 \cdot 4 \cdot 3 \cdot 2 \cdot 1 = 120$$

PROGRESS CHECK

How many different four-digit numbers can be formed using the digits 2, 4, 6, 8? (Don't repeat any of the digits.)

Answer
24

EXAMPLE 2

How many different three-letter arrangements can be made using the letters A, B, C, X, Y, Z
(a) if no letter may be repeated in an arrangement, and
(b) if letters may be repeated?

Solution

(a) We need to fill three positions. Any one of the six letters may occupy the first position; then, any one of the *remaining* five letters may occupy the second position (since repetitions are not allowed). Thus, the total number of arrangements is $6 \cdot 5 \cdot 4 = 120$.
(b) Any one of the six letters may fill any of the three positions (since repetitions are allowed). The total number of arrangements is $6 \cdot 6 \cdot 6 = 216$.

PROGRESS CHECK

The positions of president, secretary, and treasurer are to be filled from a class of 15 students. In how many ways can these positions be filled if no student may hold more than one position?

Answer
2730

Each arrangement that can be made by using all or some of the elements of a set of objects without repetition is called a **permutation.** The phrase "without repetition" indicates that no element of the set appears more than once. For example, the permutations of the letters a, b, c taken three at a time include $b\ a\ c$ but exclude $a\ a\ b$.

We will use the notation $P(n, r)$ to indicate the number of permutations of n distinct objects taken r at a time. (There are a number of other notations in common use: nPr, P_r^n, ^nP_r, $P_{n,r}$.) If $r = n$, then using the counting principle we see that

$$P(n, n) = n(n - 1)(n - 2) \ldots 2 \cdot 1$$

since any one of the n objects may fill the first position, any one of the remaining $(n - 1)$ objects may fill the second position, and so on. Using factorial notation,

$$P(n,n) = n!$$

Let's try to calculate $P(n,r)$, that is, the number of permutations of n distinct objects taken r at a time when r is less than n. We may think of this as the number of ways of filling r positions with n candidates. Once again, we may fill the first position with any one of the n candidates, the second position with any one of the remaining $(n - 1)$ candidates, and so on, so that

$$P(n,r) = \underbrace{n(n - 1)(n - 2) \ldots}_{r \text{ factors}}$$

We may write this as

$$P(n,r) = n(n - 1)(n - 2) \ldots (n - r + 1) \tag{1}$$

since $(n - r + 1)$ will be the rth factor. If we multiply the right-hand side of Equation (1) by

$$\frac{(n - r)!}{(n - r)!} = 1$$

we have

$$P(n,r) = \frac{n(n - 1)(n - 2) \ldots (n - r + 1)(n - r)(n - r - 1) \ldots 2 \cdot 1}{(n - r)!}$$

or

$$P(n,r) = \frac{n!}{(n - r)!}$$

EXAMPLE 3
Evaluate.

(a) $P(5, 5) = \dfrac{5!}{(5 - 5)!} = \dfrac{5!}{0!} = \dfrac{5 \cdot 4 \cdot 3 \cdot 2 \cdot 1}{1} = 120$

(b) $P(5,2) = \dfrac{5!}{(5 - 2)!} = \dfrac{5!}{3!} = \dfrac{5 \cdot 4 \cdot 3!}{3!} = 20$

(c) $\dfrac{P(6,2)}{3!} = \dfrac{6!}{3!(6-2)!} = \dfrac{6!}{3!4!} = \dfrac{6 \cdot 5 \cdot 4!}{3 \cdot 2 \cdot 4!} = 5$

PROGRESS CHECK
Evaluate.

(a) $P(4,4)$ (b) $P(6,3)$ (c) $\dfrac{2\,P(6,4)}{2!}$

Answers
(a) 24   *(b) 120*   *(c) 360*

EXAMPLE 4
How many different arrangements can be made by taking 5 of the letters of the word *relation?*

Solution
Since the word *relation* has eight different letters, we are seeking the number of permutations of eight objects taken five at a time or $P(8,5)$. Thus,

$$P(n, r) = \frac{n!}{(n-r)!}$$

$$P(8, 5) = \frac{8!}{(8-5)!} = \frac{8!}{3!} = 6720$$

PROGRESS CHECK
There is space on a bookshelf for displaying four books. If there are six different novels available, how many arrangements can be made?

Answer
360

EXAMPLE 5
How many arrangements can be made using all of the letters of the word *quartz* if the vowels are always to remain adjacent to each other?

Solution
If we treat the vowel pair "*ua*" as a unit, then there are five "letters" (*q, ua, r, t, z*) which can be arranged in $P(5,5)$ ways. But the vowels can themselves be arranged in $P(2,2)$ ways. By the counting principle, the total number of arrangements is

$$P(5,5) \cdot P(2,2)$$

Since $P(5,5) = 120$ and $P(2,2) = 2$, the total number of arrangements is 240.

PROGRESS CHECK
A bookshelf is to be used to display 5 new textbooks. There are 7 mathematics textbooks and 4 biology textbooks available. If we wish to put 3 mathematics books and 2 biology books on display, how many arrangements can be made if the books in each discipline must be kept together?

Answer
5040

Let's take another look at the arrangements of the letters a, b, and c taken two at a time:

$$a\ b \qquad b\ a \qquad c\ a$$
$$a\ c \qquad b\ c \qquad c\ b$$

Now let's ask a different question: In how many ways can we *select* two letters from the letters a, b, and c? We are now asked to disregard the order in which the letters are chosen. The result is then

$$a\ b \qquad a\ c \qquad b\ c$$

In general, a set of r objects chosen from a set of n objects is called a **combination.** We denote the number of combinations of r objects chosen from n objects by $C(n,r)$. $\left[\text{Other notations in common use include } nCr, C^n_r,\right.$ ^nC_r, $C_{n,r}$, and $\left.\left(\dfrac{n}{r}\right).\right]$

EXAMPLE 6
List the combinations of the letters a, b, c, d taken three at a time.

Solution
The combinations are seen to be

$$a\ b\ c \qquad a\ b\ d \qquad a\ c\ d \qquad b\ c\ d$$

Here is a rule that is helpful in determining whether a problem calls for the number of permutations or the number of combinations.

> If we are interested in calculating the number of arrangements in which different ordering of the same objects are counted, we use permutations.
>
> If we are interested in calculating the number of ways of selecting objects in which the order of the selected objects doesn't matter, we use combinations.

For example, suppose we want to determine the number of different four-card hands that can be dealt from a deck of 52 cards. Since a hand consisting of four cards is the same hand regardless of the order of the cards, we must use combinations.

Let's find a formula for $C(n,r)$. There are three combinations of the letters a, b, c taken two at a time, namely

$$ab, \quad ac, \quad \text{and} \quad bc$$

so that $C(3,2) = 3$. Now, each of these combinations can be arranged in 2! ways to yield the total list of permutations

$$ab \quad ba \quad ac \quad ca \quad bc \quad cb$$

Thus, $P(3,2) = 6 = 2!C(3,2)$. In general, each of the $C(n,r)$ combinations can be permuted in $r!$ ways so that by the counting principle the total number of permutations is $P(n,r) = r!C(n,r)$ or

$$C(n,r) = \frac{P(n,r)}{r!} = \frac{n!}{r!(n-r)!}$$

EXAMPLE 7
Evaluate.

(a) $C(5,2) = \dfrac{5!}{2!(5-2)} = \dfrac{5!}{2!3!} = \dfrac{5 \cdot 4 \cdot 3!}{2 \cdot 3!} = 10$

(b) $C(4,4) = \dfrac{4!}{4!(4-4)!} = \dfrac{4!}{4!0!} = 1$

(c) $\dfrac{P(6,3)}{C(6,3)}$

$$P(6,3) = \frac{6!}{(6-3)!} = \frac{6!}{3!} = 6 \cdot 5 \cdot 4 = 120$$

$$C(6,3) = \frac{6!}{3!(6-3)!} = \frac{6!}{3!3!} = \frac{6 \cdot 5 \cdot 4}{3 \cdot 2 \cdot 1} = 20$$

$$\frac{P(6,3)}{C(6,3)} = \frac{120}{20} = 6$$

PROGRESS CHECK
Evaluate.

(a) $C(6,2)$ (b) $C(10,10)$ (c) $\dfrac{P(3,2)}{3!C(5,4)}$

Answers

(a)   *15*   (b)   *1*   (c) $\dfrac{1}{10}$

EXAMPLE 8
In how many ways can a committee of four be selected from a group of ten people?

Solution
If A, B, C, and D constitute a committee, is the arrangement B, A, C, D a different committee? Of course not—which says that order doesn't matter. We are therefore interested in computing $C(10,4)$.

$$C(10,4) = \frac{10!}{4!6!} = \frac{10 \cdot 9 \cdot 8 \cdot 7}{4 \cdot 3 \cdot 2 \cdot 1} = 210$$

PROGRESS CHECK
In how many ways can a five-card hand be dealt from a deck of 52 cards?

Answer
2,598,960

EXAMPLE 9

In how many ways can a committee of three girls and two boys be selected from a class of eight girls and seven boys?

Solution

The girls can be selected in $C(8,3)$ ways and the boys can be selected in $C(7,2)$ ways. By the counting principle, each choice of boys can be associated with each choice of girls.

$$C(8,3) \cdot C(7,2) = \frac{8!}{3!5!} \cdot \frac{7!}{2!5!} = (56)(21) = 1176$$

PROGRESS CHECK

From five representatives of District A and eight representatives of District B, in how many ways can four persons be chosen if only one representative from District A is to be included?

Answer
280

EXAMPLE 10

A bookstore has 12 French and 9 German books. In how many ways can a group of 6 books, consisting of 4 French and 2 German books, be placed on a shelf?

Solution

The French books can be selected in $C(12,4)$ ways and the German books in $C(9,2)$ ways. The 6 books can then be selected in $C(12,4) \cdot C(9,2)$ ways. Each *selection* of 6 books can then be *arranged* on the shelf in $P(6,6)$ ways so that the total number of arrangements is

$$C(12,4) \cdot C(9,2) \cdot P(6,6) = \frac{12!}{4!8!} \cdot \frac{9!}{2!7!} \cdot \frac{6!}{(6-6)!} = 495 \cdot 36 \cdot 720 = 12{,}830{,}400$$

PROGRESS CHECK

From six different consonants and four different vowels, how many five-letter words can be made consisting of three consonants and two vowels? (Assume every arrangement is a "word".)

Answer
14,400

EXERCISE SET 9.5

1. How many different five-digit numbers can be formed using the digits 1, 3, 4, 6, 8?
2. How many different ways are there to arrange the letters in the word *study*?
3. An employee identification number consists of two letters of the alphabet followed by a sequence of three digits selected from the digits 2, 3, 5, 6, 8, and 9. If repetitions are allowed, how many different identification numbers are possible?
4. In a psychological experiment, a subject has to arrange a cube, square, triangle, and rhombus in a row. How many different arrangements are possible?

5. A coin is tossed 8 times and the result of each toss is recorded. How many different sequences of heads and tails are possible?

6. A die (from a pair of dice) is tossed 4 times and the result of each toss is recorded. How many different sequences are possible?

7. A concert is to consist of 3 guitar pieces, 2 vocal numbers, and 2 jazz selections. In how many ways can the program be arranged?

Evaluate.

8. $P(6,6)$ 9. $P(6,5)$ 10. $P(4,2)$

11. $P(8,3)$ 12. $P(5,2)$ 13. $P(10,2)$

14. $P(8,4)$ 15. $\dfrac{P(9,3)}{3!}$ 16. $\dfrac{4P(12,3)}{2!}$

17. $P(3,1)$ 18. $\dfrac{P(7,3)}{2!}$ 19. $\dfrac{P(10,4)}{4!}$

20. Find the number of ways in which five men and five women can be seated in a row
(a) if any person may sit next to any other person.
(b) if a man must always sit next to a woman.

21. Find the number of permutations of the letters in the word *money*.

22. Find the number of distinguishable permutations of the letters in the word *goose*. (*Hint:* Permutations in which the letters *oo* appear together are not distinguishable.)

23. Find the number of distinguishable permutations of the letters of the word *needed*. (See hint in Exercise 22.)

24. How many permutations of the letters a, b, e, g, h, k, and m are there when taken
(a) two at a time?
(b) three at a time?

25. How many three-letter labels of new chemical products can be formed from the letters a, b, c, d, f, g, l, and m?

26. Find the number of distinguishable permutations that can be formed from the letters of the word *mississippi* taken four at a time.

27. A family consisting of a mother, father, and three children is having a picture taken. If all five people are arranged in a row, how many different photographs can be taken?

28. List all the combinations of the numbers 4, 3, 5, 8, and 9 taken three at a time.

Evaluate.

29. $C(9,3)$ 30. $C(7,3)$ 31. $C(10,2)$

32. $C(7,1)$ 33. $C(7,7)$ 34. $C(5,4)$

35. $C(n,n-1)$ 36. $C(n,n-2)$ 37. $C(n+1,n-1)$

38. In how many ways can a committee of 2 faculty members and 3 students be selected from 8 faculty members and 10 students?

39. In how many ways can a basketball team of 5 players be selected from among 15 candidates?

40. In how many ways can a four-card hand be dealt from a deck of 52 cards?

41. How many three-letter moped plates on a local campus can be formed
(a) if no letters can be repeated?
(b) if letters can be repeated?

42. In a certain city each police car is staffed by two officers, one male and one

female. A police captain, who needs to staff 8 cars, has 15 male officers and 12 female officers available. How many different teams can be formed?

43. How many different 10-card hands with 4 aces can be dealt from a deck of 52 cards?

44. A car manufacturer makes three different models, each of which is available in five different colors with two different engines. How many cars must a dealer stock in the showroom to display the full line?

45. A penny, nickel, dime, quarter, half-dollar, and silver dollar are to be arranged in a row. How many different arrangements can be formed if the penny and dime must always be next to each other?

46. An automobile manufacturer who is planning an advertising campaign is considering seven newspapers, two magazines, three radio stations, and four television stations. In how many ways can five advertisements be placed
    (a) if all five are to be in newspapers?
    (b) if two are to be in newspapers, two on radio, and one on television?

47. In a certain police station there are 12 prisoners and 10 police officers. How many possible lineups consisting of 4 prisoners and 3 officers can be formed?

48. The notation $\binom{n}{r}$ is often used in place of $C(n,r)$. Show that $\binom{n}{r} = \binom{n}{n-r}$.

49. How many different 10-card hands with 6 red cards and 4 black cards can be dealt from a deck of 52 cards?

50. A bin contains 12 transistors, 7 of which are defective. In how many ways can four transistors be chosen so that
    (a) all four are defective?
    (b) two are good and two are defective?
    (c) all four are good?
    (d) three are defective and one is good?

9.6
PROBABILITY

There is a vast difference between the statements "It will probably rain today" and "It is equally probable that a tossed coin will result in a head or a tail." The first statement conveys an expectation but only in a vague sense; the latter statement is much more useful since it quantifies the notion of probability.

Let's take a closer look at what happens when we toss a coin. The event has only two possible outcomes: a head or a tail. Since a head represents one of two possible outcomes, we say that "the probability of a head is $\frac{1}{2}$." This leads us to define probability in the following way: If an event can occur in a total of t ways and s of these are considered successful, then the probability of success is s/t. In short,

$$\text{probability} = \frac{\text{number of successful outcomes}}{\text{total number of outcomes}}$$

EXAMPLE 1
A container holds a red ball, two white balls, and two blue balls. If one ball is drawn, what is the probability that it is white?

Solution
The selection of a ball represents a possible outcome so that there are a total of 5 possible outcomes. Since there are two ways of achieving a successful outcome of a white ball,

$$\text{probability of selecting a white ball} = \frac{\text{number of successful outcomes}}{\text{total number of outcomes}}$$

$$= \frac{2}{5}$$

PROGRESS CHECK
One card is drawn from an ordinary deck of 52 cards. What is the probability it is an ace?

Answer

$\frac{1}{13}$

EXAMPLE 2
A single die (whose faces contain the numbers 1, 2, 3, 4, 5, and 6) is tossed. What is the probability that the result is less than 5?

Solution
The four successful outcomes occur when the die shows a 1, 2, 3, or 4. Since there are six possible outcomes, we see that

$$\text{probability} = \frac{\text{number of successful outcomes}}{\text{total number of outcomes}}$$

$$= \frac{4}{6} = \frac{2}{3}$$

PROGRESS CHECK
A bag of coins contains 4 nickels, 5 dimes, and 10 quarters. If one coin is withdrawn, what is the probability that it is worth less than 25 cents?

Answer

$\frac{9}{19}$

Let's consider a bag containing three red marbles and five brown marbles. It's easy to verify that the probability of drawing a red marble in a single draw is 3/8 and that the probability of drawing a brown marble is 5/8 = 1 − 3/8. What is the probability of drawing either a red marble or a brown marble? Since any of the 8 possible outcomes is considered a success, the probability is 8/8 = 1. What is the probability of drawing a black marble? Since there are no successful outcomes, the probability is 0/8 = 0. Generalizing these results we have

A probability of 1 indicates certainty.

A probability of 0 indicates impossibility.

If p is the probability that an event will happen, then $1 - p$ is the probability that it will not happen.

EXAMPLE 3
While shuffling an ordinary deck of 52 cards, one card is dropped. What is the probability that it is not a king?

Solution
Since there are four kings in a deck, the probability of a king is $p = 4/52 = 1/13$. Then the probability that it is *not* a king is $1 - p = 12/13$.

PROGRESS CHECK
Two people throw a single die. If player A rolls a 4, what is the probability that player B does not also roll a 4?

Answer
$\dfrac{5}{6}$

The rules for computing permutations and combinations are useful in solving probability problems.

EXAMPLE 4
A bag contains three green, five white, and seven yellow balls. If three balls are drawn at random, what is the probability that they are all white?

Solution
We can select three white balls from five white balls in $C(5,3)$ ways; we can select three balls from the bag of fifteen balls in $C(15,3)$ ways. Then

$$\text{probability of selecting three white balls} = \frac{C(5,3)}{C(15,3)} = \frac{10}{455} = \frac{2}{91}$$

PROGRESS CHECK
Three cards are drawn from an ordinary deck of 52 cards. What is the probability that they are all aces?

Answer
$\dfrac{1}{5525}$

Many problems in probability involve the tossing of a pair of dice. Since the faces of the dice contain the numbers 1, 2, 3, 4, 5, and 6, the sum of the numbers on the two dice can be any of the numbers 2 through 12. The outcomes, however, are not equally probable. Table 1 displays the possible outcomes of tossing a pair of dice. In Table 2 we then summarize the

TABLE 1

Die 1	Die 2					
	1	2	3	4	5	6
1	2	3	4	5	6	7
2	3	4	5	6	7	8
3	4	5	6	7	8	9
4	5	6	7	8	9	10
5	6	7	8	9	10	11
6	7	8	9	10	11	12

number of ways in which each sum can be obtained. The probability of tossing a "3" with a pair of dice is therefore 2/36 or 1/18; the probability of tossing a "7" is 6/36 or 1/6.

TABLE 2

Sum of 2 dice	2	3	4	5	6	7	8	9	10	11	12
Number of ways	1	2	3	4	5	6	5	4	3	2	1

EXAMPLE 5

What is the probability of throwing a 10 or higher with a single throw of a pair of dice?

Solution

The favorable outcomes are 10, 11, and 12, which, by Table 2, can occur in a total of 6 ways. Then

$$\text{probability} = \frac{\text{number of successful outcomes}}{\text{total number of outcomes}}$$

$$= \frac{6}{36} = \frac{1}{6}$$

PROGRESS CHECK

What is the probability of throwing no higher than a 5 with a single throw of a pair of dice?

Answer

$\frac{5}{18}$

We conclude our introduction to probability by considering the probability of successive, independent events. For example, if a card is drawn from a deck of 52 cards, is replaced in the deck, and a second card is drawn, what is the probability that they are both aces? Note that these events are independent since the second outcome in no way depends upon the first outcome. Here is the principle that permits us to solve this type of problem.

> If p_1 is the probability that an event will occur, and p_2 is the probability that a second, independent event will occur, then $p_1 p_2$ is the probability that both events will occur.

In our example, the probability that the first card drawn is an ace is $p_1 = 4/52 = 1/13$; the probability that the second card drawn is an ace is also $1/13 = p_2$. Then the probability of drawing aces successively is $p_1 p_2 = (1/13)(1/13) = 1/169$. Of course, we can extend this principle to more than two events by forming the product of the probabilities of the independent events.

EXAMPLE 6
What is the probability of throwing a 7 twice in succession with a pair of dice?

Solution
From Table 2 we see that a 7 can occur in 6 ways so that the probability of throwing a 7 is $p_1 = 6/36 = 1/6$. The probability of throwing a 7 on the second roll is again $p_2 = 1/6$ so that the probability of throwing a 7 on both rolls is $p_1p_2 = (1/6)(1/6) = 1/36$.

PROGRESS CHECK
What is the probability of throwing an 11 twice in succession with a pair of dice?

Answer
1/324

EXAMPLE 7
What is the probability of drawing an ace three times in succession from a deck of 52 cards if the drawn cards are not replaced?

Solution
On the first draw, the probability of obtaining an ace is $p_1 = 4/52$. Since the ace is not replaced in the deck, there remain 3 aces and a total of 51 cards, so that the probability of obtaining a second ace is $p_2 = 3/51$. Arguing the same way, there now remain 2 aces and 50 cards so that the probability of drawing a third ace is $p_3 = 2/50$. Thus, the probability of drawing aces 3 times in succession without replacement is

$$p_1p_2p_3 = \frac{4}{52} \cdot \frac{3}{51} \cdot \frac{2}{50} = \frac{1}{5525}$$

PROGRESS CHECK
What is the probability of drawing a spade three times in succession from a deck of 52 cards if the drawn cards are not replaced?

Answer
11/850

EXAMPLE 8
What is the probability of throwing a 5 only on the first of two successive throws with a single die?

Solution
The probability of throwing a 5 on the first toss is $p_1 = 1/6$. A success on the second toss consists of *not* throwing a 5 and has a probability $p_2 = 1-p_1 = 5/6$. The probability of the desired result is

$$p_1p_2 = \frac{1}{6} \cdot \frac{5}{6} = \frac{5}{36}$$

EXAMPLE 9
A transistor manufacturer finds that 81 of every 1000 transistors made is defective. What is the probability that two transistors selected at random will both prove to be defective?

Solution

This is an example of **empirical probability,** that is, the probability is obtained from experience or measurement rather than by theoretical means. The probability that a transistor is defective is $p_1 = 81/1000$. The probability that a second transistor is defective is again given by $p_2 = 81/1000$. Thus, the probability that both transistors are defective is

$$p_1 p_2 = \frac{81}{1000} \cdot \frac{81}{1000} = \frac{6561}{1,000,000}$$

PROGRESS CHECK

The probability of rain in a certain town on any given day is $\frac{1}{4}$. What is the probability of having a rainy Monday, dry Tuesday, and rainy Wednesday?

Answer

$\dfrac{3}{64}$

EXERCISE SET 9.6

1. If a single die is tossed, what is the probability that an odd number occurs?
2. If two dice are tossed, what is the probability of at least one 5 showing on the top faces of the dice?
3. If a card is randomly selected from an ordinary deck of 52 cards, what is the probability that it is a
   (a) red card?   (b) spade?   (c) king?
4. Suppose that two coins are tossed. What is the probability of getting
   (a) two tails?
   (b) at least one head?
   (c) no tails?
   (d) one head and one tail?
5. If two dice are tossed, what is the probability that
   (a) at least one of the dice shows a 4 on its top face?
   (b) the sum of the numbers on the dice is 8?
   (c) neither a 3 nor a 4 appears on the top face of a die?
6. If a card is picked at random from a standard deck of 52 cards, what is the probability that it is
   (a) a club?
   (b) a 4?
   (c) not an ace?
   (d) a 4 of spades?
   (e) either an ace or a king?
   (f) neither an ace nor a king?
7. The quality control department of a calculator manufacturer determines that one percent of all calculators made are defective. What is the probability that a buyer of a calculator will get a
   (a) good calculator?
   (b) defective calculator?
8. A photography club consisting of 18 women and 12 men wishes to elect a steering committee composed of three members. If every member is equally likely to be elected, find the probability that
   (a) all three members are women?
   (b) none of the members are women?
   (c) exactly one member is a woman?
   (d) at least one member is a woman?

9. A box contains 97 good bulbs and five defective bulbs. If three bulbs are chosen at random what is the probability that
   (a) all three bulbs are defective?
   (b) exactly one of the bulbs is defective?
   (c) none of the bulbs is defective?
10. Suppose that two cards are drawn in succession from a deck of 52 cards. What is the probability that both cards will be aces if
    (a) drawn cards are replaced?
    (b) drawn cards are not replaced?
11. Suppose that four cards are selected from a deck of 52 cards without replacement. What is the probability that they are all hearts?
12. If the probability of getting an A in this course is 0.2, what is the probability of not getting an A?
13. The board of trustees of a university consists of 14 women and 12 men. Suppose that an executive committee of six persons is to be elected and that every trustee is equally likely to be elected. Find the probability that the committee will consist of three men and three women.
14. Suppose that the probability of a cloudy day in a certain town in England is 0.6.
    (a) What is the probability of a clear day?
    (b) What is the probability of two consecutive clear days?
15. A bag contains six blue marbles, five green marbles, and seven yellow marbles. If five marbles are chosen without replacement, what is the probability that two are blue, two are green, and one is yellow?
16. Suppose that two cards are chosen at random from a deck of 52 cards. What is the probability that one card is an ace and the other card is not a king?
17. If two percent of cameras made on a production line are defective, what is the probability that four cameras chosen at random will all be (a) good? (b) defective?
18. A fraternity consists of 12 seniors, 10 juniors, and 14 sophomores. A steering committee of seven members is randomly chosen. What is the probability that it consists of three seniors, two juniors, and two sophomores?

TERMS AND SYMBOLS

KEY IDEAS FOR REVIEW

■ A sequence is a function whose domain is restricted to the set of natural numbers. We generally write a sequence by using subscript notation, that is, a_n replaces $a(n)$.

■ An arithmetic progression has a common difference d between terms. We can define an arithmetic progression recursively by writing $a_n = a_{n-1} + d$ and specifying a_1.

■ A geometric progression has a common ratio r between terms. We can define a geometric progression recursively by writing $a_n = ra_{n-1}$ and specifying a_1.

■ The formulas for the nth term of arithmetic and geometric progressions are

$$a_n = a_1 + (n-1)d \qquad \text{Arithmetic}$$

$$a_n = a_1 r^{n-1} \qquad \text{Geometric}$$

■ A series is the sum of the terms of a sequence.

■ The formulas for the sum S_n of the first n terms of arithmetic and geometric progressions are

$$S_n = \frac{n}{2}(a_1 + a_n) \qquad \text{Arithmetic}$$

$$S_n = \frac{n}{2}[2a_1 + (n-1)d] \qquad \text{Arithmetic}$$

$$S_n = \frac{a_1(1 - r^n)}{1 - r} \qquad \text{Geometric}$$

■ If the common ratio r satisfies $-1 < r < 1$, then the infinite geometric series has the sum S given by

$$S = \frac{a_1}{1 - r}$$

■ Mathematical induction is a property of the natural numbers that is useful in proving certain types of theorems.

■ The notation $n!$ indicates the product of the natural numbers 1 through n.

$$n! = n(n-1)(n-2)\cdots 2\cdot 1 \quad \text{for } n \geq 1$$

$$0! = 1$$

■ The binomial formula provides the terms of the expansion of $(a + b)^n$.

$$(a + b)^n = a^n + \frac{n!}{1!(n-1)!}\, a^{n-1}b + \frac{n!}{2!(n-2)!}\, a^{n-2}b^2$$

$$+ \frac{n!}{3!(n-3)!}\, a^{n-3}b^3 + \cdots + \frac{n!}{r!(n-r)!}\, a^{n-r}b^r + \cdots + b^n.$$

■ Permutations involve arrangements or order of objects; thus, abc and bac are distinct permutations of the letters a, b, c.

■ Combinations involve selection of objects; the order is not significant. If we are selecting three letters from a box containing the letters a, b, c, d, then abc and bac are the same combination.

■ The formulas for counting permutations and combinations of n things taken r at a time are

$$P(n,r) = \frac{n!}{(n-r)!} \qquad \text{Permutations}$$

$$C(n,r) = \frac{n!}{r!(n-r)!} \qquad \text{Combinations}$$

■ Probability is a means for expressing the likelihood of the occurrence of an event. It is the ratio of the number of successful outcomes to the total number of outcomes.

■ A probability of 1 indicates that an event is certain to occur, while a probability of 0 indicates that an event cannot possibly occur.

■ If p_1 and p_2 are the probabilities of the occurrence of two independent events, then $p_1 p_2$ is the probability that both events will occur.

REVIEW EXERCISES

In Exercises 1–2 write the first three terms of a sequence where the nth term is given.

1. $a_n = n^2 + n + 1$

2. $a_n = \dfrac{n^3 - 1}{n + 1}$

In Exercises 3–4 write the first four terms of an arithmetic progression whose first term is a_1 and whose common difference is d.

3. $a_1 = 3, \quad d = 2$

4. $a_1 = -2, \quad d = -\dfrac{1}{2}$

In Exercises 5–6 find the specified term of the arithmetic progression whose first term is a_1 and whose common difference is d.

5. $a_1 = -2, \quad d = 2$; 21st term

6. $a_1 = 6, \quad d = -1$; 16th term

In Exercises 7–8, given two terms of an arithmetic progression, find the specified term.

7. $a_1 = 4, \quad a_{16} = 9$; 13th term

8. $a_1 = -4, \quad a_{23} = -15$; 26th term

In Exercises 9–10 find the sum of the first 25 terms of the arithmetic progression whose first term is a_1 and whose common difference is d.

9. $a_1 = -\dfrac{1}{3}, \quad d = \dfrac{1}{3}$

10. $a_1 = 6, \quad d = -2$

In Exercises 11–12 determine the common ratio of the given geometric progression.

11. $2, -6, 18, -54, \ldots$

12. $-\dfrac{1}{2}, \dfrac{3}{4}, -\dfrac{9}{8}, \dfrac{27}{16}, \ldots$

In Exercises 13–14 write the first four terms of the geometric progression whose first term is a_1 and whose common ratio is r.

13. $a_1 = 5, \quad r = \dfrac{1}{5}$

14. $a_1 = -2, \quad r = -1$

15. Find the sixth term of the geometric sequence $-4, 6, -9, \ldots$.

16. Find the eight term of a geometric sequence for which $a_1 = -2$ and $a_5 = -32$.

17. Insert two geometric means between 3 and 1/72.

18. Find the sum of the first six terms of the geometric progression whose first three terms are $\frac{1}{3}, \frac{1}{6}, \frac{1}{12}$.

19. Find the sum of the first six terms of the geometric progression for which $a_1 = -2$ and $r = 3$.

In Exercises 20–21 find the sum of the infinite geometric series.

20. $5 + \dfrac{5}{2} + \dfrac{5}{4} + \cdots$

21. $3 - 2 + \dfrac{4}{3} - \cdots$

22. Use the principle of mathematical induction to show that

$$3 + 6 + 9 + \cdots + 3n = \frac{3n(n + 1)}{2}$$

is true for all positive integer values of n.

In Exercises 23–25 expand and simplify.

23. $(2x - y)^4$

24. $\left(\dfrac{x}{2} - 2\right)^4$

25. $(x^2 + 1)^3$

In Exercises 26–28 evaluate the expression.

26. $6!$

27. $\dfrac{13!}{11!\,2!}$

28. $\dfrac{(n - 1)!(n + 1)!}{n!\,n!}$

In Exercises 29–32 evaluate the expression.

29. $P(6,3)$

30. $\dfrac{P(8,3)}{2!}$

31. $C(6,3)$

32. $C(n, n - 1)$

33. Four novels have been selected for display on a shelf. How many different arrangements are possible?

34. Find the number of distinguishable permutations of the letters in the word *soothe*.

35. In how many ways can a tennis team of 6 players be selected from 10 candidates?

36. In how many ways can a consonant and a vowel be chosen from the letters in the word *fouled*?

37. If two dice are tossed, what is the probability that the sum of the numbers on the dice is 7 or 11?

38. A box contains three red pens and four white pens. If two pens are selected at random what is the probability that they are both white?

39. Two cards are drawn in succession from a deck of 52 cards. What is the probability that the cards are a king and an ace if the drawn card is not replaced?

40. If ten percent of the trees in a region are found to be diseased, what is the probability that two trees chosen at random are both free of disease?

41. Six husband–wife teams volunteer for an experiment in parapsychology. If four persons are selected at random to participate in the experiment, what is the probability that they consist of two husband–wife teams?

PROGRESS TEST 9A

1. Write the first four terms of a sequence whose nth term is $a_n = n/(n + 1)^2$.

2. Write the first four terms of an arithmetic progression whose first term is -1 and whose common difference is $3/2$.

3. Find the 25th term of the arithmetic progression whose first term is -4 and whose common difference is $\frac{1}{2}$.

4. Find the 15th term of an arithmetic progression whose first and tenth terms are -1 and 26, respectively.

5. Find the sum of the first 10 terms of an arithmetic progression whose first term is -4 and whose ninth term is 8.

6. Find the common ratio of the geometric progression $12, 4, \dfrac{4}{3}, \dfrac{4}{9}, \ldots$.

7. Write the first four terms of the geometric progression whose first term is $-\frac{2}{3}$ and common ratio is 2.

8. Find the tenth term of the sequence $2, -2, 2, \ldots$.

9. Insert two geometric means between -4 and 32.

10. Find the sum of the first seven terms of the geometric progression whose first three terms are $-8, 4, -2$.

11. Find the sum of the infinite geometric series $-4 - \dfrac{4}{3} - \dfrac{4}{9} - \ldots$.

12. Use the principle of mathematical induction to show that $2 + 6 + 10 + \cdots + (4n - 2) = 2n^2$ is true for all positive integer values of n.

13. Find the first four terms in the expansion of $\left(a + \dfrac{1}{b}\right)^{10}$.

14. Evaluate $\dfrac{12!}{10!\,3!}$.

15. Evaluate $P(6,4)$.

16. Evaluate $C(n + 1, n)$.

17. Three buses arrive simultaneously at a terminal that has four parking stalls. In how many ways can the buses be parked?

18. The 1980 census staff divided a region into three districts that were to be canvassed by 10, 8, and 15 staff members, respectively. If 40 staff members were available, in how many ways could they have been assigned to the three districts?

19. The telephone company uses white, black, and green telephones which are distributed to new customers at random. If an apartment dweller requests two telephones, what is the probability that they will be the same color?

20. Four marbles are removed at random from a box containing four purple, three blue, and three red marbles. What is the probability that these are two purple and two blue marbles?

PROGRESS TEST 9B

1. Write the first four terms of a sequence whose nth term is

$$a_n = n^2 + \frac{2n}{n + 2}$$

2. Write the first four terms of an arithmetic progression whose first term is 6 and whose common difference is $-\frac{2}{3}$.

3. Find the sixth term of the arithmetic progression whose first term is -5 and whose common difference is 3.

4. Find the 30th term of an arithmetic progression whose first and 20th terms are 3 and -35, respectively.

5. The first term of an arithmetic series is -5, the last term is -2, and the sum is $-91/2$. Find the number of terms and the common difference.

6. Find the common ratio of the geometric progression $20, 4, 0.8, 0.16$.

7. Write the first four terms of the geometric progression whose first term is -1 and common ratio is $-\frac{1}{4}$.

8. Find the sixth term of a geometric progression for which $a_1 = 3$ and $a_4 = -\frac{1}{9}$.

9. Insert two geometric means between -6 and $-16/9$.

10. Find the sum of the first five terms of a geometric sequence if $a_1 = -8$ and $a_4 = -1$.

11. Find the sum of the infinite geometric progression $5 - 2 + \frac{4}{5} - \cdots$.

12. Use the principle of mathematical induction to show that

$$\frac{1}{1 \cdot 2} + \frac{1}{2 \cdot 3} + \frac{1}{3 \cdot 4} + \cdots + \frac{1}{n(n+1)} = \frac{n}{n+1}$$

is true for all positive integer values of n.

13. Find the third term in the expansion of $(2x - 1)^{10}$.

14. Evaluate $\dfrac{n \cdot n!}{(n+1)!}$.

15. Evaluate $\dfrac{P(7,2)}{6}$.

16. Evaluate $\dfrac{P(5,2)}{C(5,2)}$.

17. Four plants are to be displayed on a window shelf. If six different plants are available, how many arrangements are possible?

18. A student/faculty committee of six members is to be set up composed of three faculty members and three students. How many ways can this be done if five faculty members and six students volunteer to serve on the committee?

19. A manufacturer finds that two percent of his products are defective. If three items are selected at random what is the probability that they are all defective?

20. What is the probability that a throw of two dice results in a sum of 7 or greater?

APPENDIX/TABLES

TABLE I Exponentials and Their Reciprocals

x	e^x	e^{-x}	x	e^x	e^{-x}
0.00	1.0000	1.0000	1.4	4.0552	0.2466
0.01	1.0101	0.9900	1.5	4.4817	0.2231
0.02	1.0202	0.9802	1.6	4.9530	0.2019
0.03	1.0305	0.9704	1.7	5.4739	0.1827
0.04	1.0408	0.9608	1.8	6.0496	0.1653
0.05	1.0513	0.9512	1.9	6.6859	0.1496
0.06	1.0618	0.9418	2.0	7.3891	0.1353
0.07	1.0725	0.9324	2.1	8.1662	0.1225
0.08	1.0833	0.9231	2.2	9.0250	0.1108
0.09	1.0942	0.9139	2.3	9.9742	0.1003
0.10	1.1052	0.9048	2.4	11.023	0.0907
0.11	1.1163	0.8958	2.5	12.182	0.0821
0.12	1.1275	0.8869	2.6	13.464	0.0743
0.13	1.1388	0.8781	2.7	14.880	0.0672
0.14	1.1503	0.8694	2.8	16.445	0.0608
0.15	1.1618	0.8607	2.9	18.174	0.0550
0.16	1.1735	0.8521	3.0	20.086	0.0498
0.17	1.1853	0.8437	3.1	22.198	0.0450
0.18	1.1972	0.8353	3.2	24.533	0.0408
0.19	1.2092	0.8270	3.3	27.113	0.0369
0.20	1.2214	0.8187	3.4	29.964	0.0334
0.21	1.2337	0.8106	3.5	33.115	0.0302
0.22	1.2461	0.8025	3.6	36.598	0.0273
0.23	1.2586	0.7945	3.7	40.447	0.0247
0.24	1.2712	0.7866	3.8	44.701	0.0224
0.25	1.2840	0.7788	3.9	49.402	0.0202
0.26	1.2969	0.7711	4.0	54.598	0.0183
0.27	1.3100	0.7634	4.1	60.340	0.0166
0.28	1.3231	0.7558	4.2	66.686	0.0150
0.29	1.3364	0.7483	4.3	73.700	0.0136
0.30	1.3499	0.7408	4.4	81.451	0.0123
0.35	1.4191	0.7047	4.5	90.017	0.0111
0.40	1.4918	0.6703	4.6	99.484	0.0101
0.45	1.5683	0.6376	4.7	109.95	0.0091
0.50	1.6487	0.6065	4.8	121.51	0.0082
0.55	1.7333	0.5769	4.9	134.29	0.0074
0.60	1.8221	0.5488	5	148.41	0.0067
0.65	1.9155	0.5220	6	403.43	0.0025
0.70	2.0138	0.4966	7	1,096.6	0.0009
0.75	2.1170	0.4724	8	2,981.0	0.0003
0.80	2.2255	0.4493	9	8,103.1	0.0001
0.85	2.3396	0.4274	10	22,026	0.00005
0.90	2.4596	0.4066	11	59,874	0.00002
0.95	2.5857	0.3867	12	162,754	0.000006
1.0	2.7183	0.3679	13	442,413	0.000002
1.1	3.0042	0.3329	14	1,202,604	0.0000008
1.2	3.3201	0.3012	15	3,269,017	0.0000003
1.3	3.6693	0.2725			

TABLE II Common Logarithms

N	0	1	2	3	4	5	6	7	8	9
1.0	.0000	.0043	.0086	.0128	.0170	.0212	.0253	.0294	.0334	.0374
1.1	.0414	.0453	.0492	.0531	.0569	.0607	.0645	.0682	.0719	.0755
1.2	.0792	.0828	.0864	.0899	.0934	.0969	.1004	.1038	.1072	.1106
1.3	.1139	.1173	.1206	.1239	.1271	.1303	.1335	.1367	.1399	.1430
1.4	.1461	.1492	.1523	.1553	.1584	.1614	.1644	.1673	.1703	.1732
1.5	.1761	.1790	.1818	.1847	.1875	.1903	.1931	.1959	.1987	.2014
1.6	.2041	.2068	.2095	.2122	.2148	.2175	.2201	.2227	.2253	.2279
1.7	.2304	.2330	.2355	.2380	.2405	.2430	.2455	.2480	.2504	.2529
1.8	.2553	.2577	.2601	.2625	.2648	.2672	.2695	.2718	.2742	.2765
1.9	.2788	.2810	.2833	.2856	.2878	.2900	.2923	.2945	.2967	.2989
2.0	.3010	.3032	.3054	.3075	.3096	.3118	.3139	.3160	.3181	.3201
2.1	.3222	.3243	.3263	.3284	.3304	.3324	.3345	.3365	.3385	.3404
2.2	.3424	.3444	.3464	.3483	.3502	.3522	.3541	.3560	.3579	.3598
2.3	.3617	.3636	.3655	.3674	.3692	.3711	.3729	.3747	.3766	.3784
2.4	.3802	.3820	.3838	.3856	.3874	.3892	.3909	.3927	.3945	.3692
2.5	.3979	.3997	.4014	.4031	.4048	.4065	.4082	.4099	.4116	.4133
2.6	.4150	.4166	.4183	.4200	.4216	.4232	.4249	.4265	.4281	.4298
2.7	.4314	.4330	.4346	.4362	.4378	.4393	.4409	.4425	.4440	.4456
2.8	.4472	.4487	.4502	.4518	.4533	.4548	.4564	.4579	.4594	.4609
2.9	.4624	.4639	.4654	.4669	.4683	.4698	.4713	.4728	.4742	.4757
3.0	.4771	.4786	.4800	.4814	.4829	.4843	.4857	.4871	.4886	.4900
3.1	.4914	.4928	.4942	.4955	.4969	.4983	.4997	.5011	.5024	.5038
3.2	.5051	.5065	.5079	.5092	.5105	.5119	.5132	.5145	.5159	.5172
3.3	.5185	.5198	.5211	.5224	.5237	.5250	.5263	.5276	.5289	.5302
3.4	.5315	.5328	.5340	.5353	.5366	.5378	.5391	.5403	.5416	.5428
3.5	.5441	.5453	.5465	.5478	.5490	.5502	.5514	.5527	.5539	.5551
3.6	.5563	.5575	.5587	.5599	.5611	.5623	.5635	.5647	.5658	.5670
3.7	.5682	.5694	.5705	.5717	.5729	.5740	.5752	.5763	.5775	.5786
3.8	.5798	.5809	.5821	.5832	.5843	.5855	.5866	.5877	.5888	.5899
3.9	.5911	.5922	.5933	.5944	.5955	.5966	.5977	.5988	.5999	.6010
4.0	.6021	.6031	.6042	.6053	.6064	.6075	.6085	.6096	.6107	.6117
4.1	.6128	.6138	.6149	.6160	.6170	.6180	.6191	.6201	.6212	.6222
4.2	.6232	.6243	.6253	.6263	.6274	.6284	.6294	.6304	.6314	.6325
4.3	.6335	.6345	.6355	.6365	.6375	.6385	.6395	.6405	.6415	.6425
4.4	.6435	.6444	.6454	.6464	.6474	.6484	.6493	.6503	.6513	.6522
4.5	.6532	.6542	.6551	.6561	.6571	.6580	.6590	.6599	.6609	.6618
4.6	.6628	.6637	.6646	.6656	.6665	.6675	.6684	.6693	.6702	.6712
4.7	.6721	.6730	.6739	.6749	.6758	.6767	.6776	.6785	.6794	.6803
4.8	.6812	.6821	.6830	.6839	.6848	.6857	.6866	.6875	.6884	.6893
4.9	.6902	.6911	.6920	.6928	.6937	.6946	.6955	.6964	.6972	.6981
5.0	.6990	.6998	.7007	.7016	.7024	.7033	.7042	.7050	.7059	.7067
5.1	.7076	.7084	.7093	.7101	.7110	.7118	.7126	.7135	.7143	.7152
5.2	.7160	.7168	.7177	.7185	.7193	.7202	.7210	.7218	.7226	.7235
5.3	.7243	.7251	.7259	.7267	.7275	.7284	.7292	.7300	.7308	.7316
5.4	.7324	.7332	.7340	.7348	.7356	.7364	.7372	.7380	.7388	.7396

TABLE II (*continued*)

N	0	1	2	3	4	5	6	7	8	9
5.5	.7404	.7412	.7419	.7427	.7435	.7443	.7451	.7459	.7466	.7474
5.6	.7482	.7490	.7497	.7505	.7513	.7520	.7528	.7536	.7543	.7551
5.7	.7559	.7566	.7574	.7582	.7589	.7597	.7604	.7612	.7619	.7627
5.8	.7634	.7642	.7649	.7657	.7664	.7672	.7679	.7686	.7694	.7701
5.9	.7709	.7716	.7723	.7731	.7738	.7745	.7752	.7760	.7767	.7774
6.0	.7782	.7789	.7796	.7803	.7810	.7818	.7825	.7832	.7839	.7846
6.1	.7853	.7860	.7868	.7875	.7882	.7889	.7896	.7903	.7910	.7917
6.2	.7924	.7931	.7938	.7945	.7952	.7959	.7966	.7973	.7980	.7987
6.3	.7993	.8000	.8007	.8014	.8021	.8028	.8035	.8041	.8048	.8055
6.4	.8062	.8069	.8075	.8082	.8089	.8096	.8102	.8109	.8116	.8122
6.5	.8129	.8136	.8142	.8149	.8156	.8162	.8169	.8176	.8182	.8189
6.6	.8195	.8202	.8209	.8215	.8222	.8228	.8235	.8241	.8248	.8254
6.7	.8261	.8267	.8274	.8280	.8287	.8293	.8299	.8306	.8312	.8319
6.8	.8325	.8331	.8338	.8344	.8351	.8357	.8363	.8370	.8376	.8382
6.9	.8388	.8395	.8401	.8407	.8414	.8420	.8426	.8432	.8439	.8445
7.0	.8451	.8457	.8463	.8470	.8476	.8482	.8488	.8494	.8500	.8506
7.1	.8513	.8519	.8525	.8531	.8537	.8543	.8549	.8555	.8561	.8567
7.2	.8573	.8579	.8585	.8591	.8597	.8603	.8609	.8615	.8621	.8627
7.3	.8633	.8639	.8645	.8651	.8657	.8663	.8669	.8675	.8681	.8686
7.4	.8692	.8698	.8704	.8710	.8716	.8722	.8727	.8733	.8739	.8745
7.5	.8751	.8756	.8762	.8768	.8774	.8779	.8785	.8791	.8797	.8802
7.6	.8808	.8814	.8820	.8825	.8831	.8837	.8842	.8848	.8854	.8859
7.7	.8865	.8871	.8876	.8882	.8887	.8893	.8899	.8904	.8910	.8915
7.8	.8921	.8927	.8932	.8938	.8943	.8949	.8954	.8960	.8965	.8971
7.9	.8976	.8982	.8987	.8993	.8998	.9004	.9009	.9015	.9020	.9025
8.0	.9031	.9036	.9042	.9047	.9053	.9058	.9063	.9069	.9074	.9079
8.1	.9085	.9090	.9096	.9101	.9106	.9112	.9117	.9122	.9128	.9133
8.2	.9138	.9143	.9149	.9154	.9159	.9165	.9170	.9175	.9180	.9186
8.3	.9191	.9196	.9201	.9206	.9212	.9217	.9222	.9227	.9232	.9238
8.4	.9243	.9248	.9253	.9258	.9263	.9269	.9274	.9279	.9284	.9289
8.5	.9294	.9299	.9304	.9309	.9315	.9320	.9325	.9330	.9335	.9340
8.6	.9345	.9350	.9355	.9360	.9365	.9370	.9375	.9380	.9385	.9390
8.7	.9395	.9400	.9405	.9410	.9415	.9420	.9425	.9430	.9435	.9440
8.8	.9445	.9450	.9455	.9460	.9465	.9469	.9474	.9479	.9484	.9489
8.9	.9494	.9499	.9504	.9509	.9513	.9518	.9523	.9528	.9533	.9538
9.0	.9542	.9547	.9552	.9557	.9562	.9566	.9571	.9576	.9581	.9586
9.1	.9590	.9595	.9600	.9605	.9609	.9614	.9619	.9624	.9628	.9633
9.2	.9638	.9643	.9647	.9652	.9657	.9661	.9666	.9671	.9675	.9680
9.3	.9685	.9689	.9694	.9699	.9703	.9708	.9713	.9717	.9722	.9727
9.4	.9731	.9736	.9741	.9745	.9750	.9754	.9759	.9763	.9768	.9773
9.5	.9777	.9782	.9786	.9791	.9795	.9800	.9805	.9809	.9814	.9818
9.6	.9823	.9827	.9832	.9836	.9841	.9845	.9850	.9854	.9859	.9863
9.7	.9868	.9872	.9877	.9881	.9886	.9890	.9894	.9899	.9903	.9908
9.8	.9912	.9917	.9921	.9926	.9930	.9934	.9939	.9943	.9948	.9952
9.9	.9956	.9961	.9965	.9969	.9974	.9978	.9983	.9987	.9991	.9996

TABLE III Natural Logarithms

N	$\ln N$	N	$\ln N$	N	$\ln N$
		4.5	1.5041	9.0	2.1972
0.1	−2.3026	4.6	1.5261	9.1	2.2083
0.2	−1.6094	4.7	1.5476	9.2	2.2192
0.3	−1.2040	4.8	1.5686	9.3	2.2300
0.4	−0.9163	4.9	1.5892	9.4	2.2407
0.5	−0.6931	5.0	1.6094	9.5	2.2513
0.6	−0.5108	5.1	1.6292	9.6	2.2618
0.7	−0.3567	5.2	1.6487	9.7	2.2721
0.8	−0.2231	5.3	1.6677	9.8	2.2824
0.9	−0.1054	5.4	1.6864	9.9	2.2925
1.0	0.0000	5.5	1.7047	10	2.3026
1.1	0.0953	5.6	1.7228	11	2.3979
1.2	0.1823	5.7	1.7405	12	2.4849
1.3	0.2624	5.8	1.7579	13	2.5649
1.4	0.3365	5.9	1.7750	14	2.6391
1.5	0.4055	6.0	1.7918	15	2.7081
1.6	0.4700	6.1	1.8083	16	2.7726
1.7	0.5306	6.2	1.8245	17	2.8332
1.8	0.5878	6.3	1.8405	18	2.8904
1.9	0.6419	6.4	1.8563	19	2.9444
2.0	0.6931	6.5	1.8718	20	2.9957
2.1	0.7419	6.6	1.8871	25	3.2189
2.2	0.7885	6.7	1.9021	30	3.4012
2.3	0.8329	6.8	1.9169	35	3.5553
2.4	0.8755	6.9	1.9315	40	3.6889
2.5	0.9163	7.0	1.9459	45	3.8067
2.6	0.9555	7.1	1.9601	50	3.9120
2.7	0.9933	7.2	1.9741	55	4.0073
2.8	1.0296	7.3	1.9879	60	4.0943
2.9	1.0647	7.4	2.0015	65	4.1744
3.0	1.0986	7.5	2.0149	70	4.2485
3.1	1.1314	7.6	2.0281	75	4.3175
3.2	1.1632	7.7	2.0412	80	4.3820
3.3	1.1939	7.8	2.0541	85	4.4427
3.4	1.2238	7.9	2.0669	90	4.4998
3.5	1.2528	8.0	2.0794	95	4.5539
3.6	1.2809	8.1	2.0919	100	4.6052
3.7	1.3083	8.2	2.1041		
3.8	1.3350	8.3	2.1163		
3.9	1.3610	8.4	2.1282		
4.0	1.3863	8.5	2.1401		
4.1	1.4110	8.6	2.1518		
4.2	1.4351	8.7	2.1633		
4.3	1.4586	8.8	2.1748		
4.4	1.4816	8.9	2.1861		

TABLE IV Interest Rates

	$i = \frac{1}{2}\%$				$i = 1\%$				$i = 1\frac{1}{2}\%$		
n	$(1+i)^n$	n	$(1+i)^n$	n	$(1+i)^n$	n	$(1+i)^n$	n	$(1+i)^n$	n	$(1+i)^n$
1	1.0050 0000	51	1.2896 4194	1	1.0100 0000	51	1.6610 7814	1	1.0150 0000	51	2.1368 2106
2	1.0100 2500	52	1.2960 9015	2	1.0201 0000	52	1.6776 8892	2	1.0302 2500	52	2.1688 7337
3	1.0150 7513	53	1.3025 7060	3	1.0303 0100	53	1.6944 6581	3	1.0456 7838	53	2.2014 0647
4	1.0201 5050	54	1.3090 8346	4	1.0406 0401	54	1.7114 1047	4	1.0613 6355	54	2.2344 2757
5	1.0252 5125	55	1.3156 2887	5	1.0510 1005	55	1.7285 2457	5	1.0772 8400	55	2.2679 4398
6	1.0303 7751	56	1.3222 0702	6	1.0615 2015	56	1.7458 0982	6	1.0934 4326	56	2.3019 6314
7	1.0355 2940	57	1.3288 1805	7	1.0721 3535	57	1.7632 6792	7	1.1098 4491	57	2.3364 9259
8	1.0407 0704	58	1.3354 6214	8	1.0828 5671	58	1.7809 0060	8	1.1264 9259	58	2.3715 3998
9	1.0459 1058	59	1.3421 3946	9	1.0936 8527	59	1.7987 0960	9	1.1433 8998	59	2.4071 1308
10	1.0511 4013	60	1.3488 5015	10	1.1046 2213	60	1.8166 9670	10	1.1605 4083	60	2.4432 1978
11	1.0563 9583	61	1.3555 9440	11	1.1156 6835	61	1.8348 6367	11	1.1779 4894	61	2.4798 6807
12	1.0616 7781	62	1.3623 7238	12	1.1268 2503	62	1.8532 1230	12	1.1956 1817	62	2.5170 6609
13	1.0669 8620	63	1.3691 8424	13	1.1380 9328	63	1.8717 4443	13	1.2135 5244	63	2.5548 2208
14	1.0723 2113	64	1.3760 3016	14	1.1494 7421	64	1.8904 6187	14	1.2317 5573	64	2.5931 4442
15	1.0776 8274	65	1.3829 1031	15	1.1609 6896	65	1.9093 6649	15	1.2502 3207	65	2.6320 4158
16	1.0830 7115	66	1.3898 2486	16	1.1725 7864	66	1.9284 6015	16	1.2689 8555	66	2.6715 2221
17	1.0884 8651	67	1.3967 7399	17	1.1843 0443	67	1.9477 4475	17	1.2880 2033	67	2.7115 9504
18	1.0939 2894	68	1.4037 5785	18	1.1961 4748	68	1.9672 2220	18	1.3073 4064	68	2.7522 6896
19	1.0993 9858	69	1.4107 7664	19	1.2081 0895	69	1.9868 9442	19	1.3269 5075	69	2.7935 5300
20	1.1048 9558	70	1.4178 3053	20	1.2201 9004	70	2.0067 6337	20	1.3468 5501	70	2.8354 5629
21	1.1104 2006	71	1.4249 1968	21	1.2323 9194	71	2.0268 3100	21	1.3670 5783	71	2.8779 8814
22	1.1159 7216	72	1.4320 4428	22	1.2447 1586	72	2.0470 9931	22	1.3875 6370	72	2.9211 5796
23	1.1215 5202	73	1.4392 0450	23	1.2571 6302	73	2.0675 7031	23	1.4083 7715	73	2.9649 7533
24	1.1271 5978	74	1.4464 0052	24	1.2697 3465	74	2.0882 4601	24	1.4295 0281	74	3.0094 4996
25	1.1327 9558	75	1.4536 3252	25	1.2824 3200	75	2.1091 2847	25	1.4509 4535	75	3.0545 9171
26	1.1384 5955	76	1.4609 0069	26	1.2952 5631	76	2.1302 1975	26	1.4727 0953	76	3.1004 1059
27	1.1441 5185	77	1.4682 0519	27	1.3082 0888	77	2.1515 2195	27	1.4948 0018	77	3.1469 1674
28	1.1498 7261	78	1.4755 4622	28	1.3212 9097	78	2.1730 3717	28	1.5172 2218	78	3.1941 2050
29	1.1556 2197	79	1.4829 2395	29	1.3345 0388	79	2.1947 6754	29	1.5399 8051	79	3.2420 3230
30	1.1614 0008	80	1.4903 3857	30	1.3478 4892	80	2.2167 1522	30	1.5630 8022	80	3.2906 6279
31	1.1672 0708	81	1.4977 9026	31	1.3613 2740	81	2.2388 8237	31	1.5865 2642	81	3.3400 2273
32	1.1730 4312	82	1.5052 7921	32	1.3749 4068	82	2.2612 7119	32	1.6103 2432	82	3.3901 2307
33	1.1789 0833	83	1.5128 0561	33	1.3886 9009	83	2.2838 8390	33	1.6344 7918	83	3.4409 7492
34	1.1848 0288	84	1.5203 6964	34	1.4025 7699	84	2.3067 2274	34	1.6589 9637	84	3.4925 8954
35	1.1907 2689	85	1.5279 7148	35	1.4166 0276	85	2.3297 8997	35	1.6838 8132	85	3.5449 7838
36	1.1966 8052	86	1.5356 1134	36	1.4307 6878	86	2.3530 8787	36	1.7091 3954	86	3.5981 5306
37	1.2026 6393	87	1.5432 8940	37	1.4450 7647	87	2.3766 1875	37	1.7347 7663	87	3.6521 2535
38	1.2086 7725	88	1.5510 0585	38	1.4595 2724	88	2.4003 8494	38	1.7607 9828	88	3.7069 0723
39	1.2147 2063	89	1.5587 6087	39	1.4741 2251	89	2.4243 8879	39	1.7872 1025	89	3.7625 1084
40	1.2207 9424	90	1.5665 5468	40	1.4888 6373	90	2.4486 3267	40	1.8140 1841	90	3.8189 4851
41	1.2268 9821	91	1.5743 8745	41	1.5037 5237	91	2.4731 1900	41	1.8412 2868	91	3.8762 3273
42	1.2330 3270	92	1.5822 5939	42	1.5187 8989	92	2.4978 5019	42	1.8688 4712	92	3.9343 7622
43	1.2391 9786	93	1.5901 7069	43	1.5339 7779	93	2.5228 2869	43	1.8968 7982	93	3.9933 9187
44	1.2453 9385	94	1.5981 2154	44	1.5493 1757	94	2.5480 5698	44	1.9253 3302	94	4.0532 9275
45	1.2516 2082	95	1.6061 1215	45	1.5648 1075	95	2.5735 3755	45	1.9542 1301	95	4.1140 9214
46	1.2578 7892	96	1.6141 4271	46	1.5804 5885	96	2.5992 7293	46	1.9835 2621	96	4.1758 0352
47	1.2641 6832	97	1.6222 1342	47	1.5962 6344	97	2.6252 6565	47	2.0132 7910	97	4.2384 4057
48	1.2704 8916	98	1.6303 2449	48	1.6122 2608	98	2.6515 1831	48	2.0434 7829	98	4.3020 1718
49	1.2768 4161	99	1.6384 7611	49	1.6283 4834	99	2.6780 3349	49	2.0741 3046	99	4.3665 4744
50	1.2832 2581	100	1.6466 6849	50	1.6446 3182	100	2.7048 1383	50	2.1052 4242	100	4.4320 4565

TABLE IV (*continued*)

	$i = 2\%$				$i = 2\frac{1}{2}\%$				$i = 3\%$
n	$(1 + i)^n$	*n*	$(1 + i)^n$	*n*	$(1 + i)^n$	*n*	$(1 + i)^n$	*n*	$(1 + i)^n$
1	1.0200 0000	51	2.7454 1979	1	1.0250 0000	51	3.5230 3644	1	1.0300 0000
2	1.0404 0000	52	2.8003 2819	2	1.0506 2500	52	3.6111 1235	2	1.0609 0000
3	1.0612 0800	53	2.8563 3475	3	1.0768 9063	53	3.7013 9016	3	1.0927 2700
4	1.0824 3216	54	2.9134 6144	4	1.1038 1289	54	3.7939 2491	4	1.1255 0881
5	1.1040 8080	55	2.9717 3067	5	1.1314 0821	55	3.8887 7303	5	1.1592 7407
6	1.1261 6242	56	3.0311 6529	6	1.1596 9342	56	3.9859 9236	6	1.1940 5230
7	1.1486 8567	57	3.0917 8859	7	1.1886 8575	57	4.0856 4217	7	1.2298 7387
8	1.1716 5938	58	3.1536 2436	8	1.2184 0290	58	4.1877 8322	8	1.2667 7008
9	1.1950 9257	59	3.2166 9685	9	1.2488 6297	59	4.2924 7780	9	1.3047 7318
10	1.2189 9442	60	3.2810 3079	10	1.2800 8454	60	4.3997 8975	10	1.3439 1638
11	1.2433 7431	61	3.3466 5140	11	1.3120 8666	61	4.5097 8449	11	1.3842 3387
12	1.2682 4179	62	3.4135 8443	12	1.3448 8882	62	4.6225 2910	12	1.4257 6089
13	1.2936 0663	63	3.4818 5612	13	1.3785 1104	63	4.7380 9233	13	1.4685 3371
14	1.3194 7876	64	3.5514 9324	14	1.4129 7382	64	4.8565 4464	14	1.5125 8972
15	1.3458 6834	65	3.6225 2311	15	1.4482 9817	65	4.9779 5826	15	1.5579 6742
16	1.3727 8571	66	3.6949 7357	16	1.4845 0562	66	5.1024 0721	16	1.6047 0644
17	1.4002 4142	67	3.7688 7304	17	1.5216 1826	67	5.2299 6739	17	1.6528 4763
18	1.4282 4625	68	3.8442 5050	18	1.5596 5872	68	5.3607 1658	18	1.7024 3306
19	1.4568 1117	69	3.9211 3551	19	1.5986 5019	69	5.4947 3449	19	1.7535 0605
20	1.4859 4740	70	3.9995 5822	20	1.6386 1644	70	5.6321 0286	20	1.8061 1123
21	1.5156 6634	71	4.0795 4939	21	1.6795 8185	71	5.7729 0543	21	1.8602 9457
22	1.5459 7967	72	4.1611 4038	22	1.7215 7140	72	5.9172 2806	22	1.9161 0341
23	1.5768 9926	73	4.2443 6318	23	1.7646 1068	73	6.0651 5876	23	1.9735 8651
24	1.6084 3725	74	4.3292 5045	24	1.8087 2595	74	6.2167 8773	24	2.0327 9411
25	1.6406 0599	75	4.4158 3546	25	1.8539 4410	75	6.3722 0743	25	2.0937 7793
26	1.6734 1811	76	4.5041 5216	26	1.9002 9270	76	6.5315 1261	26	2.1565 9127
27	1.7068 8648	77	4.5942 3521	27	1.9478 0002	77	6.6948 0043	27	2.2212 8901
28	1.7410 2421	78	4.6861 1991	28	1.9964 9502	78	6.8621 7044	28	2.2879 2768
29	1.7758 4469	79	4.7798 4231	29	2.0464 0739	79	7.0337 2470	29	2.3565 6551
30	1.8113 6158	80	4.8754 3916	30	2.0975 6758	80	7.2095 6782	30	2.4272 6247
31	1.8475 8882	81	4.9729 4794	31	2.1500 0677	81	7.3898 0701	31	2.5000 8035
32	1.8845 4059	82	5.0724 0690	32	2.2037 5694	82	7.5745 5219	32	2.5750 8276
33	1.9222 3140	83	5.1738 5504	33	2.2588 5086	83	7.7639 1599	33	2.6523 3524
34	1.9606 7603	84	5.2773 3214	34	2.3153 2213	84	7.9580 1389	34	2.7319 0530
35	1.9998 8955	85	5.3828 7878	35	2.3732 0519	85	8.1569 6424	35	2.8138 6245
36	2.0398 8734	86	5.4905 3636	36	2.4325 3532	86	8.3608 8834	36	2.8982 7833
37	2.0806 8509	87	5.6003 4708	37	2.4933 4870	87	8.5699 1055	37	2.9852 2668
38	2.1222 9879	88	5.7123 5402	38	2.5556 8242	88	8.7841 5832	38	3.0747 8348
39	2.1647 4477	89	5.8266 0110	39	2.6195 7448	89	9.0037 6228	39	3.1670 2698
40	2.2080 3966	90	5.9431 3313	40	2.6850 6384	90	9.2288 5633	40	3.2620 3779
41	2.2522 0046	91	6.0619 9579	41	2.7521 9043	91	9.4595 7774	41	3.3598 9893
42	2.2972 4447	92	6.1832 3570	42	2.8209 9520	92	9.6960 6718	42	3.4606 9589
43	2.3431 8936	93	6.3069 0042	43	2.8915 2008	93	9.9384 6886	43	3.5645 1677
44	2.3900 5314	94	6.4330 3843	44	2.9638 0808	94	10.1869 3058	44	3.6714 5227
45	2.4378 5421	95	6.5616 9920	45	3.0379 0328	95	10.4416 0385	45	3.7815 9584
46	2.4866 1129	96	6.6929 3318	46	3.1138 5086	96	10.7026 4395	46	3.8950 4372
47	2.5363 4352	97	6.8267 9184	47	3.1916 9713	97	10.9702 1004	47	4.0118 9503
48	2.5870 7039	98	6.9633 2768	48	3.2714 8956	98	11.2444 6530	48	4.1322 5188
49	2.6388 1179	99	7.1025 9423	49	3.3532 7680	99	11.5255 7693	49	4.2562 1944
50	2.6915 8803	100	7.2446 4612	50	3.4371 0872	100	11.8137 1635	50	4.3839 0602

TABLE IV (*continued*)

$i = 4\%$		$i = 5\%$		$i = 6\%$		$i = 7\%$		$i = 8\%$	
n	$(1 + i)^n$	n	$(1 + i)^n$	n	$(1 + i)^n$	n	$(1 + i)^n$	n	$(1 + i)^n$
1	1.0400 0000	1	1.0500 0000	1	1.0600 0000	1	1.0700 0000	1	1.0800 0000
2	1.0816 0000	2	1.1025 0000	2	1.1236 0000	2	1.1449 0000	2	1.1664 0000
3	1.1248 6400	3	1.1576 2500	3	1.1910 1600	3	1.2250 4300	3	1.2597 1200
4	1.1698 5856	4	1.2155 0625	4	1.2624 7696	4	1.3107 9601	4	1.3604 8896
5	1.2166 5290	5	1.2762 8156	5	1.3382 2558	5	1.4025 5173	5	1.4693 2808
6	1.2653 1902	6	1.3400 9564	6	1.4185 1911	6	1.5007 3035	6	1.5868 7432
7	1.3159 3178	7	1.4071 0042	7	1.5036 3026	7	1.6057 8148	7	1.7138 2427
8	1.3685 6905	8	1.4774 5544	8	1.5938 4807	8	1.7181 8618	8	1.8509 3021
9	1.4233 1181	9	1.5513 2822	9	1.6894 7896	9	1.8384 5921	9	1.9990 0463
10	1.4802 4428	10	1.6288 9463	10	1.7908 4770	10	1.9671 5136	10	2.1589 2500
11	1.5394 5406	11	1.7103 3936	11	1.8982 9856	11	2.1048 5195	11	2.3316 3900
12	1.6010 3222	12	1.7958 5633	12	2.0121 9647	12	2.2521 9159	12	2.5181 7012
13	1.6650 7351	13	1.8856 4914	13	2.1329 2826	13	2.4098 4500	13	2.7196 2373
14	1.7316 7645	14	1.9799 3160	14	2.2609 0396	14	2.5785 3415	14	2.9371 9362
15	1.8009 4351	15	2.0789 2818	15	2.3965 5819	15	2.7590 3154	15	3.1721 6911
16	1.8729 8125	16	2.1828 7459	16	2.5403 5168	16	2.9521 6375	16	3.4259 4264
17	1.9479 0050	17	2.2920 1832	17	2.6927 7279	17	3.1588 1521	17	3.7000 1805
18	2.0258 1652	18	2.4066 1923	18	2.8543 3915	18	3.3799 3228	18	3.9960 1950
19	2.1068 4918	19	2.5269 5020	19	3.0255 9950	19	3.6165 2754	19	4.3157 0106
20	2.1911 2314	20	2.6532 9771	20	3.2071 3547	20	3.8696 8446	20	4.6609 5714
21	2.2787 6807	21	2.7859 6259	21	3.3995 6360	21	4.1405 6237	21	5.0338 3372
22	2.3699 1879	22	2.9252 6072	22	3.6035 3742	22	4.4304 0174	22	5.4365 4041
23	2.4647 1554	23	3.0715 2376	23	3.8197 4966	23	4.7405 2986	23	5.8714 6365
24	2.5633 0416	24	3.2250 9994	24	4.0489 3464	24	5.0723 6695	24	6.3411 8074
25	2.6658 3633	25	3.3863 5494	25	4.2918 7072	25	5.4274 3264	25	6.8484 7520
26	2.7724 6978	26	3.5556 7269	26	4.5493 8296	26	5.8073 5292	26	7.3963 5321
27	2.8833 6858	27	3.7334 5632	27	4.8223 4594	27	6.2138 6763	27	7.9880 6147
28	2.9987 0332	28	3.9201 2914	28	5.1116 8670	28	6.6488 3836	28	8.6271 0639
29	3.1186 5145	29	4.1161 3560	29	5.4183 8790	29	7.1142 5705	29	9.3172 7490
30	3.2433 9751	30	4.3219 4238	30	5.7434 9117	30	7.6122 5504	30	10.0626 5689
31	3.3731 3341	31	4.5380 3949	31	6.0881 0064	31	8.1451 1290	31	10.8676 6944
32	3.5080 5875	32	4.7649 4147	32	6.4533 8668	32	8.7152 7080	32	11.7370 8300
33	3.6483 8110	33	5.0031 8854	33	6.8405 8988	33	9.3253 3975	33	12.6760 4964
34	3.7943 1634	34	5.2533 4797	34	7.2510 2528	34	9.9781 1354	34	13.6901 3361
35	3.9460 8899	35	5.5160 1537	35	7.6860 8679	35	10.6765 8148	35	14.7853 4429
36	4.1039 3255	36	5.7918 1614	36	8.1472 5200	36	11.4239 4219	36	15.9681 7184
37	4.2680 8986	37	6.0814 0694	37	8.6360 8712	37	12.2236 1814	37	17.2456 2558
38	4.4388 1345	38	6.3854 7729	38	9.1542 5235	38	13.0792 7141	38	18.6252 7563
39	4.6163 6599	39	6.7047 5115	39	9.7035 0749	39	13.9948 2041	39	20.1152 9768
40	4.8010 2063	40	7.0399 8871	40	10.2857 1794	40	14.9744 5784	40	21.7245 2150
41	4.9930 6145	41	7.3919 8815	41	10.9028 6101	41	16.0226 6989	41	23.4624 8322
42	5.1927 8391	42	7.7615 8756	42	11.5570 3267	42	17.1442 5678	42	25.3394 8187
43	5.4004 9527	43	8.1496 6693	43	12.2504 5463	43	18.3443 5475	43	27.3666 4042
44	5.6165 1508	44	8.5571 5028	44	12.9854 8191	44	19.6284 5959	44	29.5559 7166
45	5.8411 7568	45	8.9850 0779	45	13.7646 1083	45	21.0024 5176	45	31.9204 4939
46	6.0748 2271	46	9.4342 5818	46	14.5904 8748	46	22.4726 2338	46	34.4740 8534
47	6.3178 1562	47	9.9059 7109	47	15.4659 1673	47	24.0457 0702	47	37.2320 1217
48	6.5705 2824	48	10.4012 6965	48	16.3938 7173	48	25.7289 0651	48	40.2105 7314
49	6.8333 4937	49	10.9213 3313	49	17.3775 0403	49	27.5299 2997	49	43.4274 1899
50	7.1066 8335	50	11.4673 9979	50	18.4201 5427	50	29.4570 2506	50	46.9016 1251

ANSWERS TO ODD-NUMBERED EXERCISES, REVIEW EXERCISES, AND PROGRESS TESTS

CHAPTER ONE

EXERCISE SET 1.1, page 8

1. $\{3, 4, 5, 6, 7\}$ 3. $\{-9\}$ 5. $\{1, 2\}$
7. $\{1, 3, 7\}$ 9. F 11. F 13. T
15. T 17. T 19. F 21. F
23. commutative (addition) 25. distributive
27. associative (addition) 29. closure (multiplication)
31. commutative (multiplication)
33. commutative, associative (multiplication)
35. multiplicative inverse 41. symmetric 43. substitution

EXERCISE SET 1.2, page 13

1.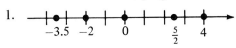

3. $A:1$, $B:2.5$, $C:-2$, $D:4$, $O:0$, $E:-3.5$
5. 4 7. -2 9. -5

11.

13. (number line from 2 to 7)   **15.** $10 > 9.99$

17. $a \geqslant 0$ **19.** $x > 0$ **21.** $b \leqslant -4$ **23.** $b \geqslant 5$
25. multiplication by negative number
27. multiplication by negative number
29. multiplication by positive number
31. 2   **33.** 1.5   **35.** -2 **37.** 1
39. 4   **41.** 2   **43.** 1/5   **45.** 3
47. 2   **49.** 8/5

EXERCISE SET 1.3, page 21

1. 12   **3.** 8/3   **5.** 5/8   **7.** 18.84
9. (a) \$2160   (b) \$2080   (c) \$2106.67   **11.** 9.37
13. -9 **15.** 3/2   **17.** 0   **19.** b^7
21. $-20y^9$ **23.** $-3x^4$ **25.** c, d **27.** 2; 3
29. 3/5; 4   **31.** 3   **33.** 4   **35.** 11
37. 176.20   **39.** $bh/2$ **41.** cost of all purchases
43. $5x + 3$
45. $2s^2t^3 - 3s^2t^2 + 2s^2t + 3st^2 + st - s + 2t - 3$
47. $-2a^2bc + ab^2c - 2ab^3 + 3$
49. $-(2x^4 - 4x^3 + x^2 - 4x + 4)$
51. $6s^3 - s^2 - 11s + 6$
53. $-4y^5 - 2y^4 + 2y^3 - 5y^2 - 3y$
55. $4a^5 - 16a^4 + 14a^3 - 3a^2 - 14a + 15$
57. $3b^4 + 3ab^3 + 2b^3 - 7ab^2 + 2b^2 - 4ab - 6a$
59. $-6x^2 + 22x - 12$ **61.** $-260x + 13y + 17z$ **63.** $x^2 + 2x - 3$
65. $4x^2 + 8x + 3$ **67.** $3x^2 - 5x + 2$ **69.** $x^2 + 2xy + y^2$
71. $9x^2 - 6x + 1$ **73.** $4x^2 - 1$ **75.** $x^4 + 2x^2y^2 + y^4$

EXERCISE SET 1.4, page 27

1. $5(x - 3)$ **3.** $-2(x + 4y)$ **5.** $5b(c + 5)$
7. $-y^2(3 + 4y^3)$ **9.** $3x^2(1 + 2y - 3z)$ **11.** $(x + 1)(x + 3)$
13. $(y - 3)(y - 5)$ **15.** $(a - 3b)(a - 4b)$ **17.** $\left(y - \dfrac{1}{3}\right)\left(y + \dfrac{1}{3}\right)$
19. $(3 - x)(3 + x)$ **21.** $(x - 7)(x + 2)$ **23.** $\left(\dfrac{1}{4} + y\right)\left(\dfrac{1}{4} - y\right)$
25. $(x - 3)^2$ **27.** $(x - 10)(x - 2)$ **29.** $(x + 3)(x + 8)$
31. $(2x + 1)(x - 2)$ **33.** $(3a - 2)(a - 3)$ **35.** $(3x + 2)(2x + 3)$
37. $(4m + 3)(2m - 3)$ **39.** $(5x + 1)(2x - 3)$
41. $(3a + 2b)(2a - 3b)$ **43.** $(5rs + 2t)(2rs + t)$
45. $(4 + 3xy)(4 - 3xy)$ **47.** $(4n + 1)(2n - 5)$
49. $2(x + 2)(x - 3)$ **51.** $5(3x - 2)(2x - 1)$
53. $3m(2x + 3)(3x + 1)$ **55.** $2x^2(3 + 2x)(2 - 5x)$
57. $(x^2 + y^2)(x + y)(x - y)$ **59.** $(b^2 + 4)(b^2 - 2)$
61. $(3b^2 - 1)(2b^2 + 3)$ **63.** $(x - 1)(2x + 7)$
65. $(2x - 1)^2(x + 2)^2[4(x + 2)(x + 1) - 3(2x - 1)^3(x + 3)]$

EXERCISE SET 1.5, page 34

1. $\dfrac{1}{x-4}$

3. $x-4$

5. $\dfrac{3x+1}{x+2}$

7. $4/9$

9. $-2b(5+a)$

11. $\dfrac{5y}{x-4}$

13. $\dfrac{(2x+1)(x-2)}{(x-1)(x+1)}$

15. $\dfrac{(x+2)(2x+3)}{x+4}$

17. $\dfrac{(x+3)(x^2+1)}{x-2}$

19. $\dfrac{x+4}{(x-5)(x+1)}$

21. xy

23. $2a$

25. $(b-1)^2$

27. $(x-2)(x+3)$

29. $x(x+1)(x-1)$

31. $\dfrac{4}{a-2}$

33. $\dfrac{x+5}{3}$

35. $\dfrac{4(a+1)}{(a-2)(a+2)}$

37. $\dfrac{4y-15}{3xy}$

39. $\dfrac{5-2x}{2(x+3)}$

41. $\dfrac{23x+24}{6(x+3)(x-3)}$

43. $\dfrac{3x^2-4x-1}{(x-1)(x-2)(x+1)}$

45. $\dfrac{5x-3}{(x+2)(x-1)}$

47. $\dfrac{x^2+x+3}{(x+1)(x+2)(x+3)}$

49. $\dfrac{3x^2+10x+1}{(x+4)(x-1)(x+1)}$

51. $\dfrac{x+2}{x-3}$

53. $\dfrac{x(x+1)}{x-1}$

55. $4x(x+4)$

57. $\dfrac{a+2}{a+1}$

59. $a-b$

61. $\dfrac{x-2}{x}$

63. $\dfrac{(y-2)(y+1)}{(y+2)(y-1)}$

67. $\dfrac{x+1}{2x+1}$

EXERCISE SET 1.6, page 39

1. x^6

3. b^4

5. $16x^4$

7. $-1/128$

9. y^{8n}

11. $-x^3/y^3$

13. x^{19}

15. $-32x^{10}$

17. x^{4n}

19. $1/x^2$

21. $30x^8$

23. 1

25. $\left(\dfrac{3}{2}\right)^n x^{2n}y^{3n}$

27. $(2x+1)^{10}$

29. $2^{2n}a^{4n}b^{6n}$

31. $4/3$

33. 3

35. 81

37. $-x^3$

39. y^6

41. 25

43. $1/3^6$

45. x^9

47. 32

49. $2x^2y$

51. $a^4b^6/9$

53. $-8y^{12}/x^9$

55. $a^9/3b^4$

57. $4a^{10}c^6/b^8$

59. $1/(a-2b^2)$

61. $(a-b)^2/(a+b)$

63. $(b+a)/(b-a)$

67. 0.074

69. 0.0113

EXERCISE SET 1.7, page 46

1. 8

3. $1/16$

5. $2x^{13/12}$

7. $x^{5/36}$

9. x^2y^{12}

11. x^9/y^6

13. $\sqrt[5]{1/16}$

15. $\sqrt[4]{a^3}$

17. $\sqrt[3]{144x^6/y^4}$

19. $8^{3/4}$

21. $(-8)^{-2/5}$

23. $(4a^3/9)^{-1/4}$

25. $2/3$

27. not real

29. 5

31. $5/4$

33. 54.82

35. $3, 4$

37. $4\sqrt{3}$

39. $3\sqrt[3]{2}$

41. $y^2 \sqrt[3]{y}$

43. $2x^2 \sqrt[4]{6x^2}$

45. $x^2 y \sqrt{xy}$

47. $2x^2 y \sqrt[4]{y}$

49. $\sqrt{5}/5$

51. $\sqrt{3y}/3y$

53. $2x\sqrt{2x}$

55. $y^2 \sqrt[3]{x^2 y}$

57. $7\sqrt{3}$

59. $7\sqrt{x}$

61. $4\sqrt{3}$

63. $11\sqrt{5} - \sqrt[3]{5}$

65. $-5\sqrt{5}$

67. $3 + 4\sqrt{3}$

69. $3xy$

71. $5 - 2\sqrt{6}$

73. $3x - 4y - \sqrt{6xy}$

75. $3(3 - \sqrt{2})/7$

77. $2(4 + \sqrt{3})/13$

79. $\dfrac{-3(3\sqrt{a} - 1)}{9a - 1}$

81. $\dfrac{-3(5 - \sqrt{5y})}{5(5 - y)}$

83. $3 + 2\sqrt{2}$

85. $2 + \sqrt{6} + 3\sqrt{2} + 2\sqrt{3}$

87. $9, 16$

EXERCISE SET 1.8, page 51

1. 1

3. $-i$

5. $-i$

7. 1

9. i

11. $-\dfrac{1}{2} + 0i$

13. $0 + 5i$

15. $0 - 6i$

17. $3 - 7i$

19. $0.3 - 7\sqrt{2}i$

21. $-2 - 4i$

23. $x = 2/3, y = -8$

25. $x = -1, y = -9/2$

27. $3 + i$

29. $5 + i$

31. $-5 - 4i$

33. $2 - 6i$

35. $-1 - i/2$

37. $5 + 0i$

39. $2 + 14i$

41. $4 - 7i$

43. 0

45. 3

53. $y \geqslant 5$

REVIEW EXERCISES, page 54

1. $\{1, 2, 3, 4\}$

2. $\{-3, -2, -1\}$

3. $\{2\}$

4. T

5. F

6. F

7. F

8. additive inverse

9. distributive

10. commutative (addition)

11. multiplicative identity

12.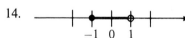

13.

14.

15. -1

16. $3/2$

17. $\$51$

18. c

19. $-0.5, 7$

20. $-7, 5$

21. $a^2 b^2 - 3a^2 b + 4b$

22. $2x^3 + 3x^2 - 2x$

23. $12x^3 + 12x^2 + 3x$

24. $2(x + 1)(x - 1)$

25. $(x + 5y)(x - 5y)$

26. $(2a + 3b)(a + 3)$

27. $(4x - 1)(x + 5)$

28. $(x + 1)(x - 1)(x^2 + 1)(x^4 + 1)$

29. $(3x^2 - 1)(2x^2 + 3)$

30. $\dfrac{-6(y - 1)}{(x - y)xy^2}$

31. $\dfrac{-3(x + 2)}{2y}$

32. $\dfrac{a - 2b}{a - b}$

33. $\dfrac{3x(x + 1)}{2x - 1}$

34. $2x^2(x + 2)(x - 2)$

35. $x(x - 1)^2$

36. $5y^2(x - 1)^2$

37. $4x^2(y + 1)^2(y - 1)$

38. $\dfrac{2(a^2 - 2)}{(a + 2)(a - 2)}$

39. $\dfrac{-2x - 5}{x^2 - 16}$

40. $\dfrac{x - 7}{(x - 1)^2(x + 2)}$

41. $\dfrac{x^3 - x^2 + 1}{x - 1}$

42. $b^9/8a^6$

43. 2
44. $1/x^4y^8$
45. x^3
46. $4\sqrt{5}$
47. $\sqrt{3}/3$
48. $x^3y^2\sqrt{xy}$
49. $2\sqrt[4]{2}x^2y\sqrt{y}$
50. $\dfrac{x - \sqrt{xy}}{x - y}$
51. $3\sqrt{xy}$
52. $8 + 2\sqrt{15}$
53. $x = -2, y = 4$
54. $-i$
55. $8 - i$
56. $3 + 4i$
57. $17 + 6i$

PROGRESS TEST 1A, page 55

1. $\{2, 4, 6, 8, 10, 12\}$
2. $\{3\}$
3. F
4. F
5. commutative (multiplication)
6. multiplicative reciprocal

7.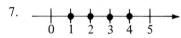
8.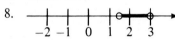

9. -1
10. 2
11. 25
12. $-7/3$
13. b
14. $-2.2, 5$
15. 14, 6
16. $2xy + 3x + 4y + 1$
17. $3a^3 + 5a^2 + 3a + 10$
18. $4a^2b(2ab^4 - 3a^3b + 4)$
19. $(2 - 3x)(2 + 3x)$
20. $6m^5/n^2$
21. $-(x - 1)/(x + 1)$
22. $4x^2(x + 1)(x - 1)(x - 2)$
23. $(11x - 15)/3(x^2 - 9)$
24. $2/(x + 1)$
25. $1/x^{17}$
26. y^{n+1}
27. -1
28. $4a^4/b^2$
29. 0
30. $32 - 10\sqrt{7}$
31. $-11\sqrt{xy}/4$
32. $x \leqslant 2$
33. $-1 + 0i$
34. $16 - 11i$

PROGRESS TEST 1B, page 56

1. $\{1, 3, 5, 7, 9\}$
2. $\{0, 12, 15, 24\}$
3. T
4. F
5. commutative (addition)
6. distributive

7.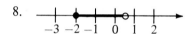
8.

9. 1
10. 7
11. 14
12. 2
13. a, d
14. 4, 5
15. 1.5, 10
16. $2s^2t^3 - 3s^2t^2 + 2s^2t + 3st^2 + st - s + 2t - 3$
17. $-3b^3 - 7b^2 + 10b + 12$
18. $5r^3s^3(s - 8rt)$
19. $(2x - 1)(x + 4)$
20. $x^2/uv(y - 1)$
21. $-1/x$
22. $4x^2(y + 1)^2(y - 1)$
23. 1
24. $(-2x^3 + 3x^2 + 3x + 2)/x(x - 1)(x + 1)$
25. $4/x$
26. b^{28}
27. x^6/y^9
28. -1
29. $x^7y^8\sqrt{y}$
30. $-2(\sqrt{x} + 1)/(x - 1)$
31. $ab\sqrt[3]{b^2}$
32. $x > 2$
33. $2 - \dfrac{3}{2}i$
34. $4 - 7i$

CHAPTER TWO

EXERCISE SET 2.1, page 62

1. T	3. T	5. -2	7. $-2/3$
9. 6	11. $-4/3$	13. $3/2$	15. $-10/3$
17. -2	19. 5	21. $-7/2$	23. 1
25. $8/(5-k)$	27. $(6+k)/5$	29. $10/3$	31. 1
33. 4	35. 4	37. 12	39. 2
41. $12/7$	43. none	45. I	47. C
49. T	51. F	53. T	

EXERCISE SET 2.2, page 70

1. $2n + 3$
3. $6n - 5 = 26$
5. 16, 28
7. 6, 7, 8
9. 68°
11. 4 meters and 8 meters
13. 10 nickels, 25 dimes
15. 300 children, 400 adults
17. 61 three-dollar tickets, 40 five-dollar tickets, 20 six-dollar tickets
19. $11,636.36 on 10-speeds, $4363.64 on 3-speeds
21. $7000
23. 20 hours
25. 50 and 54 mph
27. 40 kph, 80 kph
29. Ceylon: 2.4 ounces, Formosa: 5.6 ounces
31. 13.5 gal
33. $1/12, 1/4; -1/4, -1/12$
35. $12/5$ hours
37. $9/2$ days, 9 days
39. 8 hours
41. 140 mph

EXERCISE SET 2.3, page 77

1. $[-5, 1)$
3. $(9, \infty)$
5. $[-12, -3]$
7. $(3, 7)$
9. $(-6, -4]$
11. $5 \leqslant x \leqslant 8$
13. $x > 3$
15. $x \leqslant 5$
17. $x \geqslant 0$
19. $x < 4$
21. $x < -6$

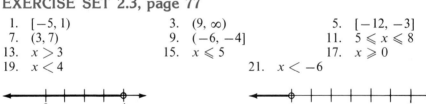

23. $x \geqslant 5$
25. $a > -1$

27. $y < -1/2$
29. $x \geqslant 0$

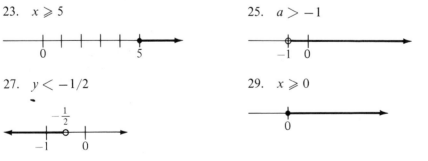

31. $r < 2$
33. $x \geqslant 1$

35. $x > 5/3$

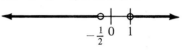

37. $(-\infty, 2]$ 39. $[2, \infty)$ 41. $(-\infty, 3/2)$ 43. $(-12, \infty)$
45. $(-\infty, 9/2)$ 47. $(-\infty, -6]$ 49. $(-\infty, 3/2)$ 51. $(-\infty, 7]$
53. $(-1/2, 5/4]$ 55. $[-3, -2]$ 57. $(-3, -1]$ 59. $[-1, 2)$
61. 98   63. 5   65. 2924   67. $L \leqslant 20$

EXERCISE SET 2.4, page 81

1. $1, -5$ 3. $3, 1$ 5. $2, -4/3$
7. $3/2, -3$ 9. $4, -2$
11. $x < -4$ or $x > 2$ 13. $x < -1/2$ or $x > 1$

15. $-1/3 < x < 1$

17. $(-\infty, -1], [7, \infty)$ 19. $(-7/2, 9/2)$ 21. no solution
23. $(-\infty, -7), (17, \infty)$ 27. $|x - 100| \leqslant 2$; $98 \leqslant x \leqslant 102$

EXERCISE SET 2.5, page 92

1. ± 3 3. $\pm \sqrt{5}$ 5. $-5/2 \pm \sqrt{2}$
7. $5/3 \pm 2\sqrt{2}/3$ 9. $\pm 8i/3$ 11. $1, 2$
13. $1, -2$ 15. $-2, -4$ 17. $0, 4$
19. $1/2, 2$ 21. ± 2 23. $1/3, 1/2$
25. $4, -2$ 27. $-1/2, 4$ 29. $1/3, -3$
31. $-1/2 \pm i/2$ 33. $1, -3/4$ 35. $-1/3 \pm \sqrt{2}i/3$
37. $0, -3/2$ 39. $2/5 \pm \sqrt{11}i/5$ 41. $2/5 \pm \sqrt{21}i/5$
43. $2/3, -1$ 45. $\pm 2\sqrt{3}/3$ 47. $0, -3/4$
49. $-1/2 \pm \sqrt{11}/2$ 51. $-2, 2/3$ 53. $-5/4 \pm \sqrt{7}i/4$
55. $\pm 1/2$ 57. $-1/4 \pm \sqrt{11}i/4, 0$ 59. $\pm \sqrt{c^2 - a^2}$
61. $\pm \sqrt{3V/\pi h}$ 63. $(-v \pm \sqrt{v^2 + 2gs})/g$
65. two complex roots   67. double real root   69. two real roots
71. two real roots   73. two complex roots   75. two complex roots
77. two real roots   79. double real root   81. 4
83. $0, -8$ 85. 4   87. 3
89. $0, 4$ 91. 5
93. $u = x^2$; $\pm i\sqrt{2}, \pm \sqrt{3}/3$ 95. $u = 1/x$; $2, -3/2$
97. $u = x^{1/5}$; $-32, -1/32$ 99. $u = 1 + 1/x$; $1/3, -2/7$

EXERCISE SET 2.6, page 96

1. A: 3 hours; B: 6 hours
3. roofer: 6 hours; assistant: 12 hours
5. $L = 12$ feet, $W = 4$ feet
7. $L = 8$ cm, $W = 6$ cm   9. 10 feet
11. 5 or 1/5   13. 3, 7; −3, −7
15. 6, 8   17. 150 shares   19. 8 days

EXERCISE SET 2.7, page 100

1. $x < -3, x > -2$ 3. $-1/2 < x < 1$ 5. $x < 0, x > 2$
7. $-5 < x < -3$ 9. $-1/2 \leqslant x < 3$ 11. $s \leqslant -2/3, s > 1/2$
13. $-2 < x < 2/3, x > 1$
15. $x < -3, x > 2$ 17. $-1 < x < 5/2$

   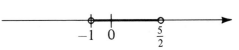

19. $-3/2 < x < 1/2$ 21. $x < -1, x \geqslant 1$

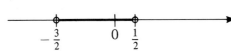   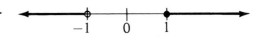

23. $x \leqslant -1, x \geqslant -1/3$ 25. $x \leqslant -2, 2 \leqslant x \leqslant 3$

   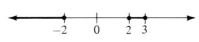

27. $-5/3 < x < -1/2, x > 3$

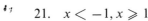

29. $x \leqslant -1, x \geqslant 2$ 31. $x \leqslant -2, x \geqslant -3/2$
33. $0 \leqslant x < 100$

REVIEW EXERCISES, page 102

1. 8/3   2. 0   3. 10/3
4. $k/2(2k + 1)$ 5. $10/3 \times 8/3$
6. 5 quarters, 14 dimes   7. 240 miles   8. 6 hours
9. F   10. F
11. $x \geqslant 1$ 12. $-9/2 \leqslant x < 5/2$

   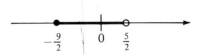

13. $(-\infty, 8)$ 14. $(5/2, \infty)$ 15. $[-9, \infty)$ 16. $5/3, -3$
17. $x = -1, x = 3/2$ 18. $x > 3$ or $x < -4$

19. $(1/5, 3/5)$
20. $(-\infty, -4/3], [8/3, \infty)$
21. $5, -4$
22. $1/2, 4/3$
23. $1 \pm i\sqrt{5}$
24. $(2 \pm i\sqrt{2})/2$
25. $-1, 1/3$
26. $\pm 3/7$
27. $\pm\sqrt{3\pi k}/k$
28. $-4, 3$
29. two real roots
30. double root
31. two complex roots
32. 4

33. 6
34. $\pm\left(\dfrac{-5 \pm \sqrt{85}}{6}\right)^{1/2}$
35. $-1/2, -1$

36. 60
37. $x \leqslant -3/2, x \geqslant 2$
38. $(-\infty, -5], [1, \infty)$
39. $(-\infty, -5), [-1/2, \infty)$
40. $(-2, -3/2), (3, \infty)$

PROGRESS TEST 2A, page 103

1. $3/4$
2. $8/13$
3. $18/5, 28/5, 29/5$ meters
4. $6000 at 6.5\%, $6200 at 7.5\%, $12,300 at 9\%
5. T
6. $-2 \leqslant x < 1$
7. $(-\infty, 15/2]$

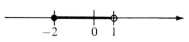
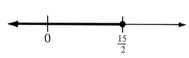

8. $[-4, 4]$

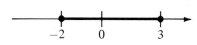

9. $5/2, -2$
10. $-2 \leqslant x \leqslant 3$

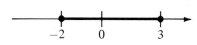

11. $(-4/3, 2)$
12. $-2, 7$
13. $(1 \pm i\sqrt{79})/10$
14. $-3/4, 1/3$
15. $(5 \pm 3i)/2$
16. $1, -3/2$
17. two real roots
18. two complex roots
19. -4
20. $\pm\sqrt{2}i, \pm\sqrt{3}/3$
21. 8×12 meters
22. $x \leqslant 1/3, x \geqslant 1$
23. $(-\infty, 1/2], [1, \infty)$
24. $[-2, 2/3], [1, \infty)$

PROGRESS TEST 2B, page 104

1. -1
2. $k^2/(k + 3)$
3. 30 ounces 60\%, 90 ounces 80\%
4. $b(d - 1)/(a - cd)$
5. F
6. $1 \leqslant x < 2$

7. $(-\infty, 4/5]$
8. $(3, \infty)$
9. $8/3, -2$

10. $x \leqslant 2, x \geqslant 6$

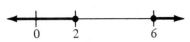

11. $(-\infty, -1), (1/5, \infty)$ 12. $-5/2, 1/3$ 13. $1/2, 2$
14. $(1 \pm i\sqrt{83})/6$ 15. $3 \pm i\sqrt{2}$ 16. $\pm\sqrt{k_2/(k_3 - k_1)}$
17.   two complex roots   18.   double real root   19.   1
20. $\pm i, \pm(-8)^{3/2}$ 21.   50¢   22. $x > 1/2$
23. $(-\infty, -5)$ 24. $(-\infty, -4), (2/3, 1)$

CHAPTER THREE

EXERCISE SET 3.1, page 114

1.

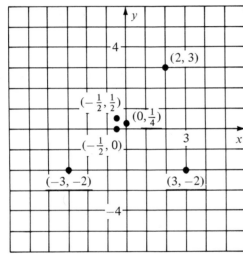

3. $3\sqrt{2}$ 5. $4\sqrt{2}$

7. $\sqrt{1345}/6$ 9. $\overline{BC} = \sqrt{37}$ 11. $\overline{RS} = \sqrt{2}/2$ 13.   no
15.   yes   21. $2\sqrt{10} + 7 + 5\sqrt{2} + \sqrt{37}$
25.   any point satisfying $x^2 + y^2 - 10y - 6x + 29 = 0$
27.   29.

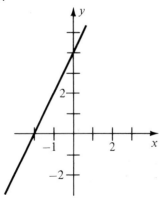

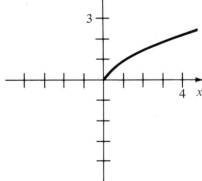

31.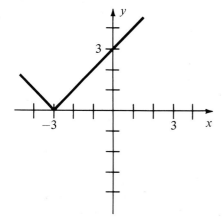

33.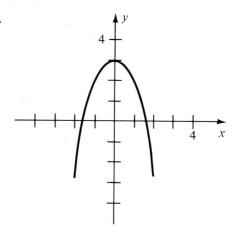

35.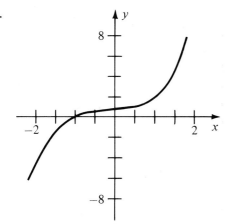

37.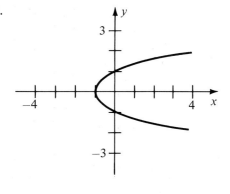

39. none   41. *x*-axis   43. *x*-axis   45. *x*-axis
47. none   49. *y*-axis   51. all   53. origin

EXERCISE SET 3.2, page 121

1. domain: all reals
   range: all reals

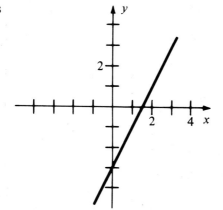

3. domain: all reals
    range: all reals

5. domain: $x \geqslant 1$
    range: $y \geqslant 0$

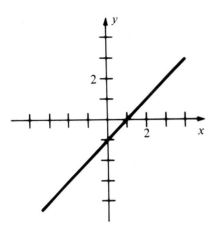

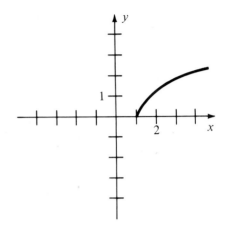

7. $x \geqslant 3/2$
9. $x > 2$
11. $x \geqslant 1, x \neq 2$
13. $7/2$
15. $3/2$
17. 5
19. $2a^2 + 5$
21. $6x^2 + 15$
23. 3
25. $1/(x^2 + 2x)$
27. $a^2 + h^2 + 2ah + 2a + 2h$
29. -0.92
31. $(3x - 1)/(x^2 + 1)$
33. $2(4x^2 + 1)/(6x - 1)$
35. -0.21
37. $2(a - 1)/(4a^2 + 4a - 3)$

EXERCISE SET 3.3, page 128

1. increasing: $(-\infty, \infty)$

3. increasing: $x \geqslant 0$
    decreasing: $x \leqslant 0$

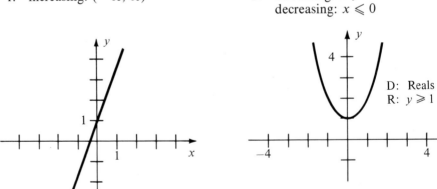

D: Reals
R: $y \geqslant 1$

5. increasing: $x \leqslant 0$
    decreasing: $x > 0$

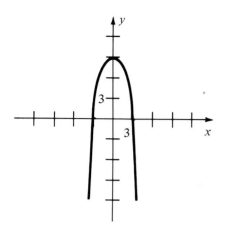

7. increasing: $x \geqslant 1/2$
    decreasing: $x \leqslant 1/2$

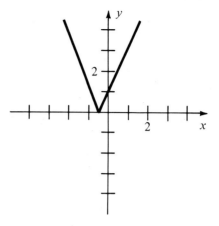

9. increasing: $x > -1$
    decreasing: $x < -1$

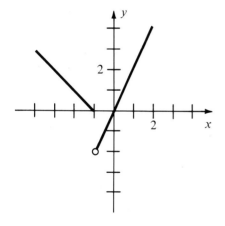

11. increasing: $x \leqslant 2$
     constant: $x \geqslant 2$

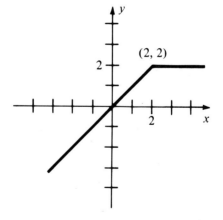

13.   increasing: $x \leqslant -2$
       decreasing: $0 \leqslant x < 1$
       constant: $1 \leqslant x \leqslant 2$

15.   constant: $x < -2, -2 \leqslant x \leqslant -1,$
       $x > -1$

increasing: $x \leqslant 0$
decreasing: $0 \leqslant x < 1, x > 2$
constant: $1 \leqslant x \leqslant 2$

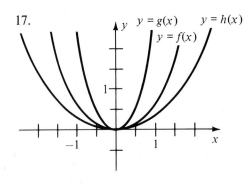

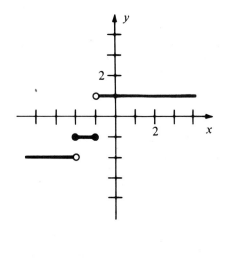

17.

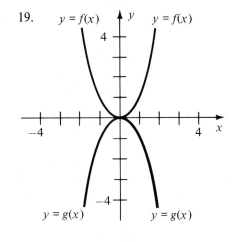

19.

21.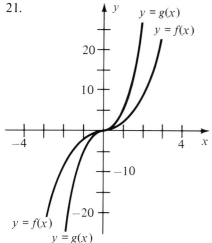

23.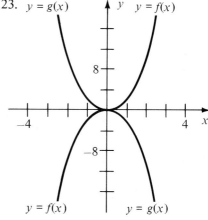

25.
$$C(u) = \begin{cases} 6.50, & 0 \leqslant u \leqslant 100 \\ 6.50 + 0.06(u - 100), & 100 < u \leqslant 200 \\ 12.50 + 0.05(u - 200), & u > 200 \end{cases}$$

27. $R(x) = \begin{cases} 30,000, & 0 \leqslant x \leqslant 100 \\ 400x - x^2, & 100 < x < 150 \end{cases}$

29. (a) $C(m) = 14 + 0.08m$
    (b) $m \geqslant 0$
    (c) $22

EXERCISE SET 3.4, page 137

1. 2; increasing   3. $-3/2$, decreasing   5. -1; decreasing
9. $2x - y + 5 = 0$ 11. $3x - y = 0$ 13. $2x - y = 0$
15. $2x - 3y = 0$ 17. $2x - y = 0$ 19. $3x - y + 2 = 0$
21. $y - 2 = 0$ 23. $x - 3y - 15 = 0$
25. $m = -3/4, b = 5/4$ 27. $m = 0, b = 4$
29. $m = -3/4, b = -1/2$ 31. (a) $y = 3$ (b) $x = -6$
33. (a) $y = 0$ (b) $x = -7$ 35. (a) $y = -9$ (b) $x = 9$
37. (a) -3 (b) $1/3$ 39. (a) $4/3$ (b) $-3/4$
41. (a) $3x + y - 6 = 0$ (b) $x - 3y + 8 = 0$
43. (a) $3x + 5y - 1 = 0$ (b) $5x - 3y + 21 = 0$

45. (a) $F = \dfrac{9}{5}C + 32$ (b) $68°F$ 47. $1,000,000

49. 5   51. $f(x) = 8x + 13$

EXERCISE SET 3.5, page 142

1. (a) 4   (b) $y = 4x$ (c)

x	2	3	4	6	8	12	20	30
y	8	12	16	24	32	48	80	120

3. (a) $-1/32$ (b) $-3/8$ 5. (a) $1/10$ (b) $5/2$
7. (a) -3 (b) $-1/4$ 9. (a) 512 (b) $512/125$
11. (a) $M = r^2/s^2$ (b) $36/25$ 13. (a) $T = 16pv^3/u^2$ (b) $2/3$
15. (a) 400 feet   (b) 5 seconds   17. $40/3$ ohms
19. (a) 800/9 candlepower   (b) 8 feet
21. 6   23. 120 candlepower/ft^2

EXERCISE SET 3.6, page 152

1. $x^2 + x - 1$ 3. $x^2 - x + 3$

5. $x^3 - 2x^2 + x - 2$ 7. $\dfrac{x^2 + 1}{x - 2}$

9. domain of f and of g: all reals   11. $4x^2 + 2x + 1$

13. 21   15. $4x^2 + 10x + 7$

17. $8x^2 - 6x + 1$ 19. $x + 6$

21. 29   23. all reals

25. $(f \circ g)(x) = x + 1; (g \circ f)(x) = x + 1$

27. $(f \circ g)(x) = (x - 1)/x, (g \circ f)(x) = -(x + 1)/x$

35. $f^{-1}(x) = (x - 3)/2$

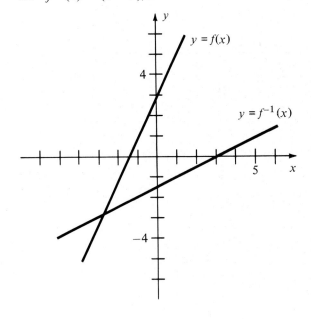

37. $f^{-1}(x) = (3 - x)/2$

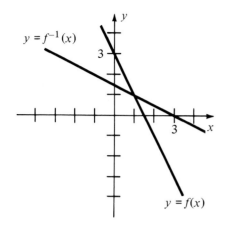

39. $f^{-1}(x) = 3x + 15$

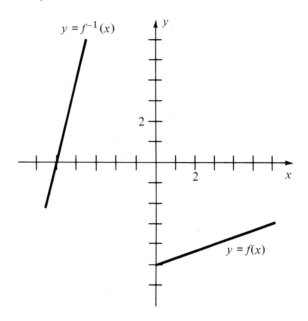

41. $f^{-1}(x) = (x - 1)^{1/3}$

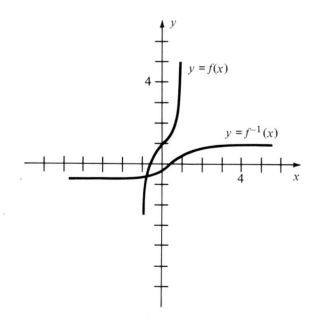

43. yes   45. no   47. yes
49. no   53. $(x - b)/a$

REVIEW EXERCISES, page 156

1. $\sqrt{61}$ 2. $\sqrt{65}$

3.   4.

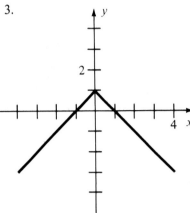

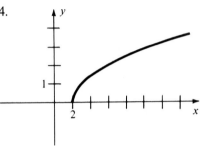

5. x-axis   6. all   7. yes   8. yes
9. $x \geqslant 5/3$ 10. $x \neq -1$ 11. 226   12. ± 3
13. 12   14. $y^2 - 3y + 2$ 15. $3 + h$

16.

17. increasing: $x \leqslant -1, 0 \leqslant x \leqslant 2$
     decreasing: $-1 < x \leqslant 0$
     constant: $x > 2$
18. -5 19. -2
20. 3   21. $y - 3x - 6 = 0$
22. $x = -4$ 23. $y = 3$
24. $y - 2x - 2 = 0$ 25. $y - 3x - 6 = 0$ 26. 160
27. 1   28. $-1/4$ 29. $x^2 + x$
30. 0   31. $(x + 1)/(x^2 - 1)$ 32. $x \neq \pm 1$

33. $x^2 + 2x$ 34. 4 35. $|x| - 2$
36. $|x| + 4 - 4\sqrt{x}$ 37. 0 38. not defined

40.

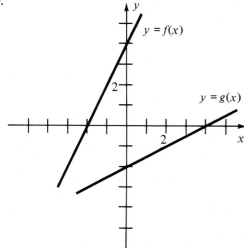

PROGRESS TEST 3A, page 157

1. $3 + \sqrt{26} + \sqrt{41}$ 2.

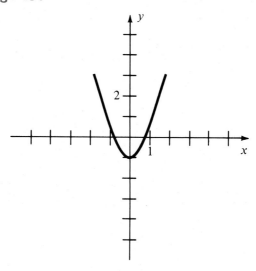

3. origin 4. $x \geqslant 0, x \neq 1$
5. 17 6. $8t^2 + 3$
7. increasing: $x > 0$
   decreasing: $-2 < x < 0$
   constant: $x < -2$
8. 0 9. 2 10. $2y - 3x - 19 = 0$
11. $x = -3$ 12. $m = 1/2; b = 2$ 13. $y = -1$
14. $3y + x - 7 = 0$ 15. -1024 16. 65,536
17. -3 18. $x^2(x - 1)$ 19. 1/4

PROGRESS TEST 3B, page 158

1. $6\sqrt{2}$

2.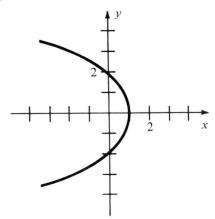

3. origin

4. $x \neq \pm 4$

5. 1

6. 1

7. increasing: $x > 3$
   decreasing: $x < -3$
   constant: $-3 < x < 3$

8. 10

9. 24

10. $-9/2$

11. $y = -5$

12. 1

13. -3

14. $x + 3y + 3 = 0$

15. $-32/9$

16. 1

17. 1

18. $\sqrt{2}/2$

19. $\sqrt{x}/x$

CHAPTER FOUR

EXERCISE SET 4.1, page 169

1.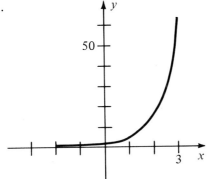

3.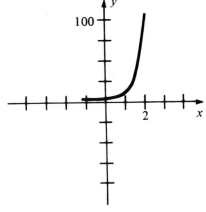

5.

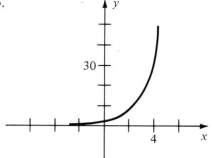

7.

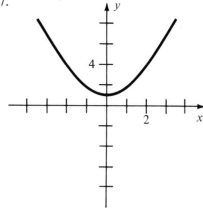

9.

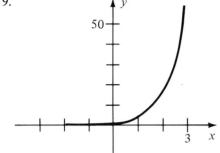

11.

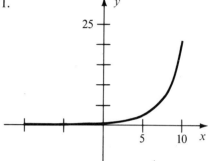

13. 3   15. 4   17. 2
19. 4   21. 2   23. 1
25. (a) 200   (b) 29,682   (c) 256.8, 543.7, 1478, 2436
27. 6.59 billion   29. 670.3 grams   31. $41,611
33. $45,417.50   35. $173.33   37. $2489

EXERCISE SET 4.2, page 176

1. $2^2 = 4$ 3. $9^{-2} = 1/81$ 5. $e^3 = 20.09$
7. $10^3 = 1000$ 9. $e^0 = 1$ 11. $3^{-3} = 1/27$
13. $\log_5 25 = 2$ 15. $\log_{10} 10,000 = 4$ 17. $\log_2 1/8 = -3$
19. $\log_2 1 = 0$ 21. $\log_{36} 6 = 1/2$ 23. $\log_{16} 64 = 3/2$
25. $\log_{27} 1/3 = -1/3$ 27. 25   29. 1/5
31. $e^2 \approx 7.39$ 33. $e^{-1/2} \approx 0.61$ 35. -2 37. 512
39. 124   41. 2   43. 3   45. 6
47. 2   49. 3   51. 1/2   53. 2
55. 1   57. 0   59. -2 61. 4
63. 2

65.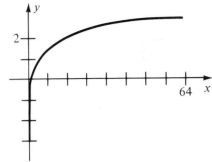

67.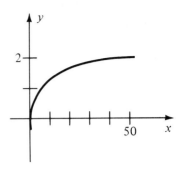

69.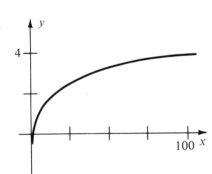

71.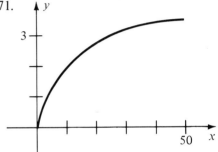

EXERCISE SET 4.3, page 180

1. $\log_{10} 120 + \log_{10} 36$

3. 4

5. $\log_a 2 + \log_a x + \log_a y$

7. $\log_a x - \log_a y - \log_a z$

9. $5 \ln x$

11. $2 \log_a x + 3 \log_a y$

13. $\frac{1}{2}(\log_a x + \log_a y)$

15. $2 \ln x + 3 \ln y + 4 \ln z$

17. $\frac{1}{2} \ln x + \frac{1}{3} \ln y$

19. $2 \log_a x + 3 \log_a y - 4 \log_a z$

21. 0.77

23. 0.94

25. 1.07

27. 0.87

29. 0.435

31. $\log x^2 \sqrt{y}$

33. $\ln \sqrt[3]{xy}$

35. $\log_a \dfrac{x^{1/3} y^2}{z^{3/2}}$

37. $\log_a \sqrt{xy}$

39. $\ln \dfrac{\sqrt[3]{x^2 y^4}}{z^3}$

41. $\log_a \dfrac{\sqrt{x-1}}{(x+1)^2}$

43. $\log_a \dfrac{x^3 (x+1)^{1/6}}{(x-1)^2}$

EXERCISE SET 4.4, page 186

1. 2.725×10^3

3. 8.4×10^{-3}

5. 7.16×10^5

7. 2.962×10^2

9. 0.5514

11. 1.5740

13. 1.5476

15. 1.8692

17. 4.6830

19. -0.4660

21. 2.520

23. 2.9

25. 7.9   27. 0.257   29. 0.000607
31. 0.0219   33. 1.028   35. 2.115
37. 103.55   39. 0.002875   41. 1.93×10^{-5}
43. 2.59×10^{-8} 45. $10,453   47. $14,660.72
49. 8.75% compounded quarterly

EXERCISE SET 4.5, page 189

1. $\log_5 18$ 3. $1 + \log_2 7$ 5. $(\log_3 46)/2$
7. $(5 + \log_5 564)/2$ 9. $(\log 2 + \log 3)/(\log 3 - 2\log 2)$
11. $-\log_2 15$ 13. $(1 - \log_4 12)/2$ 15. $\ln 18$
17. $(-3 + \ln 20)/2$ 19. 500   21. 1/2
23. 5   25. 3   27. 8
29. $-1 + \sqrt{17}$ 31. 36.62 years   33. 12.6 hours
35. 8.8 years   37. 27.47 days   39. 1.386 days

REVIEW EXERCISES, page 192

1.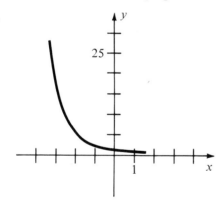

2. 3   3. 2
4. $12,750
5. $\log_9 27 = 3/2$
6. $8 = 64^{1/2}$
7. $1/8 = 2^{-3}$
8. $\log_6 1 = 0$ 9. 2
10. -2 11. e^{-4}
12. 26   13. 5
14. $-1/3$ 15. -1
16. 3

17.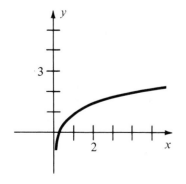

18. $\frac{1}{2}\log_a(x-1) - \log_a 2x$

19. $\log_a x + 2 \log_a(2 - x) - \dfrac{1}{2} \log_a(y + 1)$

20. $4 \ln(x + 1) + 2 \ln(y - 1)$

21. $\dfrac{2}{5} \log y + \dfrac{1}{5} \log z - \dfrac{1}{5} \log(z + 3)$

22. 1.15   23. 0.55   24. 0.4

25. −0.15   26. $\log_a \dfrac{\sqrt[3]{x}}{\sqrt{y}}$ 27. $\log(x^2 - x)^{4/3}$

28. $\ln \dfrac{3xy^2}{z}$ 29. $\log_a \dfrac{(x + 2)^2}{(x + 1)^{3/2}}$ 30. 4.765×10^2

31. 9.8×10^{-2} 32. 2.6475×10^4 33. 7.767×10^1
34. 803   35. 7.9   36. 3.5×10^{-4}
37. 11.5 hours   38. $(1 + \log_2 14)/3$ 39. $\sqrt{5000}$
40. 199/98

PROGRESS TEST 4A, page 193

1.

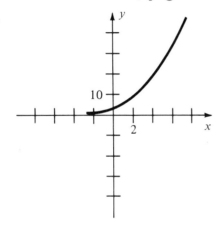

2. −2/3

3. $1/9 = 3^{-2}$

4. $\log_{16} 64 = 3/2$

5. 3   6. −1

7. 5/2   8. 1/2

9. $3 \log_a x - 2 \log_a y - \log_a z$

10. $2 \log x + \dfrac{1}{2} \log(2y - 1) - 3 \log y$

11. 0.7   12. 0.45   13. $\log \dfrac{x^2}{(y + 1)^3}$

14. $\log\left(\dfrac{x+3}{x-3}\right)^{2/3}$ 15. 2.73×10^{-4} 16. 5.972×10^0

17. 4.7×10^{-2} 18. 0.26   19. 34.6 hours

20. 200   21. 4

PROGRESS TEST 4B, page 193

1.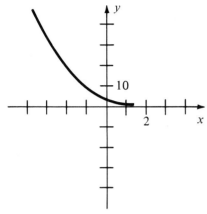

2. 8

3. $\log\dfrac{1}{1000} = -3$

4. $1 = 3^0$ 5. $3/2$

6. $3/2$ 7. $10\log_a 3$

8. 4

9. $\log_a(x-1) + \dfrac{5}{4}\log_a(y+3)$ 10. $\dfrac{1}{2}\ln x + \dfrac{1}{2}\ln y + \dfrac{1}{4}\ln 2z$

11. 0.85   12. 1.5   13. $\dfrac{1}{5}\ln\dfrac{(x-1)^3 y^2}{z}$

14. $\log\dfrac{x^2}{y^2}$ 15. 2.2684321×10^7

16. 2.97×10^{-1} 17. 9397   18. 6.2×10^{-4}

19. \$530.76   20. 3   21. 10

CHAPTER FIVE

EXERCISE SET 5.1, page 198

1. $(5/2, 5)$
3. $(1, 5/2)$
5. $(-7/2, -1)$
7. $(0, -1/2)$
9. $(-1, 9/2)$
11. $(0, 0)$

EXERCISE SET 5.2, page 202

1. $(x - 2)^2 + (y - 3)^2 = 4$
3. $(x + 2)^2 + (y + 3)^2 = 5$
5. $x^2 + y^2 = 9$
7. $(x + 1)^2 + (y - 4)^2 = 8$
9. $(h, k) = (2, 3); r = 4$
11. $(h, k) = (2, -2); r = 2$

13. $(h, k) = \left(-4, -\dfrac{3}{2}\right); r = 3\sqrt{2}$
15. no graph

17. $(x + 2)^2 + (y - 4)^4 = 16; (h, k) = (-2, 4); r = 4$

19. $\left(x - \dfrac{3}{2}\right)^2 + \left(y - \dfrac{5}{2}\right)^2 = \dfrac{11}{2}; (h, k) = \left(\dfrac{3}{2}, \dfrac{5}{2}\right); r = \dfrac{\sqrt{22}}{2}$

21. $(x - 1)^2 + y^2 = \dfrac{7}{2}; (h, k) = (1, 0); r = \dfrac{\sqrt{14}}{2}$

23. $(x - 2)^2 + (y + 3)^2 = 8; (h, k) = (2, -3); r = 2\sqrt{2}$
25. $(x - 3)^2 + (y + 4)^2 = 18; (h, k) = (3, -4); r = 3\sqrt{2}$

27. $\left(x + \dfrac{3}{2}\right)^2 + \left(y - \dfrac{5}{2}\right)^2 = \dfrac{3}{2}; (h, k) = \left(-\dfrac{3}{2}, \dfrac{5}{2}\right); r = \dfrac{\sqrt{6}}{2}$

29. $(x - 3)^2 + y^2 = 11; (h, k) = (3, 0); r = \sqrt{11}$

31. $\left(x - \dfrac{3}{2}\right)^2 + (y - 1)^2 = \dfrac{17}{4}; (h, k) = \left(\dfrac{3}{2}, 1\right); r = \dfrac{\sqrt{17}}{2}$

33. $(x + 2)^2 + \left(y - \dfrac{2}{3}\right)^2 = \dfrac{100}{9}; (h, k) = \left(-2, \dfrac{2}{3}\right); r = \dfrac{10}{3}$

35. not a circle

EXERCISE SET 5.3, page 206

1.
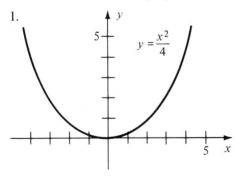

3.
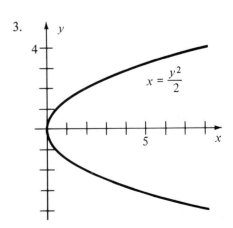

5.

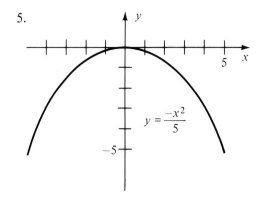

$$y = \frac{-x^2}{5}$$

7. Vertex: $(2, -1)$
    Axis: $x = 2$
    Direction: up

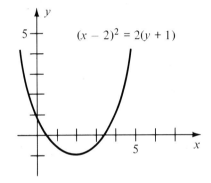

$(x - 2)^2 = 2(y + 1)$

9. Vertex: $(-4, -2)$
    Axis: $x = -4$
    Direction: down

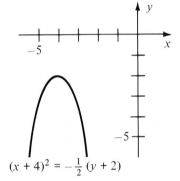

$(x + 4)^2 = -\frac{1}{2}(y + 2)$

11. Vertex: $(-1, 0)$
    Axis: $y = 0$
    Direction: left

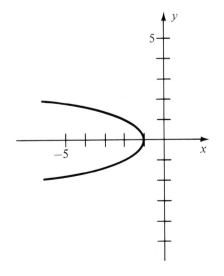

13. $V = (1, 2)$, $x = 1$, upward   15. $V = (2, 4)$, $y = 4$, opens left
17. $V = (1/2, -1/4)$, $x = 1/2$, downward
19. $V = (-1/3, 5)$, $y = 5$, opens right
21. $V = (3/2, -5/12)$, $x = 3/2$, upward
23. $V = (4, -3)$, $y = -3$, opens left
25. $V = (-1, -1)$, $x = -1$, downward

EXERCISE SET 5.4, page 213

1.

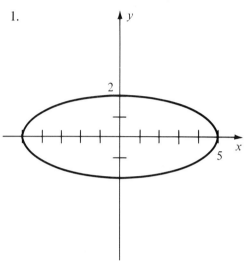

3.

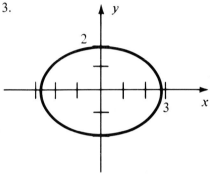

5.

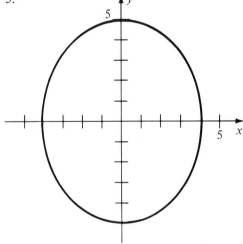

7. $\dfrac{x^2}{9} + \dfrac{y^2}{4} = 1$; $(0, \pm 2)$, $(\pm 3, 0)$

9. $\dfrac{x^2}{4} + \dfrac{y^2}{1} = 1$; $(0, \pm 1)$, $(\pm 2, 0)$

11. $\dfrac{x^2}{1} + \dfrac{y^2}{\frac{1}{4}} = 1$; $\left(0, \pm\dfrac{1}{2}\right), (\pm 1, 0)$

13. $\dfrac{x^2}{3} + \dfrac{y^2}{4} = 1$; $(0, \pm 2), (\pm\sqrt{3}, 0)$

15. $\dfrac{x^2}{\frac{1}{4}} + \dfrac{y^2}{\frac{9}{8}} = 1$; $\left(0, \pm\dfrac{3\sqrt{2}}{4}\right), \left(\pm\dfrac{1}{2}, 0\right)$

17.

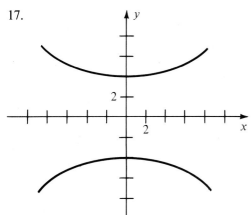

19.

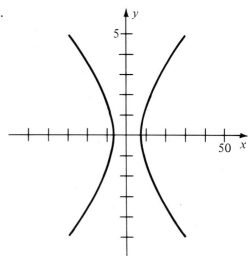

21.

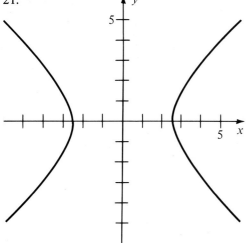

23. $\dfrac{x^2}{4} - \dfrac{y^2}{64} = 1$; $(\pm 2, 0)$

25. $\dfrac{y^2}{\frac{1}{4}} - \dfrac{x^2}{\frac{1}{4}} = 1$; $\left(0, \pm\dfrac{1}{2}\right)$

27. $\dfrac{x^2}{5} - \dfrac{y^2}{4} = 1$; $(\pm\sqrt{5}, 0)$

29.

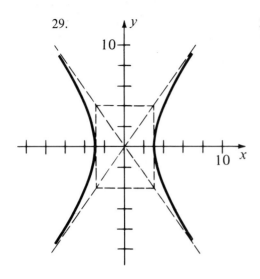

31.

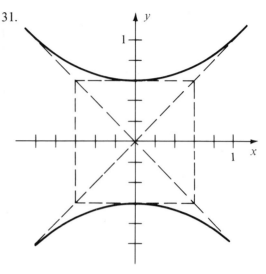

33.

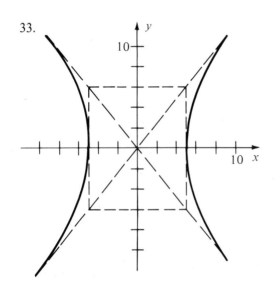

EXERCISE SET 5.5, page 216

1. parabola   3. circle   5. hyperbola   7. no graph
9. no graph   11. no graph   13. hyperbola   15. point

REVIEW EXERCISES, page 217

1. $(-1, -1)$ 2. $(-5/2, 5/2)$ 3. $(-1/2, -9/2)$
4. $(10, 7)$ 5. $\overline{P_1P_2} = \overline{P_3P_4} = \sqrt{26}, \overline{P_1P_4} = \overline{P_2P_3} = 5$
6. $\overline{AB} = \sqrt{170}, \overline{AC} = \sqrt{136}, \overline{BC} = \sqrt{34}, \overline{AB}^2 = \overline{AC}^2 + \overline{BC}^2$
7. $10x + 12y + 15 = 0$ 8. $(x + 5)^2 + (y - 2)^2 = 16$

9. $(x + 3)^2 + (y + 3)^2 = 4$
10. $(h, k) = (2, -3); r = 3$
11. $(h, k) = (-1/2, 4); r = 1/3$
12. $(h, k) = (-2, 3); r = \sqrt{3}$
13. $(h, k) = (1, -1); r = \sqrt{2}/2$
14. $(h, k) = (0, 3); r = \sqrt{6}$
15. $(h, k) = (1, 1); r = \sqrt{10}$

16. Vertex: $(3/2, -5)$
    Axis: $y = -5$

17. Vertex: $(1, 2)$
    Axis: $x = 1$

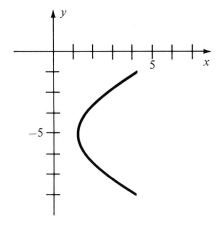

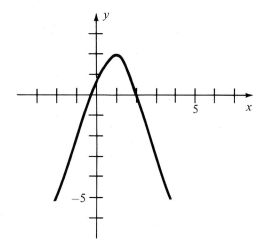

	Vertex	Axis	Direction
18.	$(-3, 0)$	$y = 0$	left
19.	$(2, -2)$	$y = -2$	left
20.	$(3, -2)$	$x = 3$	up
21.	$(-2, -1/2)$	$x = -2$	down
22.	$(0, 1)$	$y = 1$	right
23.	$(-3, 0)$	$x = -3$	down

24. $\dfrac{x^2}{4} - \dfrac{y^2}{9} = 1; (\pm 2, 0)$

25. $\dfrac{x^2}{1} + \dfrac{y^2}{9} = 1; (\pm 1, 0), (0, \pm 3)$

26. $\dfrac{x^2}{7} + \dfrac{y^2}{5} = 1; (\pm \sqrt{7}, 0), (0, \pm \sqrt{5})$

27. $\dfrac{x^2}{16} - \dfrac{y^2}{9} = 1; (\pm 4, 0)$

28. $\dfrac{x^2}{3} + \dfrac{y^2}{\frac{9}{4}} = 1; (\pm \sqrt{3}, 0), (0, \pm 3/2)$

29. $\dfrac{y^2}{\frac{20}{3}} - \dfrac{x^2}{4} = 1; (0, \pm 2\sqrt{15}/3)$

30.

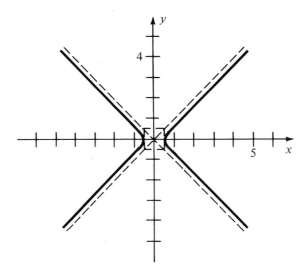

31.

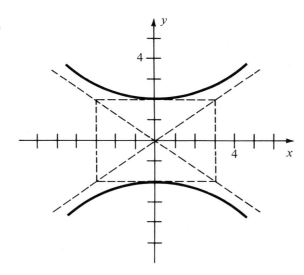

32. parabola   33. hyperbola   34. ellipse   35. no graph

PROGRESS TEST 5A, page 217

1. $(0, 4)$ 2. $(-1, 2)$ 3. slope AB = slope CD = $-5/2$;
                                                                         slope BC = slope AD = $2/7$

4. $(x - 2)^2 + (y + 3)^2 = 36$ 5. $(h, k) = (1, -2)$; $r = 2$
6. $(h, k) = (2, 0)$; $r = \sqrt{5}$

7. Vertex: $(-3, 1)$
   Axis: $x = -3$

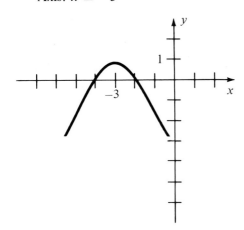

8. Vertex: $(1, 2)$
   Axis: $y = 2$

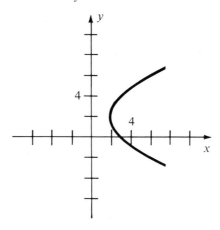

9. vertex: $(3, 2)$, axis: $x = 3$, direction: down
10. vertex: $(-2, -4)$, axis: $y = -4$, direction: right

11. $\dfrac{x^2}{4} + \dfrac{y^2}{1} = 1$; $(\pm 2, 0), (0, \pm 1)$

12. $\dfrac{y^2}{9} - \dfrac{x^2}{4} = 1$; $(0, \pm 3)$

13. $\dfrac{x^2}{\frac{1}{4}} - \dfrac{y^2}{\frac{1}{4}} = 1$; $(\pm 1/2, 0)$

14.

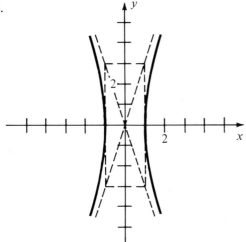

15. circle
16. ellipse

PROGRESS TEST 5B, page 218

1. $(-1/2, -1)$

2. $(-7, 1)$

3. $PR = QS = \sqrt{65}$

4. $(x + 2)^2 + (y + 5)^2 = 25$

5. $(h, k) = (-3, 2); r = 3$

6. $(h, k) = (1/2, 1); r = \sqrt{10}$

7. Vertex: $(22, 2)$
   Axis: $y = 2$
   Direction: right

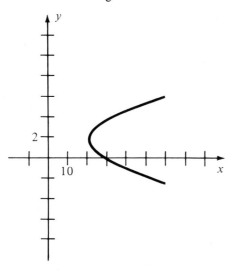

8. Vertex: $(1/3, -1/3)$
   Axis: $x = 1/3$
   Direction: down

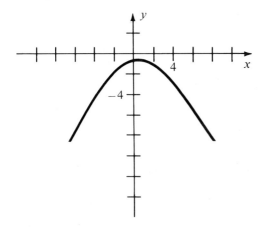

9. vertex: $(-1, 2)$, axis: $y = 2$, direction: right

10. vertex: $(1/2, -3)$, axis: $x = 1/2$, direction: up

11. $\dfrac{x^2}{5} + \dfrac{y^2}{\frac{25}{9}} = 1$; $(\pm\sqrt{5}, 0)$, $(0, \pm 5/3)$

12. $\dfrac{x^2}{3} + \dfrac{y^2}{\frac{7}{2}} = 1$; $(\pm\sqrt{3}, 0)$, $(0, \pm\sqrt{14}/2)$ 13. $\dfrac{y^2}{9} - \dfrac{x^2}{3} = 1$; $(0, \pm 3)$

14.

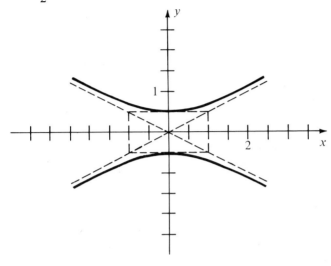

15. hyperbola
16. parabola

CHAPTER SIX

EXERCISE SET 6.1, page 224

1. $x = 2, y = -1$ 3. none   5. $x = 1, y = -1$
7. all points on the line $2x + 5y = -6$
9. $x = 1, y = -3$ 11. $x = 3/2, y = 1/2$
13. $x = 1, y = 1/2$ 15. none
17. all points on the line $3x - y = 18$

EXERCISE SET 6.2, page 228

1. $x = 1, y = -2$ 3. $x = 1, y = -1/2$
5. all points on the line $x + 2y = 6$
7. $x = -4/7, y = 6/7$
9. no solution   11. $x = 3, y = -1$
13. $x = 39/4, y = 1/4$ 15. 22 nickels, 12 quarters
17. 6/5 pounds nuts, 4/5 pounds raisins
19. 125 type-A, 75 type-B

EXERCISE SET 6.3, page 234

1. 25 nickels, 15 dimes
3. color: $2.50, black and white: $1.50
5. $4000 in bond A, $2000 in bond B
7. 10 rolls of 12″, 4 rolls of 15″
9. 8 pounds of $1.20 coffee, 16 pounds of $1.80 coffee
11. speed of bicycle: 105/8 mph, wind speed: 15/8 mph
13. 34   15.   30 pounds of nuts, 20 pounds of raisins
17. $6000 in type A, $12,000 in type B
19. 5 units Epiline I, 4 units Epiline II
21. (a) $R = 95x$

(b)

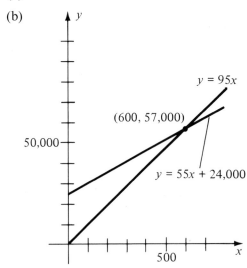

(c) $57,000
23. (a) $p = 4$ (b) 18

EXERCISE SET 6.4, page 240

1. $x = 2, y = -1, z = -2$ 3. $x = 1, y = 2/3, z = -2/3$
5. no solution   7. $x = 1, y = 2, z = 2$
9. $x = 1, y = 1, z = 0$ 11. $x = 1, y = 27/2, z = -5/2$
13. no solution   15. no solution
17. $x = 5, y = -5, z = -20$ 19. A: 2, B: 3, C: 3
21. three 12″-sets, eight 16″-sets, five 19″-sets

EXERCISE SET 6.5, page 244

1. $x = 3, y = 2; x = 1/5, y = -18/5$
3. $x = 1, y = 1; x = 9/16, y = -3/4$
5. $x = 1, y = 2; x = 13/5, y = -6/5$
7. $x = \dfrac{-1 + \sqrt{5}}{2}, y = \dfrac{1 + \sqrt{5}}{2}; x = \dfrac{-1 - \sqrt{5}}{2}, y = \dfrac{1 - \sqrt{5}}{2}$
9. $x = 3, y = 2; x = 3, y = -2$
11. $x = 3, y = 2; x = -3, y = 2; x = 3, y = -2; x = -3, y = -2$
13. no solution

15. $x = \sqrt{2}, y = 5; x = -\sqrt{2}, y = 5; x = \sqrt{2}, y = -5; x = -\sqrt{2}, y = -5$
17. $x = 1, y = -1; x = 5/2, y = 1/2$
19. 6 and 8   21. 4 and 5

EXERCISE SET 6.6, page 252

1.
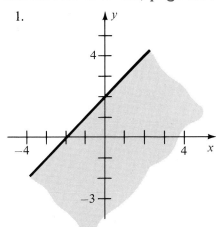

3.
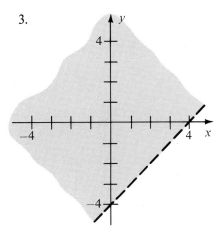

5.
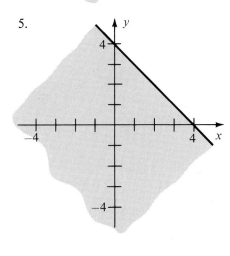

7.
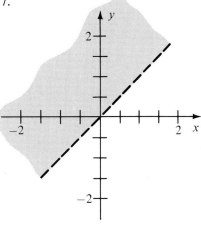

9.
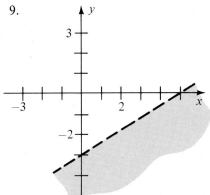

11.
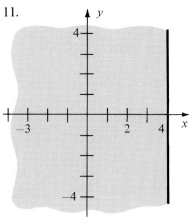

13.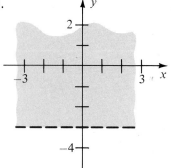

15.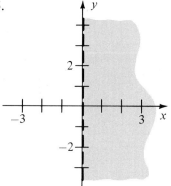

17.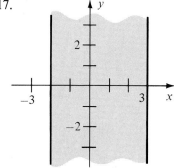

19. $2x + 5y \leqslant 15$; $x \geqslant 0$; $y \geqslant 0$

21.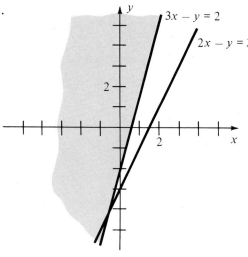

23.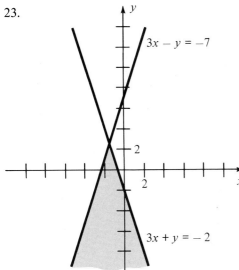

25.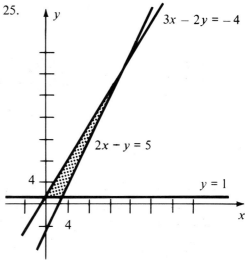

27.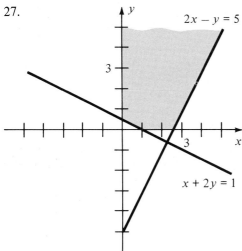

29. no solution

31.
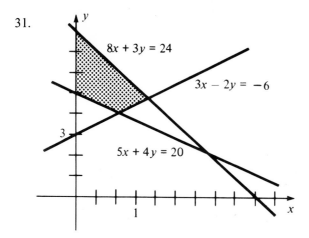

33.
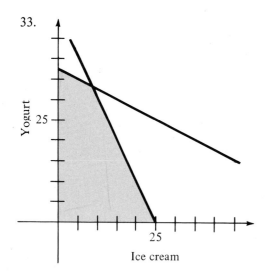

35.
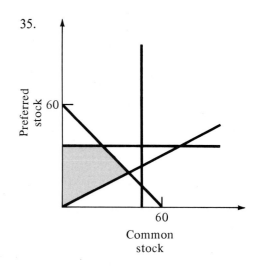

REVIEW EXERCISES, page 255

1. $x = -1/2, y = 1$
2. $x = 1/2, y = -2$
3. $x = 5, y = -1$
4. $x = -4, y = 3/2$
5. $x = -7/20, y = -7/10$
6. $x = -3, y = 1/2$
7. $x = 0, y = -2$
8. $x = -5, y = 1$
9. $x = -3, y = 5$
10. $x = 2, y = -2$
11. $x = 4, y = -1$
12. $x = -2, y = 3$
13. $x = 1/4, y = -1/2$
14. $x = 3, y = -4$
15. 45
16. 72
17. steak: 13/4 lb, hamburger: 9/5 lb
18. 600 kph
19. 3, 11
20. 575, \$9200
21. $x = -3, y = 1, z = 4$
22. $x = -2, y = 1/2, z = 3$
23. $x = 1, y = -1, z = 2$
24. $x = 3, y = 1/4, z = -1/3$
25. $x = -3, y = 4$
26. $x = -5/3, y = 5/6$
27. $x = -2, y = -1, z = -3$
28. $x = 1/2, y = -1, z = 1$
29. $x = 5, y = 0; x = -4, y = 3$
30. $x = 1, y = -1; x = 5, y = 3$

31. none

32. $x = 0, y = 3$

33. $x = 4, y = 4; x = 36/25, y = -12/5$ 34. $x = 5, y = 2; x = 10, y = -3$

35. $x = 0, y = -4$

36.

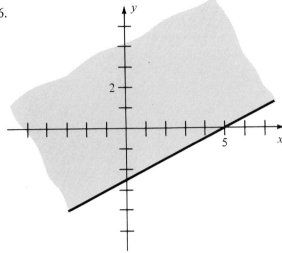

37.

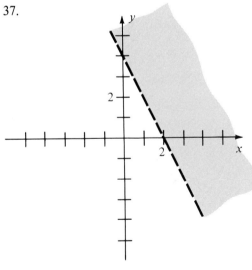

38.

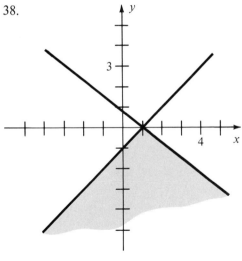

39.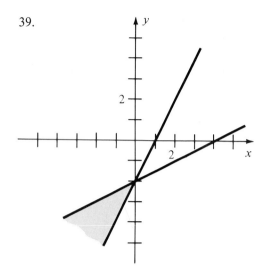

40.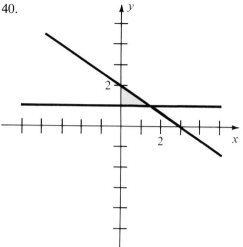

41.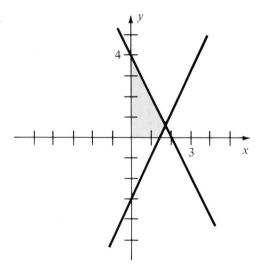

PROGRESS TEST 6A, page 257

1. $x = -5, y = 2$
2. $x = -1, y = 6$
3. $x = 4, y = 0$
4. $x = 1, y = -3$
5. $x = 2, y = -1$
6. 38
7. shirts: $15, ties: $10
8. 1100
9. $x = -2, y = 4, z = 6$
10. $x = -1/3, y = -1$
11. $x = 2/3, y = 2, z = -2$
12. $x = 3, y = \pm 4; \; x = -3, y = \pm 4$
13. $x = 2, y = \pm \sqrt{10}; \; x = 3, y = \pm \sqrt{15}$

14.
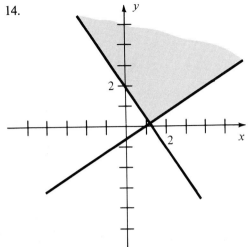

15.
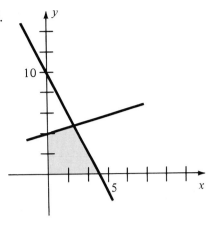

PROGRESS TEST 6B, page 258

1. $x = -3, y = 2/3$
2. $x = 2/3, y = -1$
3. $x = 1/5, y = -1/5$
4. $x = -7/2, y = -1/2$
5. $x = 1, y = -3$
6. 86
7. 5/2 kph
8. 6, 31
9. $x = -3, y = -10, z = 5$
10. $x = 1/2, y = -1/4$
11. $x = 2, y = -3, z = 1$
12. $x = 0, y = 2; x = 3, y = 1$
13. $x = 5, y = \pm 4; x = -5, y = \pm 4$

14.
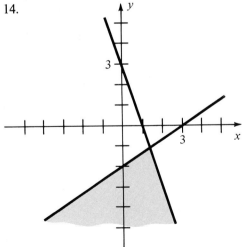

15.
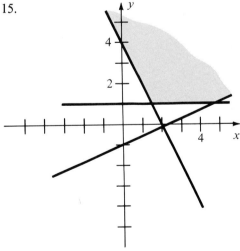

CHAPTER SEVEN

EXERCISE SET 7.1, page 267

1. 2×2 3. 4×3 5. 3×3

7. (a) -4 (b) 7 (c) 6 (d) -3

9. $\begin{bmatrix} 3 & -2 \\ 5 & 1 \end{bmatrix}$, $\left[\begin{array}{cc|c} 3 & -2 & 12 \\ 5 & 1 & -8 \end{array}\right]$

11. $\begin{bmatrix} \frac{1}{2} & 1 & 1 \\ 2 & -1 & -4 \\ 4 & 2 & -3 \end{bmatrix}$, $\left[\begin{array}{ccc|c} \frac{1}{2} & 1 & 1 & 4 \\ 2 & -1 & -4 & 6 \\ 4 & 2 & -3 & 8 \end{array}\right]$

13. $\frac{3}{2}x + 6y = -1$
    $4x + 5y = 3$

15. $x + y + 3z = -4$
    $-3x + 4y = 8$
    $2x + 7z = 6$

17. $x = -13, y = 8, z = 2$
19. $x = 35, y = 14, z = -4$
21. $x = 2, y = 3$
23. $x = 2, y = -1, z = 3$
25. $x = 3, y = 2, z = -1$
27. $x = -5, y = 2, z = 3$
29. $x = -5/7, y = -2/7, z = -3/7, w = 2/7$

EXERCISE SET 7.2, page 274

1. $a = 3, b = -4, c = 6, d = -2$

3. $\begin{bmatrix} 2 & -1 & 5 \\ 7 & 1 & 6 \\ 4 & 3 & 7 \end{bmatrix}$

5. $\begin{bmatrix} -2 & 18 & 8 \\ -3 & 8 & 11 \end{bmatrix}$

7. not possible

9. $\begin{bmatrix} 17 & 5 \\ 10 & 12 \end{bmatrix}$

11. not possible

13. $\begin{bmatrix} 10 & 4 \\ 12 & 28 \end{bmatrix}$

15. not possible

17. $\begin{bmatrix} 18 & 23 & 29 \\ 17 & -12 & 13 \end{bmatrix}$

19. $AB = \begin{bmatrix} 8 & -6 \\ -8 & 6 \end{bmatrix}$ $AC = \begin{bmatrix} 8 & -6 \\ -8 & 6 \end{bmatrix}$

25. The amount of pesticide 2 eaten by herbivore 3.

27. $A = \begin{bmatrix} 3 & 4 \\ 3 & -1 \end{bmatrix}$, $X = \begin{bmatrix} x \\ y \end{bmatrix}$, $B = \begin{bmatrix} -3 \\ 5 \end{bmatrix}$

29. $A = \begin{bmatrix} 3 & -1 & 4 \\ 2 & 2 & 3/4 \\ 1 & -1/4 & 1 \end{bmatrix}$, $X = \begin{bmatrix} x \\ y \\ z \end{bmatrix}$, $B = \begin{bmatrix} 5 \\ -1 \\ 1/2 \end{bmatrix}$

31. $x_1 - 5x_2 = 0$
    $4x_1 + 3x_2 = 2$

33. $4x_1 + 5x_2 - 2x_3 = 2$
    $3x_2 - x_3 = -5$
    $2x_3 = 4$

EXERCISE SET 7.3, page 284

1. no   3. yes   5. $\begin{bmatrix} \frac{2}{3} & \frac{5}{6} \\ \frac{1}{3} & \frac{1}{6} \end{bmatrix}$

7. $\begin{bmatrix} 1 & -1 \\ 2 & -1 \end{bmatrix}$ 9. $\begin{bmatrix} 8 & 7 & -1 \\ -4 & -4 & 1 \\ -5 & -5 & 1 \end{bmatrix}$ 11. $\begin{bmatrix} \frac{4}{7} & -\frac{3}{7} \\ \frac{1}{7} & \frac{1}{7} \end{bmatrix}$

13. none   15. $\begin{bmatrix} \frac{1}{2} & 0 \\ 0 & -\frac{1}{3} \end{bmatrix}$ 17. $\begin{bmatrix} -1 & 1 & -1 \\ 2 & -1 & 2 \\ -2 & 1 & -1 \end{bmatrix}$

19. $x = 3, y = -1$ 21. $x = 2, y = -3, z = 1$
23. $x = 0, y = 2, z = -3$ 35. $x = \begin{bmatrix} 25 \\ 3 \\ 1 \end{bmatrix}, \begin{bmatrix} -2 \\ 5 \\ 10 \end{bmatrix}$

EXERCISE SET 7.4, page 294

1. 22   3. -8 5. 0
7. (a) -6 (b) -1 (c) 1   (d) 7
9. (a) -6 (b) 1   (c) 1   (d) 7
11. 52   13. -3 15. 0
17. -12 19. 0
21. $x = 1, y = -2, z = -1$
23. $x = 3, y = 2, z = -1$
25. $x = -3, y = 0, z = 2$
27. $x = -5/7, y = -2/7, z = -3/7, w = 2/7$

REVIEW EXERCISES, page 297

1. 3×5 2. -1 3. 4   4. 8

5. $\begin{bmatrix} 3 & -7 \\ 1 & 4 \end{bmatrix}$ 6. $\begin{bmatrix} 3 & -7 & | & 14 \\ 1 & 4 & | & 6 \end{bmatrix}$

7. $4x - y = 3$ 8. $-2x + 4y + 5z = 0$
   $2x + 5y = 0$ $6x - 9y + 4z = 0$
                            $3x + 2y - z = 0$

9. $x = -1, y = -4$ 10. $x = 1/2, y = 5$
11. $x = -4, y = 3, z = -1$ 12. $x = -1, y = 1, z = -3$
13. $x = 1/2, y = 3/2$ 14. $x = -5, y = 2$
15. $x = 3, y = 1/3, z = -2$
16. $x = 3 + 5t/4, y = 3 + t/2, z = t$

17. -3 18. -3 19. $\begin{bmatrix} 1 & 4 \\ 7 & -1 \end{bmatrix}$

20. $\begin{bmatrix} -3 & 6 \\ 1 & -5 \end{bmatrix}$ 21. not possible   22. $\begin{bmatrix} 5 & 15 & 20 \\ -5 & 0 & -30 \end{bmatrix}$

23. $\begin{bmatrix} -1 & -3 & -4 \\ -4 & 0 & -24 \\ 4 & 6 & 20 \end{bmatrix}$

24. $\begin{bmatrix} 7 & 4 \\ -11 & 12 \end{bmatrix}$

25. not possible

26. $\begin{bmatrix} 1 & -5 \\ 16 & -12 \\ -10 & 16 \end{bmatrix}$

27. $\begin{bmatrix} 0 & 9 \\ 11 & -4 \end{bmatrix}$

28. $\begin{bmatrix} 6 & -13 \\ -5 & -9 \end{bmatrix}$

29. $\begin{bmatrix} -\frac{4}{11} & \frac{3}{11} \\ \frac{1}{11} & \frac{2}{11} \end{bmatrix}$

30. $\begin{bmatrix} \frac{2}{5} & -\frac{7}{5} & -\frac{8}{5} \\ -1 & 3 & 4 \\ -\frac{2}{5} & \frac{2}{5} & \frac{3}{5} \end{bmatrix}$

31. $x = 2, y = 3$

32. $x = -1, y = -1, z = 1/2$

33. 10

34. -6

35. 0

36. 12

37. 0

38. -3

39. $x = 1/2, y = 4$

40. $x = 1, y = -4$

41. $x = 10, y = -4$

42. $x = -4, y = 2, z = 1$

43. $x = 1/3, y = 2/3, z = -1$

44. $x = 1/4, y = -2, z = 1/2$

PROGRESS TEST 7A, page 299

1. 3×2

2. 0

3. $\begin{bmatrix} -7 & 0 & 6 & | & 3 \\ 0 & 2 & -1 & | & 10 \\ 1 & -1 & 1 & | & 5 \end{bmatrix}$

4. $-5x + 2y = 4$
   $3x - 4y = 4$

5. $x = -1/2, y = 1/2$

6. $x = -34/3, y = -2/3$

7. $x = 1/2, y = 1/2, z = 1/2$

8. 3

9. $\begin{bmatrix} 2 & 14 \\ -2 & -4 \\ -5 & 1 \end{bmatrix}$

10. $\begin{bmatrix} -7 & -11 \\ 11 & 15 \end{bmatrix}$

11. $\begin{bmatrix} -10 \\ 2 \\ 0 \end{bmatrix}$

12. not possible

13. $\begin{bmatrix} \frac{1}{27} & \frac{12}{27} & \frac{4}{27} \\ \frac{5}{27} & \frac{6}{27} & -\frac{7}{27} \\ \frac{7}{27} & \frac{3}{27} & \frac{1}{27} \end{bmatrix}$

14. $x = -2, y = 1$

15. -2

16. 27

17. $x = 4, y = -3$

PROGRESS TEST 7B, page 300

1. 2×4

2. 1

3. $\begin{bmatrix} 2 & -6 & | & 5 \\ 1 & 3 & | & -2 \end{bmatrix}$

4. $16x \qquad + 6z = 10$
   $-4x - 2y + 5z = 8$
   $2x + 3y - z = -6$

5. $x = 0, y = -1, z = -3$

6. $x = -1, y = -3$

7. $x = 5, y = -2, z = 1/2$

8. -5

9. 10, 0

10. $\begin{bmatrix} 6 & -17 \\ 3 & 2 \\ 2 & 13 \end{bmatrix}$

11. not possible

12. not possible

13. $\begin{bmatrix} -\frac{3}{2} & -\frac{5}{2} & -\frac{3}{4} \\ -\frac{1}{2} & -\frac{1}{2} & -\frac{1}{4} \\ -\frac{3}{4} & -\frac{5}{4} & -\frac{1}{8} \end{bmatrix}$

14. $x = -1, y = 1/2$

15. -22

16. -1

17. $x = -3/2, y = 1/2$

CHAPTER EIGHT

EXERCISE SET 8.1, page 306

1. $Q(x) = x - 2, R(x) = 2/(x - 5)$
3. $Q(x) = 2x - 4, R(x) = (8x - 4)/(x^2 + 2x - 1)$
5. $Q(x) = 3x^3 - 9x^2 + 25x - 75, R(x) = 226/(x + 3)$
7. $Q(x) = 2x - 3, R(x) = (-4x + 6)/(x^2 + 2)$
9. $Q(x) = x^2 - x + 1, R(x) = 0$
11. $Q(x) = x^2 - 3x, R(x) = 5(x + 2)$
13. $Q(x) = x^3 + 3x^2 + 9x + 27, R(x) = 0$
15. $Q(x) = 3x^2 - 4x + 4, R(x) = 4/(x + 1)$
17. $Q(x) = x^4 - 2x^3 + 4x^2 - 8x + 16, R(x) = 0$
19. $Q(x) = 6x^3 + 18x^2 + 53x + 159, R(x) = 481/(x - 3)$

EXERCISE SET 8.2, page 310

1. -7 3. -34 5. 0   7. -1
9. 0   11. -62

13.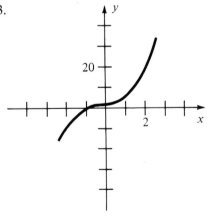

15.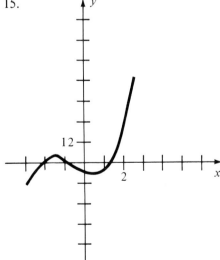

17.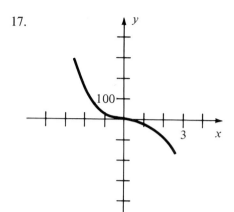

19. yes   21. no
23. yes   25. yes
27. $r = 3, -1$ 29. $5/2$

EXERCISE SET 8.3, page 318

1. 5
3. 25
5. 20
7. $-13/10 + 11i/10$
9. $-7/25 - 24i/25$
11. $8/5 - i/5$
13. $5/3 - 2i/3$
15. $4/5 + 8i/5$
17. $4/25 - 3i/25$
19. $9/10 + 3i/10$
21. $0 + i/5$
25. $x^3 - 2x^2 - 16x + 32$
27. $x^3 + 6x^2 + 11x + 6$
29. $x^3 - 6x^2 + 6x + 8$
31. $x^3/3 + x^2/3 - 7x/12 + 1/6$
33. $x^3 - 4x^2 - 2x + 8$
35. $3, -1, 2$
37. $-2, 4, -4$
39. $-2, -1, 0, -1/2$
41. $5, 5, 5, -5, -5$
43. $x^3 + 6x^2 + 12x + 8$
45. $4x^4 + 4x^3 - 3x^2 - 2x + 1$
47. $2, -1$
49. $(3 \pm i\sqrt{3})/2$
51. $-1, -2, 4$
53. $x^2 + (1 - 3i)x - (2 + 6i)$
55. $x^2 - 3x + (3 + i)$
57. $x^3 + (1 + 2i)x^2 + (-8 + 8i)x + (-12 + 8i)$
59. $(x^2 - 6x + 10)(x - 1)$
61. $(x^2 + 2x + 5)(x^2 + 2x + 4)$
63. $(x - 2)(x - 3)(x + 2)(x^2 + 6x + 10)$
65. $x - (a + bi)$

EXERCISE SET 8.4, page 324

	positive roots	negative roots	complex roots
1.	3	1	0
	1	1	2
3.	0	0	6
5.	3	2	0
	1	2	2
	3	0	2
	1	0	4
7.	1	2	0
	1	0	2
9.	2	0	2
	0	0	4
11.	1	1	6

13. 1, −2, 3
15. 2, −1, −1/2, 2/3
17. 1, −1, −1, 1/5
19. 1, −3/4
21. 3, 3, 1/2
23. −1, 3/4, ±i
25. 3/5, ±2, ±i√2
27. 0, 1/2, 2/3, −1
29. 1/2, −4, 2 ± √2
31. $k = 3, r = -2$
33. $k = 7, r = 1$

EXERCISE SET 8.5, page 335

1. $x \neq 1$

3. $x \neq 0, 2$

5. all real numbers

7. $x = 4, y = 0$

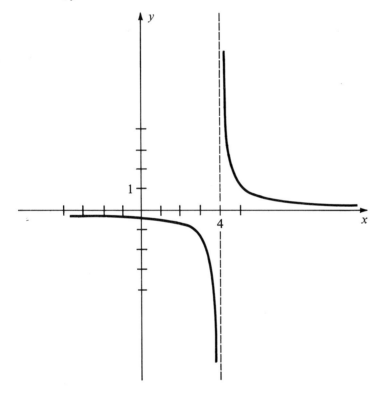

9. $x = -2, y = 0$

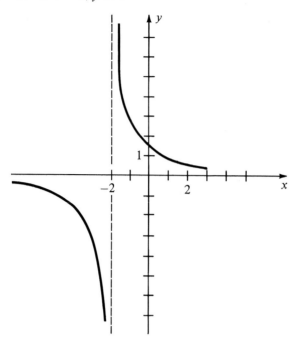

11. $x = -1, y = 0$

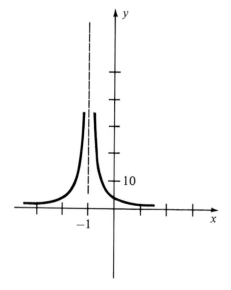

13. $x = 2, y = 1$

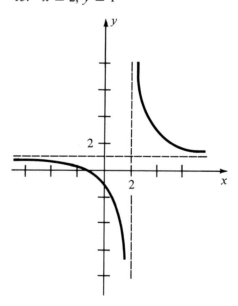

15. $x = 2, x = -2, y = 2$

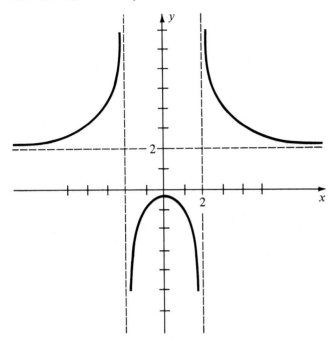

17. $x = 2, x = -3/2, y = 1/2$

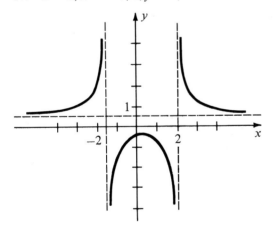

19. $x = 1$

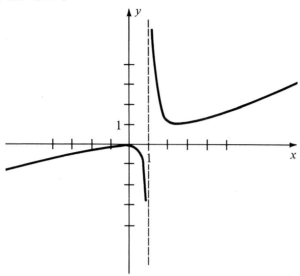

21. $x = 5, x = -5$

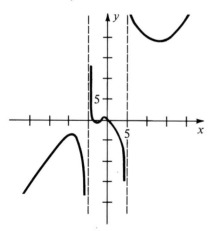

23. $x \neq -2$

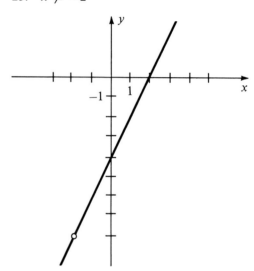

25. $x \neq 2$

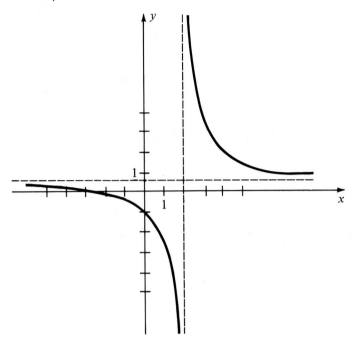

27. $x \neq 0, x \neq -1$

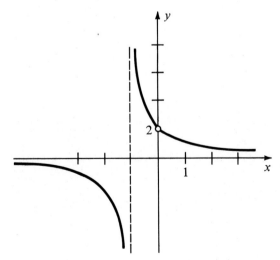

REVIEW EXERCISES, page 337

1. $Q(x) = 2x^2 + 2x + 8$, $R(x) = 4/(x - 1)$
2. $Q(x) = x^3 - 5x^2 + 10x - 18$, $R(x) = 31/(x + 2)$
3. $46, -8$ 4. $4, 1$ 7. $6/25 - 17i/25$
8. $-1/5 + 2i/5$ 9. $-5/2 + 5i/2$ 10. $1/10 - 3i/10$
11. $i/4$ 12. $2/29 + 5i/29$
13. $x^3 + 6x^2 + 11x + 6$ 14. $x^3 - 3x^2 + 3x - 9$
15. $x^4 + x^3 - 5x^2 - 3x + 6$ 16. $4x^4 + 4x^3 - 3x^2 - 2x + 1$
17. $x^4 + 2x^2 + 1$ 18. $x^4 - 6x^2 - 8x - 3$
19. $-\frac{1}{2}, 3$ 20. $-1 \pm \sqrt{2}$
21. $4, -1/2, 2 - i$ 22. 1 positive, 1 negative
23. 5 positive, 0 negative   24. 1 positive, 0 negative
25. 2 positive, 2 negative   26. $3, -2/3, -3/2$
27. $1, -2, 2/3, 3/2$ 28. none
29. $-1, (-9 \pm \sqrt{321})/12$ 30. $2, 3/2, -1 \pm \sqrt{2}$

31.   32.

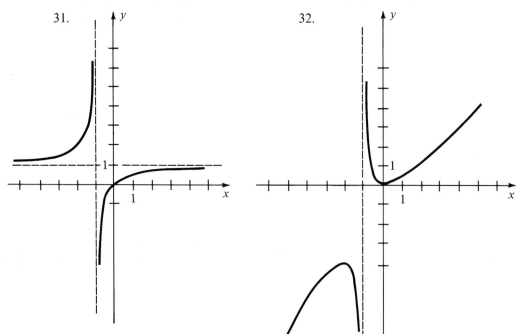

PROGRESS TEST 8A, page 338

1. $Q(x) = 2x^2 - 5$, $R(x) = 11/(x^2 + 2)$
2. $Q(x) = 3x^3 - 7x^2 + 14x - 28$, $R(x) = 54/(x + 2)$
3. -25 4. -165
6. $x^3 - 2x^2 - 5x + 6$ 7. $x^4 - 6x^3 + 6x^2 + 6x - 7$
8. $2, \pm i$ 9. $-1, -1, (3 \pm \sqrt{17})/2$
10. $x^5 + 3x^4 - 6x^3 - 10x^2 + 21x - 9$

11. $16x^5 - 8x^4 + 9x^3 - 9x^2 - 7x - 1$
12. $x^2 - (1 + 2i)x + (-1 + i)$
13. $1/2, 1/2$
14. $1, 1 \pm i$
15. $(x^2 - 4x + 5)(x - 2)$
16. 2
17. 1
18. none
19. $1, 1, -1, -1, 1/2$
20. $2/3, -3, \pm i$

21.

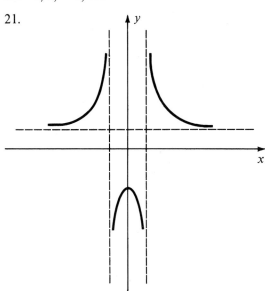

PROGRESS TEST 8B, page 339

1. $Q(x) = 3x^3/2 + 3x^2/4 + 17x/8 + 15/16, R(x) = (49x - 17)/16(2x^2 - x - 1)$

2. $Q(x) = -2x^2 + x + 1, R(x) = 0$

3. -1

4. 24

6. $2x^4 - x^3 - 3x^2 + x + 1$

7. $x^3 - 4x^2 + 2x + 4$

8. $1, 2, 2, 2$

9. $(-3 \pm \sqrt{13})/2, -3, -3$

10. $8x^4 + 4x^3 - 18x^2 + 11x - 2$

11. $x^4 + 4x^3 - x^2 - 6x + 18$

12. $x^4 - 4x^3 - x^2 + 14x + 10$

13. $(3 \pm \sqrt{17})/2$

14. $-2 \pm 2\sqrt{2}$

15. $(x^2 - 2x + 2)(2x^2 + 3x - 2)$

16. 1

17. 2

18. $-1, 2/3, -2$

19. $-1/2, 3/2, \pm i$

20. $0, 1/2, \pm\sqrt{2}$

21.

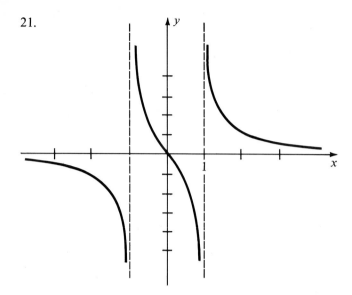

CHAPTER NINE

EXERCISE SET 9.1, page 346

1. 2, 4, 6, 8
3. 5, 7, 9, 11
5. 1/3, 4/5, 9/7, 16/9
7. yes
9. no
11. yes
13. yes
15. 2, 6, 10, 14
17. 3, 5/2, 2, 3/2
19. 1/3, 0, −1/3, −2/3
21. 25
23. −8
25. −2
27. 19/3
29. 821/160
31. 440
33. −126
35. 1720
37. 30
39. $n = 30$, $d = 3$
41. $n = 6$, $d = 1/4$
43. −2

EXERCISE SET 9.2, page 354

1. $r = 2$
3. $r = -3/4$
5. $r = 1/5$
7. 3, 9, 27, 81
9. 4, 2, 1, 1/2
11. −3, −6, −12, −24
13. −384
15. 1/4
17. 1/243
19. 27/8
21. 2
23. 7
25. 1, 3
27. 1/4, 1/16
29. 1093/243
31. −1353/625
33. 1020
35. 55/8
37. $10,235
39. 58,594
41. 2
43. 3/4
45. 8/3
47. 1
49. 1/5
51. 364/990
53. 325/999

EXERCISE SET 9.4, page 363

1. $243x^5 + 810x^4y + 1080x^3y^2 + 720x^2y^3 + 240xy^4 + 32y^5$
3. $256x^4 - 256x^3y + 96x^2y^2 - 16xy^3 + y^4$
5. $32 - 80xy + 80x^2y^2 - 40x^3y^3 + 10x^4y^4 - x^5y^5$

7. $a^8b^4 + 12a^6b^3 + 54a^4b^2 + 108a^2b + 81$
9. $a^8 - 16a^7b + 112a^6b^2 - 448a^5b^3 + 1120a^4b^4 - 1792a^3b^5 + 1792a^2b^6$
   $- 1024ab^7 + 256b^8$
11. $\frac{1}{27}x^3 + \frac{2}{3}x^2 + 4x + 8$
13. $1024 + 5120x + 11{,}520x^2 + 15{,}360x^3$
15. $19{,}683 - 118{,}098a + 314{,}928a^2 - 489{,}888a^3$
17. $16{,}384x^{14} - 344{,}064x^{13}y + 3{,}354{,}624x^{12}y^2 - 20{,}127{,}744x^{11}y^3$
19. $8192x^{13} - 53{,}248x^{12}yz + 159{,}744x^{11}y^2z^2 - 292{,}864x^{10}y^3z^3$
21. 120   23. 12   25. 990   27. 5040
29. 120   31. 210   33. $-35{,}840x^4$ 35. $\frac{495}{256}x^8y^4$
37. $2016x^{-5}$ 39. $-540x^3y^3$ 41. $181{,}440x^4y^3$ 43. $-144x^6$
45. $\frac{35}{8}x^{12}$ 47. 4.8268

EXERCISE SET 9.5, page 371

1. 120   3. 146,016   5. 256   7. 5040
9. 720   11. 336   13. 90   15. 84
17. 3   19. 210   21. 120   23. 60
25. 336   27. 120   29. 84   31. 45

33. 1   35. n 37. $\dfrac{n^2 + n}{2}$ 39. 3003

41. (a) 15,600   (b) 17,576   43. 12,271,512
45. 240   47. 59,400

49. $\dfrac{(26!)^2}{22!20!6!4!}$

EXERCISE SET 9.6, page 378

1. 1/2
3. (a) 1/2   (b) 1/4   (c) 1/13
5. (a) 11/36   (b) 5/36   (c) 4/9
7. (a) 99/100   (b) 1/100
9. (a) 1/17170   (b) 1164/8585   (c) 7372/8585
11. 11/4165   13. 8008/23023   15. 75/612
17. (a) 92.2   (b) $1.6 \times 10^{-7}\%$

REVIEW EXERCISES, page 381

1. 3, 7, 13   2. 0, 7/3, 13/2   3. 3, 5, 7, 9
4. $-2, -5/2, -3, -7/2$ 5. 38
6. -9 7. 8   8. $-33/2$
9. 275/3   10. -450 11. -3
12. $-3/2$ 13. 5, 1, 1/5, 1/25   14. $-2, 2, -2, 2$
15. 243/8   16. ±256 17. 1/2, 1/12
18. 21/32   19. -728 20. 10
21. 9/5   23. $16x^4 - 32x^3y + 24x^2y^2 - 8xy^3 + y^4$
24. $x^4/16 - x^3 + 6x^2 - 16x + 16$ 25. $x^6 + 3x^4 + 3x^2 + 1$
26. 720   27. 78   28. $(n + 1)/n$ 29. 120
30. 168   31. 20   32. n 33. 24
34. 360   35. 210   36. 9   37. 2/9
38. 2/7   39. 8/663   40. 81   41. 1/33

PROGRESS TEST 9A, page 382

1. 1/4, 2/9, 3/16, 4/25
2. -1, 1/2, 2, 7/2
3. 8
4. 41
5. 55/2
6. 1/3
7. $-2/3$, $-4/3$, $-8/3$, $-16/3$
8. -2
9. 8, -16
10. $-43/8$
11. -6
13. $a^{10} + 10a^9/b + 45a^8/b^2 + 120a^7/b^3$
14. 22
15. 360
16. $n + 1$
17. 24
18. 8.46×10^{20}
19. 1/3
20. 3/35

PROGRESS TEST 9B, page 383

1. 5/3, 5, 51/5, 52/3
2. 6, 16/3, 14/3, 4
3. 10
4. -55
5. 13, 1/4
6. 0.2
7. -1, 1/4, $-1/16$, 1/64
8. $-1/81$
9. -4, $-8/3$
10. $-31/2$
11. 25/7
13. $11520x^8$
14. $n/(n + 1)$
15. 7
16. 2
17. 360
18. 200
19. 0.8×10^{-5}
20. 7/12

INDEX

C 3
D 4
E 5
F 6
G 7
H 8
I 9